Solar Hydrogen Generation

Solar Hydrogen Generation

Toward a Renewable Energy Future

Edited by

Krishnan Rajeshwar
University of Texas at Arlington, TX, USA

Robert McConnell
Amonix, Inc., Torrance, CA, USA

Stuart Licht
University of Massachusetts, Boston, USA

 Springer

Editors

Krishnan Rajeshwar
Department of Chemistry & Biochemistry
University of Texas, Arlington
Arlington TX 76019-0065
USA
rajeshwar@uta.edu

Robert McConnell
Amonix, Inc.
3425 Fujita St.
Torrance, CA 90505
bob@amonix.com

Stuart Licht
Chemistry Division
National Science Foundation
4201 Wilson Blvd.
Arlington, VA 022230
USA
slicht@nsf.gov

Department of Chemistry
University of Massachusetts, Boston
100 Morrissey Blvd.
Boston MA 02135
USA
stuart.licht@umb.edu

ISBN: 978-0-387-72809-4   e-ISBN: 978-0-387-72810-0

Library of Congress Control Number: 2007943478

Dedication

Krishnan Rajeshwar

To the three girls in my life, Rohini, Reena, and Rebecca: I could not have done this
without your love and support

Robert McConnell

To my wife Suzie Star whose love and support made this possible. To my children and
especially my grandson Tharyn. My hope for them is to live in a cleaner
world powered by renewable energy and hydrogen.

Stuart Licht

To my children: Reeva, Gadi, Ariel, Jacob and Dov; I hope to open a
path to a sustainable energy future for them. To my wife Bregt, this is here
because you are here.

Preface

This book examines ways to generate hydrogen from sunlight and water. It largely arose out of a desire to bring all the disparate ways to accomplish this goal within the confines of a single edited volume. Thus we are aware of many books and reports discussing the pros and cons of a hydrogen economy but none, that we are aware of, that focus on the science and technology of generating hydrogen from sunlight and water. While renewable hydrogen currently remains an elusive goal, at least from a cost perspective, the *scientific* principles behind its generation are well understood. Thus over and above reviewing this substantial fundamental database, part of the incentive for creating this book was to hopefully inspire future generations of scientists and engineers to respond to the grand challenge of translating the impressive laboratory advances and prototype demonstrations to a practical renewable energy economy. Much of this daunting hurdle has to do with optimizing the efficiency and hence the cost-effectiveness of hydrogen producing solar energy systems.

History certainly is on our side in meeting this challenge. Many early civilizations used the sun, water, and the wind to meet basic needs. Even geothermal heat was used by North American Indians some 10,000 years ago for cooking. The ancient Greeks used hydro power to grind flour and the Persians used windmills to pump water in the first millennium. The human race is very good at solving technological problems and we can certainly wean ourselves from fossil fuels if we collectively put our minds to it. But cost is certainly going to be a driver and no amount of civic sense is going to render the hydrogen economy practically realizable if a gallon of gasoline continues to be substantially cheaper than a kilogram of hydrogen. Unfortunately however we can only give short shrift to the issue of economics in this book because of the rapidly shifting landscape of assumptions that an evolving technology brings with it. Nonetheless, the concluding chapter of this book examines investments, levelized hydrogen prices, and fuel cycle greenhouse gas emissions of a centralized electrolytic hydrogen production and distribution system powered by photovoltaic electricity.

Another important and related topic, not specifically addressed in this book, concerns the issue of how to store hydrogen, especially in a mobile transportation application. We felt that this topic was specialized and wide ranging enough to warrant a separate volume to be created by scientists and engineers far more qualified and knowledgeable than us. While fuel cells are briefly introduced in Chapter 1, how hydrogen is to be utilized to generate power is again left to many other excellent treatises in the literature; some of these are cited in what follows.

Every effort was made to remove redundancy and add homogeneity to the material in this multi-author volume. Indeed, the more authoritative level of discussion afforded by having specialists write each chapter will have hopefully overridden any "rough edges" that remain from chapter to chapter. Undoubtedly, many flaws remain for which we as editors are wholly responsible; we would welcome feedback on these.

A project of this magnitude could not have been completed without the collective contributions of many people, some of whom we wish to acknowledge at this juncture. First, Ken Howell deserves special thanks for his many useful suggestions. His patience as this book production went through countless delays is also much appreciated. Don Gwinner, Al Hicks and their production team at NREL managed to create quality illustrations from the drawings and graphs (many in primitive form) that were furnished to them. Maria Gamboa is thanked for very capably doing the pre-print lay-out of the various manuscripts. Finally we offer simple thanks to our families for their love and support and for putting up with the many weekends away spent in putting this volume together.

Krishnan Rajeshwar
Arlington, Texas

Robert McConnell
Torrance, CA

Stuart Licht
Washington, DC

Contents

Contributors

Maria L. Ghiradi,
National Renewable Energy Laboratory
1617 Cole Blvd., Golden, CO 80401
marie_ghiradi@nrel.gov

Kevin Harrison
National Renewable Energy Laboratory
NREL MS3911
1617 Cole Blvd., Golden, CO 80401
Ph: 303-384-7091, F:303-384-7055, Kevin_Harrison@nrel.gov

Johanna Ivy Levene
National Renewable Energy Laboratory
NREL MS3911
1617 Cole Blvd., Golden, CO 80401
johanna_levene@nrel.gov

Stuart Licht
Chemistry Division
National Science Foundation
4201 Wilson Blvd., Arlington, VA 022230
Ph: 703-292-4952, slicht@nsf.gov

Chemistry Department
100 Morrissey Boulevard
University of Massachusetts, Boston, MA 02135-3395
Ph: 617-287-6156, stuart.licht@umb.edu

Frederick M. MacDonnell
Department of Chemistry and Biochemistry
The University of Texas at Arlington, Arlington, TX 76019-0065
Ph: 817-272-2972, F:817-272-3808, macdonn@uta.edu

Pin Ching Maness
National Renewable Energy Laboratory
1617 Cole Blvd., Golden, CO 80401
pinching_maness@nrel.gov

James Mason
Hydrogen Research Institute
52 Columbia St., Farmingdale, NY 11735
Ph: 516-694-0759, E: hydrogenresearch@verizon.net

Robert McConnell
Amonix, Inc.
3425 Fujita St., Torrance, CA 90505
Ph: 310-325-8091, F: 310-325-0771, E: bob@amonix.com

Daryl Myers
Electric System Center
NREL MS3411
1617 Cole Blvd., Golden, CO 80401
Ph: 303-384-6768, F:303-384-6391, E: daryl_myers@nrel.gov,
W:http://www.nrel.gov/srrl

Krishnan Rajeshwar
College of Science, Box 19065
The University of Texas at Arlington, Arlington, TX 76019
Ph: 817-272-3492, F:817-272-3511, E: rajeshwar@uta.edu,

Michael Seibert
National Renewable Energy Laboratory
1617 Cole Blvd., Golden, CO 80401
Ph: 303-384-6279, F: 303-384-6150, mike_seibert@nrel.gov

Ken Zweibel
Primestar Solar Co., Longmont, CO
ken.zweibel@primestarsolar.com

Biographical Sketches of Authors

Maria L. Ghirardi is a Senior Scientist at NREL and a Research Associate Professor at the Colorado School of Mines. She has a B.S., an M.S. and a Ph.D degree in Comparative Biochemistry from the University of California at Berkeley and has extensive experience working with photosynthetic organisms. Her research at NREL involves photobiological H_2 production and covers metabolic, biochemical and genetic aspects of algal metabolism, generating over 60 articles and several patents.

Kevin W. Harrison is a Senior Engineer in the Electrical Systems Center at NREL. He received his Ph.D. at the University of North Dakota and leveraging management, automated equipment design and quality control experience, gained while working for Xerox Corporation, he joined NREL in 2006. At NREL he leads all aspects of the renewable hydrogen production task whose objective is to improve the efficiency and reduce the capital costs of a closely coupled wind to hydrogen demonstration project. Generally speaking his research interests are in reducing the environmental impact of the world's energy use by integrating and utilizing renewable energy for electricity and transportation fuels.

Johanna Ivy Levene is a Senior Chemical Applications Engineer at NREL. She specializes in the technical and economic analysis of electrolysis systems, and her current focus is the production of fuels from renewable resources. Prior to her work at NREL, Johanna has worked as a process control engineer, a database administrator, a systems administrator and a programmer. Results from her work have been published in Solar Today and Science.

Stuart Licht is a Program Director in the Chemistry Division of the National Science Foundation (NSF) and Professor of Chemistry at the University of Massachusetts, Boston. His research interests include solar and hydrogen energy, energy storage, unusual analytical methodologies, and fundamental physical chemistry. Prof. Licht received his doctorate in 1986 from the Weizmann Institute of Science, followed by a Postdoctoral Fellowship at MIT. In 1988 he became the first Carlson Professor of Chemistry at Clark University, and in 1995 a Gustella

Professor at the Technion Israel Institute of Science, in 2003 became Chair of the Department of Chemistry at the University of Massachusetts Boston, and in 2007 a Program Director at the NSF. He has contributed 270 peer reviewed papers and patents ranging from novel efficient solar semiconductor/electrochemical processes, to unusual batteries, to elucidation of complex equilibria and quantum electron correlation theory.

F. M. MacDonnell is Professor of Inorganic Chemistry at the University of Texas at Arlington (UTA). He received his PhD at Northwestern University in 1993. After a postdoctoral stint at the Chemistry Department of Harvard University, he joined the Chemistry and Biochemistry Department at UTA in 1995. His research interests are in the design of photocatalysts for light harvesting and energy conversion and has published over 100 articles in these areas.

Pin Ching Maness is a Senior Scientist at NREL. She received her Masters Degree in 1976 at Indiana State University, Terre Haute, IN. She worked as a Research Specialist at the University of California, Berkeley, CA from 1976 to 1980, before joining NREL in 1981. Her research interests are in studies of the physiology, biochemistry, and molecular biology of various biological H_2-production reactions in cyanobacteria, photosynthetic bacteria, and cellulolytic fermentative bacteria.

James M. Mason is Director of the Hydrogen Research Institute in Farmingdale, New York. He received his PhD at Cornell University in 1996. His research interests are the economic modelling of centralized hydrogen production and distribution systems using renewable energy sources.

Robert D. McConnell recently joined Amonix, Inc., a concentrator photovolatics (PV) company located in Torrance, CA as Director of Government Affairs and Contracts. He earned his PhD at Rutgers University in Solid State Physics following a Bachelor's degree in Physics at Reed College in Portland, Oregon. After a postdoctoral stint at the University of Montreal and employment at the research institute of the electric utility, Hydro Quebec, he joined NREL in 1978. He has authored numerous papers and edited or co-edited five books and chaired four international conferences on centrator PV. His technology interests include concentrator PV, future generation PV concepts, hydrogen, superconductivity, and wind energy. He has served as Chairman of the Energy Technology Division of the Electrochemical Society and is presently Convener of the international working group developing concentrator PV standards under the aegis of the International Electrotechnical Commission located in Geneva, Switzerland.

Daryl R. Myers is a Senior Scientist at NREl. In 1970 He received a Bachelor of Science in Applied Mathematics from the University of Colorado, Boulder, School of Engineering. Prior to joining NREL in 1978, he worked for four years at the Smithsonian Institution Radiation Biology Laboratory in Rockville Maryland, and is a Cold War veteran, serving as a Russian linguist in the United States Army from 1970 to 1974. He has over 32 years of experience in terrestrial broadband and spectral solar radiation physics, measurement instrumentation, metrology (calibration), and modelling radiative transfer through the atmosphere. Daryl is active in International Lighting Commission (CIE) Division 2 on Physical Measurement of Light and Radiation, the American Society for Testing and Materials (ASTM) committees E44 on Solar, Geothermal, and Other Alternative Energy Sources and G03 on Weathering and Durability, and the Council for Optical Radiation Measurements (CORM).

Krishnan Rajeshwar is a Distinguished Professor in the Department of Chemistry and Biochemistry and Associate Dean in the College of Science at the University of Texas at Arlington. He is the author of over 450 refereed publications, several invited reviews, book chapters, a monograph, and has edited books, special issues of journals, and conference proceedings in the areas of materials chemistry, solar energy conversion, and environmental electrochemistry. Dr. Rajeshwar is the Editor of the Electrochemical Society Interface magazine and is on the Editorial Advisory Board of the Journal of Applied Electrochemistry. Dr. Rajeshwar has won many Society and University awards and is a Fellow of the Electrochemical Society.

Michael Seibert is a Fellow at the National Renewable Energy Laboratory in Golden, CO, USA. He received his Ph.D. in the Johnson Research Foundation at the University of Pennsylvania and then worked at GTE Laboratories before joining NREL (formerly the Solar Energy Research Institute) in 1977. His research has resulted in over 180 publications and several patents in the areas of materials development for electronic microcircuits, primary processes of bacterial and plant photosynthesis, cryopreservation and photomorphogenesis of plant tissue culture, water oxidation by photosystem II in plants and algae, microbial H_2 production, hydrogenase structure and function, genomics of Chlamydomonas, and computational approaches for improving H_2 metabolism in algae. He also holds a concurrent position as Research Professor at the Colorado School of Mines and is a Fellow of the AAAS.

Ken Zweibel is President of PrimeStar Solar, a CdTe PV company located in Colorado, USA. He graduated in Physics from the University of Chicago in 1970. He was employed for 27 years at SERI and then NREL in Golden, CO, where he worked on the development of CdTe, copper indium diselenide, and amorphous and thin film silicon. When he left in December 2006, he was manager of the Thin Film PV Partnership Program. He has published numerous papers and articles, and two books on PV, the most recent being, "Harnessing Solar Power: The PV Challenge." Besides the success of PrimeStar Solar, he is interested in solar policy and solutions to climate change and rising energy prices.

1

Renewable Energy and the Hydrogen Economy

Krishnan Rajeshwar,[1] Robert McConnell,[2] Kevin Harrison,[2] and Stuart Licht[3]

[1] University of Texas at Arlington, Arlington, TX
[2] NREL, Golden, CO
[3] University of Massachussetts, Boston, MA

1 Renewable Energy and the Terawatt Challenge

Technological advancement and a growing world economy during the past few decades have led to major improvements in the living conditions of people in the developed world. However, these improvements have come at a steep environmental price. Air quality concerns and global climate impact constitute two major problems with our reliance on fossil energy sources. Global warming as a result of the accumulation of greenhouse gases such as CO_2 is not a new concept. More than a century ago, Arrhenius put forth the idea that CO_2 from fossil fuel combustion could cause the earth to warm as the infrared opacity of its atmosphere continued to rise.[1] The links between fossil fuel burning, climate change, and environmental impacts are becoming better understood.[2] Atmospheric CO_2 has increased from ~275 ppm to ~370 ppm (Figure 1); unchecked, it will pass 550 ppm this century. Climate models indicate that 550 ppm CO_2 accumulation, if sustained, could eventually produce global warming comparable in magnitude but opposite in sign to the global cooling of the last Ice Age.[3] The consequences of this lurking *time bomb* could be unpredictably catastrophic and disastrous as recent hurricanes and tsunamis indicate.

Every year, a larger percentage of the 6.5 billion global population seeks to improve their standard of living by burning ever-increasing quantities of carbon-rich fossil fuels. Based on United Nations forecasts, another 2.5 billion people are expected by 2050 with the preponderance of them residing in poor countries.[4] Coupled with this growing population's desire to improve their quality of life are the developed countries already high and rising per capita energy use which promises to add to the environmental pressure.

Oil, coal, and natural gas have powered cars, trucks, power plants, and factories, causing a relatively recent and dramatic buildup of greenhouse gases in the atmos-

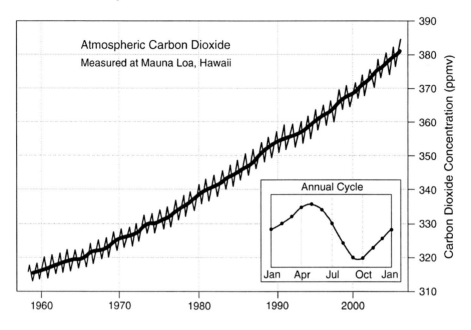

Fig. 1. Atmospheric carbon dioxide record from Mauna Loa. Data courtesy of C. D. Keeling and T. P. Whorf.

phere, most notably CO_2. The anthropogenic buildup of heat-trapping gases is intensifying the earth's natural greenhouse effect, causing average global temperatures to rise at an increasing rate. We appear to be entering into a period of abrupt swings in climate partially due to buildup of human-released CO_2 in the atmosphere. Most alarming is not the fact that the climate is changing but rather the rate at which the buildup of CO_2 is occurring.

Ice core samples from Vostok, Antarctic, look back over 400,000 years before present at atmospheric CO_2 levels by examining the composition of air bubbles trapped in the polar ice buried over 3623 m (11,886 ft) deep.[5] These data show that the range of CO_2 concentrations over this time period have been relatively stable, cycling between about 180 and 300 parts per million by volume (ppmv). According to the World Meteorological Organization the CO_2 concentration in 2005 reached an unprecedented 379.1 ppmv.[6]

This environmental imperative requires us to quickly come to terms with the actual costs, including environmental externalities, of all of our energy use. Only then will the economic reality of energy consumption be realized and renewable sources expand through true market forces. That is not to say that fossil fuels like oil, natural gas, and coal do not have a future in helping to meet this growing demand. However, it should go without saying that all new sources of CO_2 should be captured and stored (i.e., sequestered). Although integrating the systems required to safely and economically storing CO_2 deep underground have not been realized. More than ever, CO_2 released into the atmosphere by coal-fired power plants must be addressed to effectively deal with global climate change. In addition to greenhouse gas emissions, destructive extraction and processing of the fuel, fine particulates of 2.5 micrometers (μm) released from coal-fired power plants are responsible for the deaths of roughly 30,000 Americans every year.[7]

Even notwithstanding this climate change and global warming concern are issues with the supply side of a fossil-derived energy economy. Gasoline and natural gas supplies will be under increasing stress as the economies of heavily-populated developing countries (such as India and China) heat up and become more energy intensive. It is pertinent to note that this supply problem is exacerbated because the United States alone consumes a disproportionately higher fraction (more than the next five highest energy-consuming nations, Ref. 8) of the available fossil fuel supply. There are no signs that the insatiable energy appetite of the U. S. and other advanced parts of the world are beginning to wane. While there is considerable debate about when global oil and natural gas production is likely to peak,[9] there is no debate that fossil fuels constitute a non-renewable, finite resource. We are already seeing a trend in some parts of the world (e.g., Alberta, Canada) of a switch to "dirtier" fossil fuels, namely, coal, heavy oil or tar sand as petroleum substitutes. This switch would mean an increase in CO_2 emissions (note that the carbon content of these sources is higher than gasoline or natural gas), a greater temperature rise than is now being forecast, and even more devastating effects on the earth's biosphere than have already been envisioned.[10]

Currently, renewable energy only constitutes a very small fraction of the total energy mix in the U. S. and in other parts of the world (Figure 2). For example, in 2000, only about 6.6 quads (one quad is about 10^{18} J) of the primary energy in the U. S. came from renewables out of a total of 98.5 quads.[11] Of this small fraction supplied by renewable energy, about 3.3 quads were from biomass, 2.8 from hydroelectric generation, 0.32 from geothermal sources, 0.07 from solar thermal energy and 0.05 quads from wind turbines.[8] This profile would have to switch to an energy mix that resembles the right-side panel in Figure 2 if the CO_2 emissions are to be capped at environmentally safe levels. This is what the late Professor Rick Smalley, winner of the Nobel Prize in Chemistry, referred to as the *Terawatt Challenge*. Recent analyses[12] have posited that researching, developing, and commercializing carbon-free primary power to the required level of 10-30 TW (one terawatt = 10^{12} W) by 2050 will require efforts of the urgency and scale of the Manhattan Project and the Apollo Space Program.

This book examines the salient aspects of a hydrogen economy, particularly within the context of a renewable, sustainable energy system.

2 Hydrogen as a Fuel of the Future

Jules Verne appears to be one of the earliest people to recognize, or at least articulate, the idea of splitting water to produce hydrogen (H_2) and oxygen (O_2) in order to satisfy the energy requirements of society. As early as 1874 in *The Mysterious Island*, Jules Verne alluded to clean hydrogen fuels, writing:

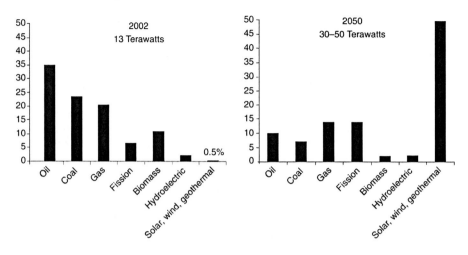

Fig. 2. The terawatt renewable energy challenge; the energy mix has to switch from the panel on the left to the panel on the right to cap CO_2 levels at safe limits. Data from the International-al Energy Agency.

"Yes, my friends, I believe that water will someday be employed as fuel, that hydrogen and oxygen, which constitute it, used singly or together, will furnish an inexhaustible source of heat and light....I believe, then, that when the deposits of coal are exhausted, we shall heat and warm ourselves with water. Water will be the coal of the future."

Remarkable words indeed from a prophetic visionary who foresaw also the technological development of spacecraft and submarines. Hydrogen gas was first isolated by Henry Cavendish in 1766 and later recognized as a constituent of water by Lavoisier in 1783.[13] The production of hydrogen and oxygen by the electrolytic decomposition of water has been practiced since the year 1800, when the process was first discovered by Nicholson and Carlisle.[14] Since then, the idea of society using hydrogen as a primary energy carrier has been explored and refined.

In the late 1920s and the early 1930s a German inventor, Rudolf A. Erren, recognized and worked towards producing hydrogen from off-peak electricity and modifying the internal combustion engine to run on hydrogen.[15] Erren's primary objective was to eliminate pollution from the automobile and reduce oil imports. In the 1970s Derek Gregory appears to have been one of the leading advocates in creating the case for a hydrogen-based economy.[13,15,16]

The literature suggests that the term *hydrogen economy* may have been coined by H. R. Linden, one of Gregory's colleagues at the Institute of Gas Technology, in 1971.[13] Gregory points to hydrogen's environmental benefits and recognizes that, while fossil fuels are inexpensive, requiring the atmosphere to assimilate the by-products of their combustion is not without consequence.

The water electrolyzer industry grew substantially during the 1920s and 1930s, as elaborated later in Chapter 3. This included products from companies such as Oerlikon, Norsk Hydro, and Cominco in multi-megawatt sizes.[14,17,18] Most of these installations were near hydroelectric plants that supplied an inexpensive source of electricity. As more hydrogen was needed for industries, steam reforming of methane gradually took over as the hydrogen production process of choice because it was less expensive.

Hydrogen is often blamed for the 1937 Hindenburg disaster. The shell of the German airship was a mixture of two major components of rocket fuel, aluminum and iron oxide, and a doping solution which was stretched to waterproof the outer hull. Researchers concluded that the coating of the Hindenburg airship was ignited by an electrical discharge and the ensuing explosion to be inconsistent with a hydrogen fire.[19] It turns out that 35 of the 37 people who died in the disaster, perished from jumping or falling from the airship to the ground. Only two of the victims died of burns, and these were from the burning airship coating and on-board diesel fuel.[20] Modern laboratory tests confirmed that the 1930s fabric samples to still be combustible.

"Although the benefits of the hydrogen economy are still years away, our biggest challenges from a sustainability standpoint are here today,"

said Mike Nicklas, Past Chair of the American Solar Energy Society, during his opening comments at the first Renewable Hydrogen Forum in Washington, D.C., in April 2003.[21]

Hydrogen (H) is the simplest of atoms, consisting of one proton and one electron also called a protium. As atoms, hydrogen is very reactive and prefers to join into molecular pairs (H_2) and when mixed in sufficient quantities with an oxidant (i.e., air, O_2, Cl, F, N_2O_4, etc.) becomes a combustible mixture. Like all other fuels, H_2 requires proper understanding and handling to avoid unwanted flammable or explosive environments. Hydrogen is not a primary source of energy; rather it is an energy carrier much like electricity. Therefore, energy is required to extract hydrogen from substances like natural gas, water, coal, or any other hydrocarbon.

At 25 °C and atmospheric pressure the density of air is 1.225 kg m^{-3} while hydrogen is 0.0838 kg m^{-3}, making it 14.6 times lighter than air. This is an important safety consideration in that a hydrogen leak will dissipate quickly. Hydrogen's positive buoyancy significantly limits the horizontal spreading of hydrogen that could lead to combustible mixtures. Hydrogen is the lightest (molecular weight 2.016) and smallest of all gases requiring special considerations for containing and sensing a leak.

Figure 3 shows the two types of molecular hydrogen distinguished by the spin, ortho- and para-hydrogen. They differ in the magnetic interactions as ortho-hydrogen atoms are both spinning in the same direction and in para-hydrogen the protons are spinning anti-parallel. At 300 K, the majority (75%) is ortho-hydrogen, while at 20 K 99.8% of the hydrogen molecules are para-hydrogen. As the gas transitions from gas to liquid at 20 K heat is released and ortho-hydrogen becomes unstable.[22] Hydrogen becomes a liquid below its boiling point of −253 °C (20 K) at atmospheric pres-

a)   b)

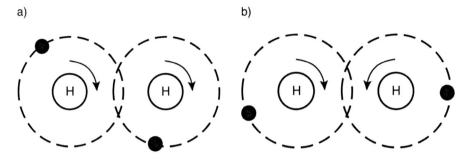

Fig. 3. Ortho- (left) and para-hydrogen (right).

sure. Pressurization of the hydrogen to 195 pisg (13 barg) increases the boiling point to −240 °C (−400 °F), pressures above that don't return a significant improvement.[22]

At ambient temperature and pressure hydrogen is colorless, odorless, tasteless and nontoxic. However, leaks of hydrogen (or any gas for that matter) can displace oxygen and act as an asphyxiant. Any atmosphere with less than 19.5% oxygen by volume in considered oxygen deficient and asphyxiation can lead to physiological hazards.

The primary hazard associated with gaseous hydrogen is the unintentional mixing of the fuel with an oxidant (typically air) in the presence of an ignition source. Hydrogen fires and deflagrations have resulted when concentrations within the flammability limit were ignited by seemingly harmless ignition sources. Ignition sources include electrical, mechanical, thermal and chemical. For example; sparks from valves, electrostatic discharges, sparks from electrical equipment, mechanical impact, welding and cutting, open flame, personnel smoking, catalyst particles and lightning strikes in the proximity of hydrogen vent stacks.[23]

With the exception of helium, hydrogen has the lowest boiling point at atmospheric pressure of where it becomes a transparent and odorless liquid. Liquid hydrogen has a specific gravity of 0.071, which is roughly 1/14th the density of water and is neither corrosive nor reactive. The low specific gravity of liquid hydrogen further reveals hydrogen's low volumetric energy density in that a cubic meter of water contains more hydrogen (111 kg) than a cubic meter of pure hydrogen in liquid state (71 kg). The values of the main physical properties of gaseous hydrogen are shown in Table 1.

Leaking hydrogen gas and (once ignited) its flame are nearly invisible. The pale blue flame of a hydrogen fire is barely visible and is often detected by placing a standard household wicker broom in the path of the suspected hydrogen flame. The hydrogen flame temperature in air (2045 C, 3713 F) releases most of its energy in the ultraviolet (UV) region requiring UV sensors for detecting the presence of a flare or fire. The UV radiation from a flaring hydrogen fire can also cause burns akin to over-exposure to the sun's damaging UV radiation.

Table 1. Selected properties of gaseous hydrogen at 20 °C and 1 atm.

Physical Property		Units
Molecular weight	2.016	
Density	0.0838	kg/m^3
Specific gravity	0.0696	(Air = 1)
Viscosity	8.813 x 10^{-5}	g/cm sec
Diffusivity	1.697	m^2/hr
Thermal conductivity	0.1825	W/m K
Expansion ratio	1:848	Liquid to gas
Boiling point (1 atm)	−253 (−423)	°C (°F)
Specific heat, constant pressure	14.29	J/g K
Specific heat, constant volume	10.16	J/g K
Specific volume	11.93	m^3/kg
Diffusion coefficient in air	6.10	cm^2/sec
Enthalpy	4098	kJ/kg
Entropy	64.44	J/g K

The amount of thermal radiation (heat) emitted from a hydrogen flame is low and is hard to detect by feeling (low emissivity). Most commercially available combustible gas detectors can be calibrated for hydrogen detection. Typically alarms from these sensors are set by the manufacturer between 10%–50% of the lower flammability limit (LFL) of hydrogen to avoid the presence of an unwanted flammable environment.

Table 2 compares the same fuels as above and reports their volumetric energy density in kg m^{-3}. Hydrogen has the highest energy content per unit mass than any fuel making it especially valuable when traveling into space. As mentioned earlier, hydrogen suffers volumetrically when compared with traditional fuels making storing sufficient on-board terrestrial vehicles an engineering challenge.

The LFL of hydrogen represents the minimum concentration required below which the mixture is too lean to support combustion.[24] Hydrogen has a wide flammability range of (4%–75%) while gasoline is (1.5%–7%) when mixed with air at standard temperature (25 °C) and pressure (1 atm). Hydrogen in oxygen has a slightly wider flammability range (4%–95%). Table 3 summarizes a selected number of important combustion properties of hydrogen.

Table 2. Comparing hydrogen properties with other fuels. Based on LHV and 1 atm, 25 °C for gases.

	Hydrogen	Methane	Gasoline	Diesel	Methanol
Density, kg m^{-3}	0.0838	0.71	702	855	799
Energy density, MJ m^{-3}	10.8	32.6	31,240	36,340	14,500
Energy density, kWh m^{-3}	3.0	9.1	8680	10,090	4030
Energy, kWh kg^{-1}	33.3	12.8	12.4	11.8	5.0

*Energy density = LHV * density (□), and the conversion factor is 1 kWh = 3.6 MJ.

Table 3. Selected combustion properties of hydrogen at 20 °C and 1 atm.[a]

Combustion Property		Units
Flammability limits in air	4 – 75	vol%
Flammability limits in oxygen	4 – 95	vol%
Detonability limits in air	18 – 59	vol%
Detonability limits in oxygen	15 – 90	vol%
Minimum ignition energy in air	17	αJ
Auto ignition temperature	585 (1085)	°C (°F)
Quenching gap in air	0.064	cm
Diffusion coefficient in air	0.061	cm^2/sec
Flame velocity	2.7 – 3.5	m/s
Flame emissivity	0.1	
Flame temperature	2045 (3713)	°C (°F)

[a]From Ref. 19.

Each fuel is limited to a fixed amount of energy it can release when it reacts with an oxidant. Every fuel has been experimentally tested to determine the amount of energy it can release and is reported as the fuel's higher heating value (HHV) and lower heating value (LHV). The difference between the two values is the latent heat of vaporization of water, and the LHV assumes this energy is not recovered.[22] In other words, LHVs neglect the energy in the water vapor formed by the combustion of hydrogen in the fuel because it may be impractical to recover the energy released when water condenses. This heat of vaporization typically represents about 10% of the energy content.

It is often confusing to know which heating value to use when dealing with similar processes such as electrolysis and fuel cells. The appropriate heating value depends on the phase of the water in the reaction products. When water is in liquid form, the HHV is used; if water vapor (or steam) is formed in the reaction, then the LHV would be appropriate. An important distinction is that water is produced in the form of vapor in a fuel cell as well as in a combustion reaction and, therefore, the LHV represents the amount of energy available to do work. Table 4 shows both the LHV and the HHV for common fuels.

Obviously, the most important virtue of using hydrogen as a fuel is its pollution-free nature. When burned in air, the main combustion product is water with O_2 in a fuel cell to directly produce electricity; the only emission is water vapor. Indeed this

Table 4. HHVs and LHVs at 25 °C and 1 atm of common fuels, kJ g^{-1} [a]

Fuel	HHV	LHV
Hydrogen	141.9	119.9
Methane	55.5	50.0
Gasoline	47.5	44.5
Diesel	44.8	42.5
Methanol	20.0	18.1

[a]From Ref. 22.

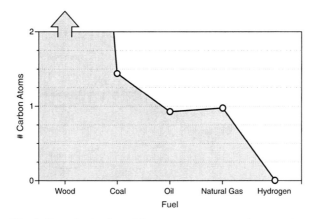

Fig. 4. Decarbonization of the energy source over the centuries.

fuel cell product is clean enough to furnish drinking water to the crews in spacecraft! Crucially, the use of hydrogen completes the *decarbonization* trend that has accompanied the evolution of energy sources for mankind over the centuries (Figure 4). The combustion of H_2, unlike fossil fuels, generates no CO_2. Unlike fossil fuels, however, hydrogen is not an energy *source* but is an energy *carrier* since it almost never occurs by itself in nature, at least terrestrially. (The atmospheres of other planets, e.g., Mars, are rich in hydrogen. Should space travel prove to be economical and accessible in the future, we may have a viable means to "mine" H_2 as we are doing for petroleum and coal these days!)

In the interim timeframe: Where is the H_2 to come from? Historically, H_2 has been used for energy since the 1800s. It is a major constituent (up to ~50% by volume) of *syngas* generated from the gasification of coal, wood, or municipal wastes. Indeed, syngas was used in urban homes in the U. S. for heating and cooking purposes from the mid-1800s until the 1940s and is still used in parts of Europe, Latin America and China where natural gas is unavailable or too expensive. Most of the H_2 manufactured these days comes from the steam reforming of methane (see above). Other processes for making H_2 from fossil fuel sources include the water gas shift reactions. Neither of these approaches is carbon-neutral in that significant amounts of CO_2 are generated in the H2 manufacture process itself.

The ultimate goal would be to produce H_2 with little or no greenhouse gas emissions. One option is to combine H_2 production from fossil fuels with CO_2 sequestration. Carbon sequestration, however, is as yet an unproven technology. Another approach is biomass gasification—heating organic materials such as wood and crop wastes so that they release H_2 and carbon monoxide. This technique is carbon-neutral because any carbon emissions are offset by the CO_2 absorbed by the plants during their growth. A third possibility is the electrolysis of water using power generated by renewable energy sources such as wind turbines and solar cells. This approach is discussed in Chapters 3 and 4.

Although electrolysis and biomass gasification involve no major technical hurdles, they are cost-prohibitive, at least at present: $6–10 per kilogram of H_2 pro-

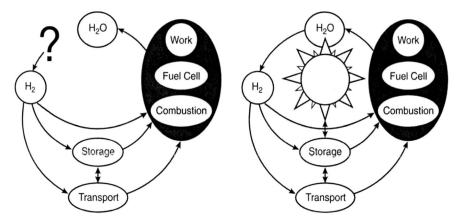

Fig. 5. The water splitting/hydrogen fuel cycle without (left panel) or with (right panel) inclusion of solar energy input.

duced.The goal is to be able to develop and scale-up technologies to afford a pump price for H₂ of $2–4 per kilogram. In such a scenario, hydrogen in a fuel cell powered car would cost less per kilometer than gasoline in a conventional car today.

Clearly, water would be the ideal and most sustainable source for H₂ and this H₂ generation concept dates back two centuries. Table 5 summarizes various schemes for generating H₂ via splitting of water and Figure 5 depicts the water splitting/hydrogen fuel cycle without (left panel) or with (right panel) inclusion of solar energy input. The approaches considered in Table 5 and Figure 5 form the topics of discussion in Chapters 4 through 7 of this book.

The power needed for water electrolysis could come from nuclear energy although producing H₂ this way would not be significantly cheaper than using renewable power sources. Nuclear plants can generate H₂ in a non-electrolytic, thermal mode because of the intense heat generated in a thermonuclear reaction. This ap-

Table 5. The ability of nuclear and various renewable energy sources to meet the 14-20 TW demand of carbon-free power by 2050.[a]

Source	Power available TW	Comments
Biomass	7–10	Entire arable land mass of the planet must be used excluding the area needed to house 9 billion people
Wind on land	2.1	Would saturate the entire Class 3 (wind speed at 5.1 m/s at 10 m above ground) global land mass with windmills
Nuclear	8	Requires the construction of 8000 new nuclear power plants
Hydroelectric	1.5	Would require damming of all available rivers

[a] From Ref. 26

proach, while potentially cost-effective, has not been demonstrated yet. It must be noted that any option involving nuclear power has the same hurdles that have dogged the nuclear electric power industry for decades, namely those of waste disposal problems, proliferation concerns and lack of public acceptance. (This contrasts with the success of the nuclear power industry in some countries, e.g., France.) Producing 10 TW of nuclear power would require the construction of a new 1-GW$_e$ nuclear fission plant somewhere in the world every other day for the next 50 years![25]

3 Solar Energy and the Hydrogen Economy

Solar energy is a virtually inexhaustible and freely available energy source. More sunlight (~1.2×10^5 TW) falls on the earth's surface in 1 h than is used by all human activities in 1 year globally. The sun is earth's natural power source, driving the circulation of global wind and ocean currents, the cycle of water evaporation and condensation that creates rivers and lakes, and the biological cycles of photosynthesis and life. It is however a dilute energy source (1 kW/m^2 at noon, Chapter 2); about 600–1000 TW strikes the earth's terrestrial surfaces at practical sites suitable for solar energy harvesting.[27] Covering 0.16% of the land on earth with 10% efficient solar conversion systems would provide 20 TW of power,[28] nearly twice the world's consumption rate of fossil energy and an equivalent 20,000 1-GW$_e$ nuclear fission plants. Clearly, solar energy is the largest renewable carbon-free resource amongst the other renewable energy options.

Consider the total amounts possible for each in the light of the 14–20 TW of carbon-free power needed by 2050. Table 5 provides a summary;[26] clearly the additional energy needed per year over the 12.8 TW fossil fuel energy base is simply not attainable from biomass, wind, nuclear and hydroelectric options. The answer to this supply dilemma must lie with solar energy. Chapter 2 provides an overview of the solar energy resource with particular emphasis on the solar spectrum.

Solar energy can be harnessed in many ways[25] but three routes of particular relevance to the theme of this book rely on *electrical*, *chemical*, and *thermal* conversion. Thus the energy content of the solar radiation can be captured as excited electron-hole pairs in a semiconductor, a dye, or a chromophore, or as heat in a thermal storage medium. Excited electrons and holes can be tapped off for immediate conversion to electrical power, or transferred to biological or chemical molecules for conversion to fuel. Solar energy is "fixed" in plants via the photosynthetic growth process. These plants are then available as biomass for combustion as primary fuels or for conversion to secondary fuels such as ethanol or hydrogen. All of these possibilities are addressed in more detail in the Chapters that follow.

While there is tremendous potential for solar energy to contribute substantially to the future carbon-free power needs, none of the routes listed above are currently competitive with fossil fuels from cost, reliability, and performance perspectives. Photovoltaic solar cells have been around for decades and have been widely deployed in space vehicles. Terrestrially, their utilization thus far has been limited to niche applications or remote locales where less expensive electricity is not available. Costs for turnkey installations were 6–10 times more expensive in 1999 for solar electrical energy than for electricity derived from coal or oil. The present cost of photovoltaic (PV) modules is ~$3.50/peak watt. Considering the additional balance of system costs (land, maintenance, etc.) this translates to an energy cost of ~$0.35/kWh. The target at present is ~$0.40/peak watt corresponding to electricity at $0.02/kWh or H_2 produced by PV hybrid water electrolyzers at $0.11/kWh. Major advances in electrolyzer technology could bring this hydrogen cost to $0.04/kWh,[29] which is about the present cost of H_2 from steam reforming of natural gas. These issues are further elaborated in Chapters 2, 3, and 9. A cost goal of $0.40/peak watt requires solar photovoltaic conversion at a total cost of $125/m² combined with a cell energy conversion efficiency of ~50%. Such combinations of cost and efficiency require truly disruptive photovoltaic technologies. Many such approaches are being actively pursued in research laboratories around the world. A critical discussion of outstanding issues, including dispelling the *seven myths of solar electricity* may be found in Refs.25, 29, and 30.

The economic outlook for the other two solar approaches is not much rosier, at least at present. Solar fuels in the form of biomass produce electricity and heat at costs that are within the range of fossil fuels, but their production capacity is limited. The low efficiency with which plants convert sunlight to stored energy means large land areas are required. To produce the full 13 TW of power used by the planet, nearly all the arable land on earth would need to be planted with switchgrass, the fastest growing energy crop. Artificial photosynthetic systems, however, are more promising (see next Section) and these are discussed in Chapter 6. Solar thermal systems provide the lowest-cost electricity at the present time, but require large areas in the Sun Belt in the U. S. and continuing advances in materials science/engineering.

4 Water Splitting and Photosynthesis

The decomposition of liquid water to form gaseous hydrogen and oxygen:

$$H_2O_{(\ell)} \rightarrow H_{2(g)} + \frac{1}{2}O_{2(g)} \tag{1}$$

is a highly endothermic and endergonic process with $\Delta H° = 285.9$ kJ/mol and $\Delta G° = 237.2$ kJ/mol. This reaction may be driven either electrochemically or thermally via the use of solar energy.

The standard potential $\Delta E°$ for Reaction 1 corresponding to the transfer of two electrons is given by:

$$\Delta E° = -\Delta G°/2F = -1.23 \text{ V} \tag{2}$$

In Eq. 2, F is the Faraday constant (96485 C mol⁻¹) and the negative sign denotes the thermodynamically non-spontaneous nature of the water splitting process. The actual *voltage* required for electrolysis will depend on the fugacities of the gaseous products in Reaction 1 as well as on the electrode reaction kinetics (overpotentials)

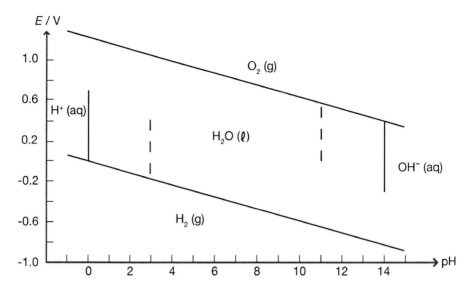

Fig. 6. Pourbaix diagram of water.[31]

along with the Ohmic resistance losses in the cell. In practice, steady-state electroly-sis of water at 298 K requires the application of ~1.50 V.

Figure 6 contains a Pourbaix diagram for water;[31] the zones in this diagram are labeled by the formulas for the predominant species at the electrode potential and pH indicated on the axes. Thus the threshold (thermodynamic) potentials for the decom-position of water via:

$$H_2O(\ell) \rightarrow 4\,e^- + O_{2(g)} + 4\,H^+(aq) \tag{2a}$$

or

$$4\,OH^-(aq) \rightarrow 4\,e^- + O_{2(g)} + 2\,H_2O(\ell) \tag{2b}$$

and

$$2\,H^+(aq) + 2\,e^- \rightarrow H_{2(g)} \tag{3a}$$

or

$$2\,H_2O(\ell) + 2\,e^- \rightarrow H_{2(g)} + 2\,OH^-(aq) \tag{3b}$$

clearly depend on solution pH and they vary at a Nernstian rate of −0.059 V/pH at 298 K.

Optimizing the rates of the electrochemical processes (Reactions 2 and 3) consti-tute much of the R&D focus in electrochemical or photoelectrochemical splitting of water. Two-compartment cells are also employed to spatially separate the evolved gases with special attention being paid to the proton transport membranes (e.g., Na-fion[R]). Chapter 3 provides a summary of the progress made in water electrolyzer technologies.

Water is transparent to the wavelengths constituting the solar spectrum. Therefore, photocatalytic or photoelectrochemical splitting of water requires an agent (semiconductor, dye, or chromophore) capable of first absorbing sunlight and generating electron-hole pairs. Molecular approaches are discussed in Chapter 6 and semiconductor-based approaches are described in Chapter 7.

Thermochemical splitting of water involves heating water to a high temperature and separating the hydrogen from the equilibrium mixture. Unfortunately the decomposition of water does not proceed until temperatures around 2500 K are reached. This and other thermal routes are discussed in Chapter 5. Solar thermal processes are handicapped by the Carnot efficiency limits. On the other hand, solar *photonic* processes are limited by fundamental considerations associated with bandgap excitation; these have been reviewed in Refs.32 and 33.

The water splitting reaction, Eq. 1, have been stated here as the Holy Grail[21] of hydrogen generation using solar energy. However other chemical reactions have been investigated and include, for example:[34,35]

$$2\,HBr \xrightarrow{\ h\upsilon\ } H_2 + Br_2 \tag{4}$$

$$2\,H_2O + 2\,Cl - \xrightarrow{\ h\upsilon\ } 2\,OH^- + Cl_2 + H_2 \tag{5}$$

However, these alternative schemes are fraught with problems associated with the generation and handling of toxic or hazardous by-products such as Br_2 and Cl_2.

Turning to photobiological schemes for producing H_2 (Chapter 8), a complex reaction scheme uses solar energy to convert H_2O into O_2 and reducing equivalents which appear as NADPH. In photosystem 1, the reducing equivalents in NADPH are used to reduce CO_2 to carbohydrates:

$$6\,H_2O + 6\,CO_2 + 48\,h\upsilon \rightarrow C_8H_{12}O_6 + 6\,O_2 \tag{6}$$

or in bacteria, used directly as a reductive energy source.[36,37] In artificial photosynthesis, the goal is to harness solar energy to drive high-energy, small-molecule reactions such as water splitting (Reaction 1) or CO_2 reduction, Reaction 7:[38]

$$2\,H_2O + 2\,CO_2 + 4\,h\upsilon \rightarrow 2\,HCOOH + O_2 \tag{7}$$

Photobiological processes for H_2 production are considered in Chapter 8.

5 Completing the Loop: Fuel Cells

The high-energy chemicals such as H_2 that form in the reactions considered in the preceding reaction, can be recombined in fuel cells to extract the stored chemical energy as electricity. A fuel cell is an electrochemical device that converts the chemical energy in a fuel (such as hydrogen) and an oxidant (oxygen, pure or in air) directly to electricity, water, and heat. Fuel cells are classified according to the electrolyte that they use (Table 6). For automobile applications, the polymer-electrolyte-membrane (PEM) type of fuel cell is the leading candidate for developing zero-emission vehicles. Other types of fuel cells (e.g., solid oxide fuel cells or SOFCs)

Table 6. Types of fuel cells.[a]

Type	Electrolyte	Operating temperature, °C
Polymer-electrolyte membrane (PEM)	Sulfuric acid impregnated in membrane	60–80
Alkaline	KOH	70–120
Phosphoric acid	Phosphoric acid	160–200
Molten carbonate	Lithium/potassium carbonate	650
Solid oxide	Yttria-stabilized zirconia	1,000

[a]Adapted from Ref. 39.

have been considered for stationary power needs. Figure 7 contains the schematic diagram of a PEM fuel cell.[39]

The major virtue of a fuel cell, other than its clean emissions, is its high electrical conversion efficiency. This is not Carnot-limited (unlike in heat engines) and for an ideal hydrogen-oxygen fuel cells, can approach an impressive 83%.[40] In practical devices, up to 60% of the energy content in H_2 can be converted to electricity, the remainder being dissipated as heat. For comparison, practical internal combustion engines using H_2 fuel achieve efficiencies of only 45%.[40]

The principle of fuel cells has been known since 1838 thanks to William Grove. However, widespread deployment did not begin till the 1960s and 70s when fuel cells were used in space and for military (e.g., submarine) applications. Nowadays, fuel cells are being considered for low-polluting co-generation of heat and power in buildings and for transportation applications.

As with the technologies considered earlier, the main deterrent is cost. Today's fuel cell demonstration cars and buses are custom-made prototypes that cost about $1 million apiece.[41] Economies of scale in mass manufacture would bring this cost to a more reasonable $6,000-10,000 range. This translates to about $125 per kilowatt of engine power, which is about four times as high as the $30 per kilowatt cost of a comparable gasoline-powered internal combustion engine.[41] A major cost component in the PEM fuel cell is the noble metal (usually Pt) electrocatalyst. Efforts are underway in many laboratories to find less expensive substitutes (see for example, Refs. 42–44).

Other technical hurdles must be overcome to make fuel cells more appealing to automakers and consumers. Durability is a key issue and performance degradation is usually traceable to the proton exchange membrane component of the device. Depending on the application, 5,000–40,000 h of fuel cell lifetime is needed. Chemical attack of the membrane and electrocatalyst deactivation (due to gradual poisoning by impurities such as CO in the feed gases) are critical roadblocks that must be overcome.

High temperature membranes, that can operate at temperatures above 100 °C, are desirable to promote heat rejection, speed up electrode reaction rates, and to improve tolerance to impurities. This is an active area of materials research. Unfortunately, space constraints preclude a detailed description of fuel cell technologies and the underlying issues. Instead, the reader is referred to excellent reviews and books that exist on this topic.[45-47]

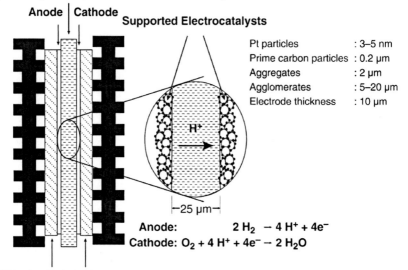

Proton Exchange Membrane
(25–185 μm thick, sulfonic acid clusters: 5 nm)

Anode | Cathode

Supported Electrocatalysts

Pt particles	: 3–5 nm
Prime carbon particles	: 0.2 μm
Aggregates	: 2 μm
Agglomerates	: 5–20 μm
Electrode thickness	: 10 μm

H^+

|−25 μm−|

Anode: $2 H_2 \rightarrow 4 H^+ + 4e^-$

Cathode: $O_2 + 4 H^+ + 4e^- \rightarrow 2 H_2O$

Gas Diffusion Substrates
(~350 μm thick, pore size : 100 nm in microporous layer, 5 μm in base carbon layer)

Fig. 7. Schematic diagram of a PEM fuel cell. Reproduced from Ref. 39. Copyright (2004), by permission of The Electrochemical Society.

6 Concluding Remarks

If renewable energy economy based on hydrogen were to become a reality, a nexus of three technologies, namely solar energy (thermal and photovoltaic), hydrogen production, and fuel cells will have to occur. However, many grand challenges remain in overcoming the technical and cost hurdles associated with each of these technologies. Many of these have been outlined above but are also elaborated in the Chapters that follow. Nonetheless, it is interesting to note, at this juncture, that all three technologies are poised at a very interesting stage of development in their translation from the R&D laboratory to the commercial world. How soon will they reach the marketplace will depend on many factors, some more tangible than others.

References

1. S. Arrhenius, *Phil. Mag.* **41** 237 (1896).
2. M. I. Hoffert et al., *Science* **298** 981 (2002).
3. M. I. Hoffert and C. Covey, *Nature* **360** 573 (1992).

4. R. D. McConnell, J. A. Turner, J. B. Lasich, and D. Holland, Concentrated solar energy for the electrolytic production of hydrogen., in *International Solar Concentrator Conference for the Generation of Electricity or Hydrogen*, Alice Springs, Australia, 2004, p. 24.

5. J. R. Petit, et al., *Vostok Ice Core Data for 420,000 Years*, NOAA/NGDC Paleoclimatology Program, IGBP PAGES/World Data Center for Paleoclimatology Data Contribution Series #2001-076, Boulder, CO, 2001.

6. *WMO Greenhouse Gas Bulletin 2005: Atmospheric Carbon Dioxide Levels Highest On Record*, World Meteorological Organization, Retrieved on November 5, 2006, from http://www.wmo.int/web/arep/gaw/ghg/PR_762_E.doc, 2006.

7. C. G. Schneider, *Death Disease and Dirty Power: Mortality and Health Damage Due to Air Pollution from Power Plants*, Clean Air Task Force,

8. Sources: Energy Information Administration, U. S. Department of Energy, Rep. DOE/EIA-0384 (2000), Wash. D. C., August, 2001. International Energy Agency, http://omrpublic.iea.org.

9. J. Rifkin, *The Hydrogen Economy*, Jeremy P. Tarcher/Putnam, New York, 2002.

10. D. L. Albritton et al., *Summary for Policy Makers: Climate Changes 2001: A Report of Working Group I of the Intergovernmental Panel on Climate Change*, IPCC, 2001. www.earth.usgcrp.gov/ipcc/wg1spm.pdf.

11. S. F. Baldwin, *Physics Today*, April 2002, p. 62.

12. M. I. Hoffert et al., *Nature* **395**, 881 (1998).

13. D. Gregory, *A Brief History of the Hydrogen Energy Movement*, Symposium Papers: Hydrogen for Energy Distribution, Institute of Gas Technology, Chicago, 1978.

14. D. H. Smith, Industrial water electrolysis, in *Industrial Electrochemical Processes*, edited by A. T. Kuhn, Elsevier Publishing Company, 1971, pp. 127–157.

15. D. Gregory, The hydrogen economy, in *Scientific American*. **228** (1) 13 (1973).

16. D. Gregory and J. B. Pangborn, Hydrogen energy, in *Hydrogen for Energy Distribution*, Institute of Gas Technology, 1978, pp. 279–310.

17. K. E. Cox, J. K.D. Wiliamson, Hydrogen: Its technology and implications, *Production Technology,* CRC Press, **1** (1977).

18. A. Konopka, D. Gregory, Hydrogen production by electrolysis: Present and future, in *10th Intersociety Energy Conversion Engineering Conference*, IEEE Cat. No. 75CHO 983-7 TAB, 1975.

19. *Safe Use of Hydrogen and Hydrogen Systems*, NASA Training Center, 2006.

20. J. C. Bokow, *Fabric, Not Filling, to Blame Hydrogen Exonerated in Hindenburg Disaster*, National Hydrogen Association, http://www.hydrogenus.com/advocate/ad22zepp.htm,, 1997.

21. R. W. Larson, The right future? ASES and the renewables community examine renewable hydrogen's potential benefits — and weigh growing concerns, in *Solar Today*, 2004.

22. Hydrogen Properties, College of the Desert, http://www.eere.energy.gov/hydrogenand fuelcells/techvalidation/pdfs/fcm01r0.pdf, December 2001.

23. P. M. Ordin, *Safety Standard for Hydrogen and Hydrogen Systems*, I. NSS 1740.16: NASA, Office of Safety and Mission Assurance, 1997.

24. *OSH Answers: Compressed Gases-Hazards*, Canadian Centre for Occupational Health & Safety (CCOHS), Retrieved on March 25, 2006, from http://www.ccohs.ca/oshanswers/ chemicals/compressed/compress.html, 2005.

25. *Basic Research Needs for Solar Energy Utilization*, Report of the Basic Energy Sciences Workshop on Solar Energy Utilization, April 18-21, 2005, Office of Science, U. S. Department of Energy, Wash. D. C. See also http://www.sc.doc.gov/bes/reports/files/ SEU_rpt.pdf.

26. Prof. Nate Lewis' website: http://nsl.caltech.edu/energy.html.

27. A. J. Nozik, *Inorg. Chem.* **44**, 6893 (2005).

28. J. A. Turner, M. C. Williams, K. Rajeshwar, The Electrochemical Society *Interface*, Fall 2004, p. 24.
29. *The Hydrogen Economy: Opportunities, Costs, Barriers, and R&D Needs*, The National Academies Press, Washington, D. C., 2004.
30. *Solar Electricity: The Power of Choice*, Second Quarter, 2001; www.nrel.gov/ncpv.
31. K. B. Oldham, J. C. Myland, *Fundamentals of Electrochemical Science*, Academic Press, San Diego, 1994, p. 129.
32. M. D. Archer, J. R. Bolton, *J. Phys. Chem.* **94** 8028 (1990).
33. J. R. Bolton, *Solar Energy* **57** 37 (1996).
34. A. J. Bard, M. A. Fox, *Acc. Chem. Res.* **28** 141 (1995).
35. J. S. Kilby, J. W. Lathrop, W. A. Porter, U. S. Patents 4 021 323 (1977); 4 100 051 (1978); 4 136 436 (1979).
36. R. E. Blankenship, *Molecular Mechanisms of Photosynthesis*, Blackwell Science, Oxford, U. K., 2002.
37. R. D. Britt, ed., *Oxygenic Photosynthesis: The Light Reactions*, Kluwer Academic Publishers, Dordrecht, The Netherlands, 1996.
38. J. H. Alstrum-Acevedo, M. K. Brennaman and T. J. Meyer, *Inorg. Chem.* **44**, 6802 (2005).
39. V. Ramani, H. R. Kunz and J. M. Fenton, The Electrochemical Society *Interface*, Fall 2004, p. 17.
40. J. M. Ogden, *Physics Today*, **April** 69 (2002).
41. J. M. Ogden, *Sci. Amer.* **295** 94 (2006).
42. D. Berger, *Science* **286** 49 (1999).
43. R. Bashyam and P. Zelenay, *Nature* **443** 63 (2006).
44. C. He, S. Desai, G. Brown, and S. Bollepalli, The Electrochemical Society *Interface* **Fall** 41 (2005).
45. M. F. Mathias, R. Makharia, H. A. Gasteiger, J. J. Conley, T. J. Fuller, C. J. Gittleman, S. S. Kocha, D. P. Miller, C. K. Mittelsteadt, T. Xie, S. G. Yan, and P. T. Yu, The Electrochemical Society *Interface* **Fall** 24 (2005).
46. W. Vielstich, A. Lamm, and H. A. Gasteiger, Ed., *Handbook of Fuel Cells – Fundamentals, Technology, and Applications*, John Wiley & Sons, Chicester, U. K., 2003.
47. H. A. Liebhafsky and E. J. Cairns, *Fuel Cells and Fuel Batteries: A Guide to Their Research and Development*, John Wiley & Sons, New York, 1969.

2

The Solar Resource

Daryl R. Myers

NREL, Golden, CO

1 Introduction: Basic Properties of the Sun

The sun is class G2-V yellow dwarf star of radius 6.95508 x 10^7 km and surface area of 6.087 x 10^{22} cm². It emits radiation produced by the internal conversion of matter into radiation into the entire 4-pi steradian solid angle (sphere), with the sun at the center. The mean radiation intensity, or radiance of the solar surface is 2.009x10^7 watts per square meter per steradian (Wm^{-2} sr^{-1}), or a total of 2.845x10^{26} watts. The Earth's orbit is elliptical with an eccentricity of 0.0167 (1.4710x10^9 km at perihelion, 1.5210x10^9 km at aphelion). At the mean Earth-Sun distance, the sun subtends a solid angle of 9.24 milliradians or 0.529°.[1] Thus the sun is not truly a point source, and the rays from the sun are not truly parallel, but diverge into a cone with half angle of about 0.529°. At the mean distance of the Earth from the sun of 1.495979 x 10^9 km, the solar radiation reaching the top of the Earth's atmosphere is 1366.1 Wm^{-2} ± 7.0 Wm^{-2} or 1.959 calories cm^{-2} minute^{-1}.[2]

The Earth's elliptical orbit causes the distance between the Earth and the Sun (the Earth's radius vector) to vary by 3.39% from perihelion (closest) to aphelion (farthest). These variations in distance cause the intensity of solar radiation at the top of the atmosphere to vary as $1/R^2$, where R is the radius vector. Thus the solar input at the top of the atmosphere varies from 1414 Wm^{-2} (in December) to 1321 Wm^{-2} (in July). Additional variations in solar intensity, or brightness, result from the solar sunspot cycle, and even solar oscillations. These slight variations in the solar output are usually accounted for in the calculation of solar energy available at the top of the atmosphere, or the total extraterrestrial solar radiation, referred to as ETR. The ETR has only been monitored from space since the early 1970's, or almost three solar sunspot cycles. Excellent histories of ETR measurements and analysis are provided in Frohlich[3] and Gueymard.[4]

2 The Spectral Distribution of the Sun as a Radiation Source

In this chapter we briefly describe the solar spectral distribution, or distribution of energy with respect to wavelength, over the region of the electromagnetic spectrum of use to renewable energy systems. The sun radiates energy at wavelengths ranging from the X-ray and gamma ray spectral region out into the very long wavelength radio spectral region. We will restrict our discussion, for the most part, to solar energy in the wavelength region between the ultraviolet (UV) of wavelength 250 nanometers (nm) and the near infrared (NIR) with wavelength of 4000 nm or 4.0 micrometers.

The Planck theory of blackbody radiation provides a first approximation to the spectral distribution, or intensity as a function of wavelength, for the sun. The blackbody theory is based upon a "perfect" radiator with a uniform composition, and states that the spectral distribution of energy is a strong function of wavelength and is proportional to the temperature (in units of absolute temperature, or Kelvin), and several fundamental constants. Spectral radiant exitance (radiant flux per unit area) is defined as:

$$M(\lambda) = \frac{2\pi c^2 h}{\lambda^5 (e^{hc/\lambda kT} - 1)} \qquad (1)$$

where λ is wavelength (in meters), h is Planck's constant = 6.626196×10^{-34} Joule seconds (J s), c is the velocity of light in vacuum = 2.9979250×10^8 meter per second (ms^{-1}), k is Boltzman's constant = 1.3806×10^{-23} Joule per Kelvin (J K^{-1}), and T is absolute temperature in Kelvins.

The sun is not a "perfect" radiator, nor does it have uniform composition. The sun is composed of about 92% hydrogen, 7.8% helium. The remaining 0.2% of the sun is made up of about 60 other elements, mainly metals such as iron, magnesium, and chromium. Carbon, silicon, and most other elements are present as well.[1] The interaction of the atoms and ions of these elements with the radiation created by the annihilation of matter deep within the sun modifies and adds structure to the solar spectral distribution of energy. Astrophysicists such as Kurucz have used quantum calculations and the relative abundance of elements in the sun to compute the theoretical spectral distribution from first principles.[5] Figure 1 shows a plot of the Kurucz computed spectral distribution at very high resolution (0.005 nanometer at UV) as well as an inset showing much lower resolution (0.5 nanometer in UV to 5 nm in IR) plot.

Figure 2 is a plot of the low resolution ETR spectrum compared with the Planck function for a blackbody with a temperature of 6000 Kelvin. The differences in the infrared, beyond 1000 nanometers are small. The larger differences in the shortwavelength region are due to the absorption of radiation by the constituents of the solar composition, resulting in the "lines" observed by Fraunhofer and named after him.

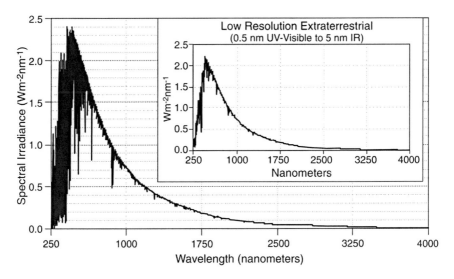

Fig. 1. The theoretical extraterrestrial solar spectral distribution (at the top of the Earth's atmosphere at the mean Earth-Sun distance of one astronomical unit) of Kurucz at high and low spectral resolution.

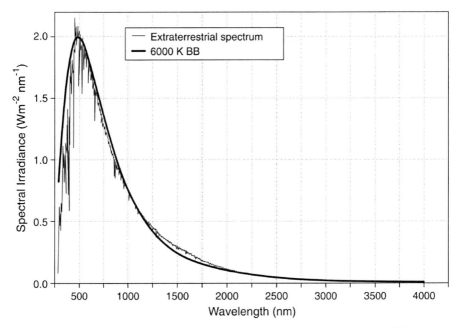

Fig. 2. The low-resolution ETR spectral distribution (gray jagged curve) and the 6000-Kelvin blackbody spectral distribution.

3 The Earth's Atmosphere as a Filter

Above, we described the solar resource at the top of the atmosphere. The atmosphere acts as a continuously variable filter for the ETR radiation. The atmosphere has stable components of 78% nitrogen, 21% oxygen, and 1% argon and other "noble" gases. There are also variable components in the atmosphere, such as water vapor (0% to 2% of the total composition), and gases dumped into the atmosphere by manmade and natural process, such as carbon dioxide, (0.035%), methane, and nitrous oxides.[6]

At a concentration of only 0.3 parts per million, Ozone absorbs and attenuates the dangerous UV radiation below 280 nm. Water vapor absorbs mainly in the infrared, contributing to the heating of the atmosphere. Similarly, small concentrations of "greenhouse" gases such as carbon dioxide and methane absorb in the infrared, but with such strength that their increasing concentration may pose a threat to the stability of the Earth's climate. Suspended particulates such as aerosols, dust and smoke, as well as condensed water vapor (clouds) also strongly modify the solar resource as the atmosphere is traversed by photons from the top of the atmosphere to the surface.

As the photons propagate through the atmosphere, radiation in the narrow cone of light from the solar disk (the *direct beam*) interacts with the atmosphere by being absorbed or scattered out of the beam. Molecules of atmospheric gases, aerosols, dust particles, and so on, do the absorption and scattering. Scattered radiation contributes to the bright blue of the clear sky dome, or the dull gray of overcast skies. The scattered radiation also illuminates clouds, which reflect most of the wavelengths of light in the visible region, making them appear white. Figure 3 schematically shows this process of atmospheric sorting of the radiation into three components: the direct beam, the scattered radiation (called diffuse radiation) and the combination of the direct and diffuse radiation, called the total hemispherical, or *global* radiation from the entire sky dome.

As a result of the absorption and scattering processes in the atmosphere, the ETR spectral distribution is significantly modified. Figure 4 shows the effect of the atmosphere on the ETR spectral distribution for a very specific solar geometry, referred to as *Air Mass 1.5*. As indicated in Fig. 3, Air Mass 1.0 occurs when the sun is directly overhead. The angle between the horizon and the observer (solar elevation) is then 90°, and the angle between the zenith (overhead point) and the observer is 0°. The relative position of the sun with respect to the horizon or zenith is specified as the elevation angle, ε, or zenith angle, z, respectively. The term Air Mass refers to the relative path length through the atmosphere, with respect the minimum path length of 1 for zenith angle 0°. Geometrically, Air Mass M is:

$$M = 1/\cos(z) \quad \text{or} \quad M = 1/\sin(\varepsilon) \tag{2}$$

Figures 5 and 6 show how the direct beam and global horizontal spectral distributions are modified on a clear day as a function of increasing Air Mass. Note how, as the Air Mass increases, the direct beam spectra change more than the global sky

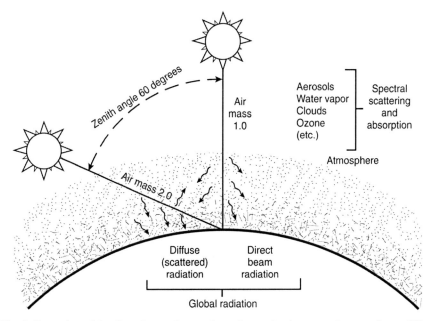

Fig. 3. Scattering of the direct beam photons from the sun by the atmosphere produces diffuse and global sky irradiance.

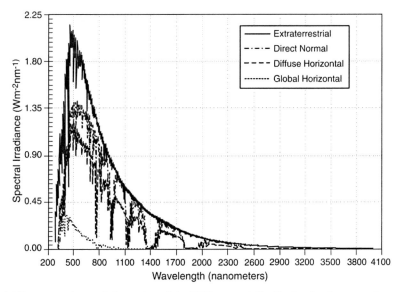

Fig. 4. The attenuation and absorption of the ETR spectral distribution by the atmosphere. Top curve is the ETR spectral distribution. In decreasing order, the global, direct, and diffuse spectral distributions at the bottom of the atmosphere (at sea level) for the sun at zenith angle 48.2° (Air Mass 1.5) are shown.

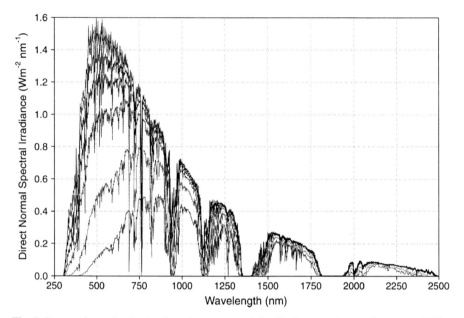

Fig. 5. Progressive reduction in direct beam spectral distributions as air mass in increased. The plots representing uniformly increasing 10° steps in zenith angle from 0° (top curve) to 80° (bottom curve).

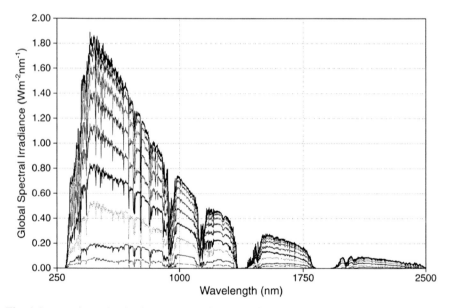

Fig. 6. Progressive reduction in global total hemispherical spectral distributions as Air Mass is increased, as in Fig. 5.

spectra. This is because the energy scattered out of the direct beam is "transferred" into the global spectra, as increasing contributions to the diffuse sky radiation.

Figure 5 in particular illustrates the shift of the spectral peak to longer (red) wavelengths associated with red skies at sunset and sunrise. Figure 7 schematically shows the relationship of absolute temperatures, and color as perceived by the human eye for various outdoor natural conditions, and artificial indoor sources. Our eyes adapted to take advantage of the energy peak in solar spectral distribution between 400 nm and 700 nm, and evolved under the spectral distribution of our sun. Therefore, the match (or mismatch) of artificial source spectral distributions with the solar spectral distribution is important for both natural and artificial lighting applications.

As mentioned above, other elements besides the gases in the atmosphere interact with the ETR spectral distribution. One of the most important of these other atmospheric constituents are small particles called aerosols. Particles scatter radiation most efficiently when the wavelength of the radiation is smaller than the particle size. Many of the particles that work their way into our atmosphere (dust, decaying organic material, smoke from fires, etc.) have diameters that scatter shortwave (UV) and visible light very efficiently. This removes a great deal of energy from the direct beam, and redistributes the energy over the sky dome.[7]

A measure of the scattering power of aerosols is the amount of energy removed from (or the attenuation of) the beam radiation. The Beer-Bouger-Lambert law for the attenuation of a beam of intensity I_o to intensity I, resulting from the amount, x, of a material is given by: $I/I_o = e^{-\tau x}$ or $I = I_o\, e^{-\tau x}$ where τ is the *attenuation coefficient*, called the aerosol optical depth , or AOD. The attenuation by aerosols is both exponential, and a strong function of wavelength, implying a large impact on the direct beam solar spectral distribution. Anders Angstrom first proposed a relation between the wavelength and τ, dependent on two parameters and a reference point at a wavelength of 1000 nm as $\tau = \beta\,\lambda^{-\alpha}$. β is related to the size of the particles, and ranges from about 1 to 2. α is related to the scattering properties of the particles, and ranges from about 0.001 to 0.5.

Most often, AOD is referred to a specific wavelength, usually 500 nm, or occasionally 550 nm. An AOD of 0.01 at 500 nm represents very clean, pristine, clean atmosphere. Values of AOD at 500 nm of 0.1 to 0.2 are quite typical of average conditions. Values of 0.4 or greater represent a heavy aerosol load and very hazy skies. Figure 8 shows the clear sky direct beam spectral distribution at Air Mass 1.5 for a range of AOD from 0.05 to 0.40

From the discussion above and Figs. 5 to 7, the fluctuations in the solar spectrum at the Earth's surface are dependent on many factors. These variations can be characterized with measurements and models to account for their impact on solar renewable energy technologies.

4 Utilization of Solar Spectral Regions: Spectral Response of Materials

Why this extended discussion of the solar spectral distribution? The primary reason is provided by the example discussing the sensitivity of our eyes, in the previous

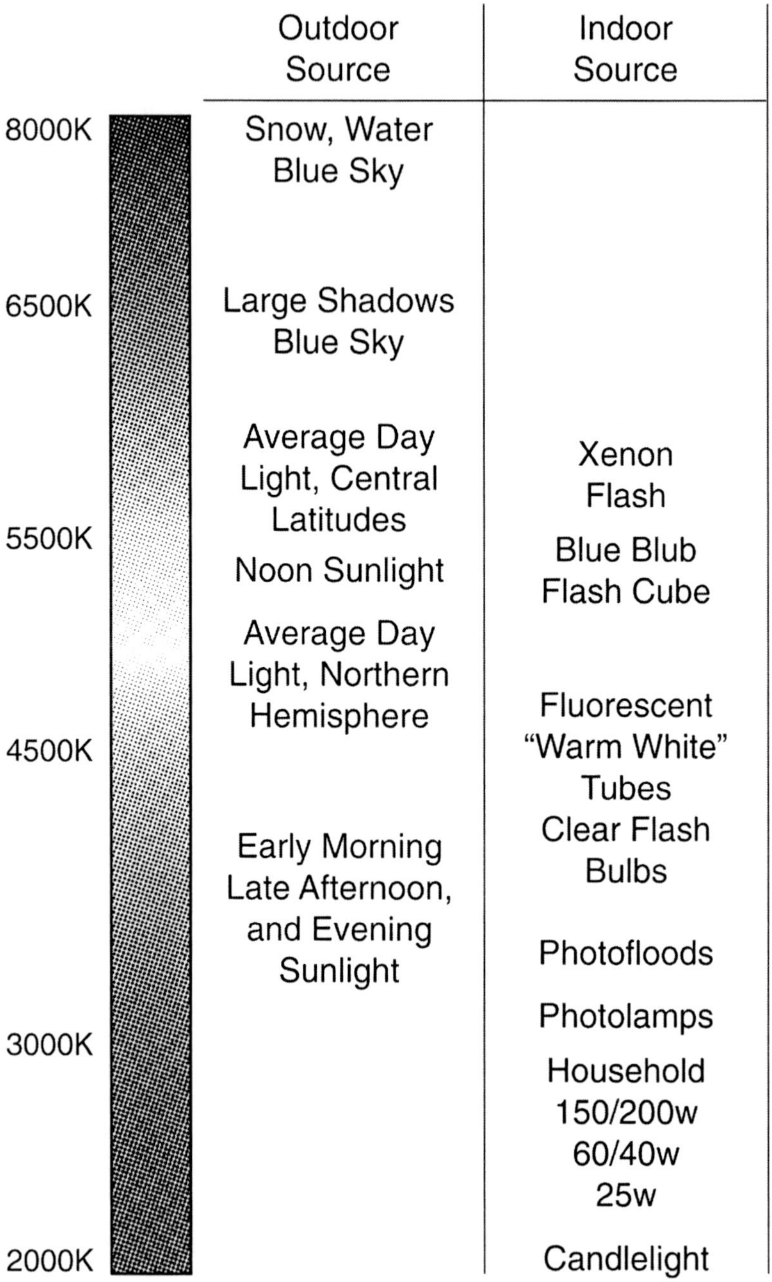

	Outdoor Source	Indoor Source
8000K	Snow, Water Blue Sky	
6500K	Large Shadows Blue Sky	
5500K	Average Day Light, Central Latitudes Noon Sunlight	Xenon Flash Blue Blub Flash Cube
4500K	Average Day Light, Northern Hemisphere	Fluorescent "Warm White" Tubes Clear Flash Bulbs
	Early Morning Late Afternoon, and Evening Sunlight	Photofloods
3000K		Photolamps Household 150/200w 60/40w 25w
2000K		Candlelight

Fig. 7. Temperature and color relationships for various natural and artificial sources of optical radiation.

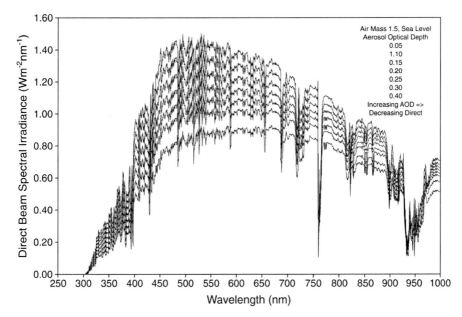

Fig. 8. Decreasing direct beam spectral irradiance as aerosol optical depth at 500 nm increases from 0.05 (top curve) to 0.04 (bottom curve).

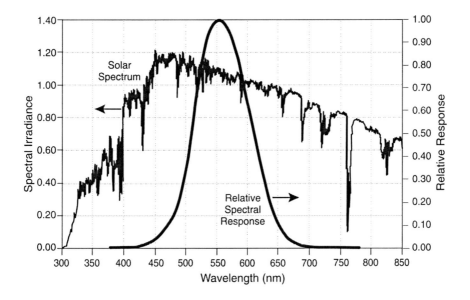

Fig. 9. Human eye daytime (photopic) relative spectral response (right axis) and solar spectrum (jagged curve).

Section. Figure 9 shows the wavelength region where the human eye responds, the spectral response of the eye, overlaying an Air Mass 1.5 global solar spectrum. The peak of this response is at 555 nm, corresponding to the color green.

Many of the materials used in renewable, and specifically, solar energy system applications have a significant response to the solar spectrum over a limited spectral interval.

The green color of most plant leaves is the result of absorption of blue and red light by chlorophyll, which reflects almost all of the light in the region around 550 nm, as shown in Fig. 10.

Our skin contains compounds that absorb ultraviolet light with wavelengths shorter than 400 nm. These compounds react with the UV photons to produce suntans, sunburns, or even cancer.

Similarly, many semiconductors, such as silicon, germanium, etc. produce a flow of electrons (the photovoltaic effect) when photons of a certain wavelength interact with the materials. Figure 11 shows Air Mass 1.5 direct normal, diffuse sky, and total global solar spectral distributions, with indications of the spectral regions where our vision, plants, and various photovoltaic materials interact with the solar spectral distribution.

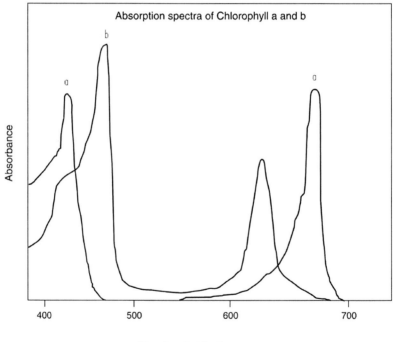

Fig. 10. Blue and red light absorbed by chlorophyll (types a and b shown) used in photosynthesis. Green (around 550 nm) is reflected and not used.

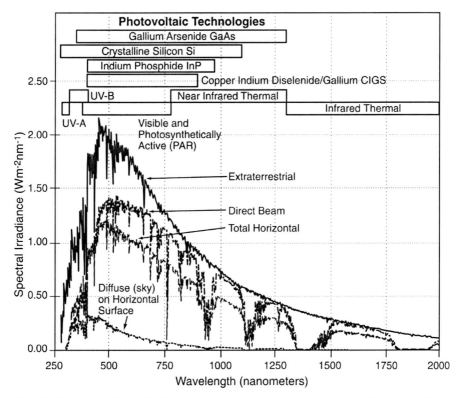

Fig. 11. Solar spectral distributions and the various regions of spectral sensitivity for vision, plant photosynthesis, and photovoltaic conversion technologies.

Figure 12 shows that various combinations of photovoltaic materials can be constructed to respond over different spectral ranges, utilizing more or less of the solar spectral distribution.[8]

The different response regions shown in Fig. 12 are the result of the *band gap* between bound electrons in the material and the conduction band for electrons (and holes) in terms of energy (in electron volts, eV). The energy, E, of a photon is related to the wavelength as $E = (h\,c)\,/\,\lambda$. In semiconductors suitable for Photovoltaic applications, the band gaps are relatively close together, so photons with relatively low energy (0.5 to 1.5 eV) can stimulate electrons to enter the conduction band. We can convert the power versus wavelength plot to one of number of photons versus energy, as on the left of Fig. 13. On the right of Fig. 13 we show the relationship between the available solar energy and projected conversion efficiencies of some PV materials can approach 50%, if concentrated (focused by lens or mirrors) solar radiation is used in conjunction with future generation materials.

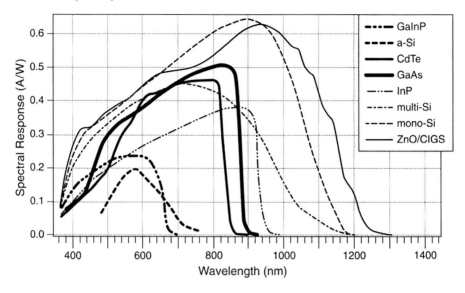

Fig. 12. Spectral response regions for various photovoltaic technologies. CIG stands for cadmium indium gallium selenide.

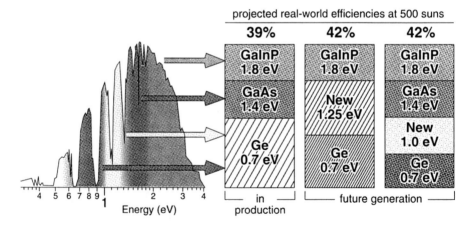

Fig. 13. Correlation of available solar spectral photon energies with band gap of present and future generation photovoltaic materials. *New* represents some new combination of materials optimized or tailored for a specific band gap.

As Fig. 13 shows, future generation materials with "designer" band gaps can produce higher efficiency devices to generate more electricity with the same solar spectrum. By stacking the available materials, additional components of the solar spectrum contribute to the overall production of conduction electrons, or electric current.

Optimization of the performance of solar energy systems, as well as building thermal performance (heating and cooling loads), and daylighting (window performance) all require knowledge of the terrestrial solar spectral distribution. Optical properties of materials such as transmittance, reflectance, and absorption are always dependent on the wavelength of the incident radiation. For example, Fig. 14 shows the properties of several components of a window for building applications. The top left panel shows the properties of a single pane of glass which permits thermal infrared radiation into a room; high *inside* reflectance keeps the thermal energy trapped, offsetting the need for more heating energy in a cold climate.

In the lower panels of the figure, properties of each pane of a double pane structure are shown. Low IR transmittance (lower left) of the outer pane keeps thermal infrared solar radiation from entering the building. The broad transmittance band of the inner layer and low *inside* reflectance of the inner pane allows thermal infrared energy to escape. This structure reduces the cooling load in a sunny environment.

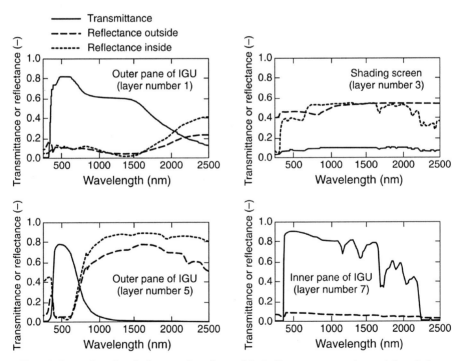

Fig. 14. Examples of optical properties of materials (reflectance, transmittance) for window structures. When used in conjunction with solar spectral distributions, energy savings can be computed.

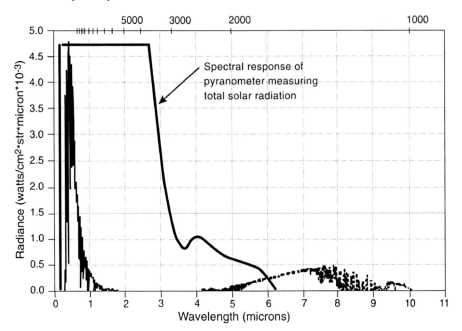

Fig. 15. Spectral response of thermopile pyranometer measuring *total* solar radiation is shown with thick black line. Spectral radiance (brightness) of the sky dome (blue line). The cut-off at 3000 nm means the radiometer will not respond to the infrared sky radiation that peaks at 7000 nm (7 micrometers).

Similar principles can be used for designing selective absorbers, where the goal is absorb as much of the solar spectrum as possible and convert that absorbed energy into heat, or thermal energy. Conversely, reflective materials can be designed to select only visible (*cold mirror*) or infrared (*hot mirrors*) to isolate and direct selected portions of the solar spectrum for various applications.

Knowledge of the optical properties of materials in relation to the solar spectrum is also important in measuring broadband solar radiation. For instance, a pyranometer used to monitor *total* solar radiation for a renewable energy system has a spectral response (due to the special glass dome protecting the detector) that does not respond to the thermal infrared radiation of the sky beyond 3000 nm, as shown in Fig. 15. However, there will be thermal infrared radiation exchanged between the radiometer and the sky dome, which will influence the measurement performance of the pyranometer.[9]

5 Reference Spectral Distributions

From the discussion in the preceding Sections, even without addressing the influence of clouds, it is clear that the terrestrial solar spectrum is highly variable. So, how can we relate the spectral responses and spectral optical properties of materials to each other when this variability is present? The answer is to establish a *standard* spectral distribution with which to compute performance. Then comparisons can be made based on standard set of conditions. Furthermore, if measurements are made under conditions, deviations from the standard conditions can be computed, documented, and in most cases corrected or scaled to the reference conditions.

Several national and international consensus standards organizations, such as the American Society for Testing and Materials (ASTM) and the International Standards Organization (ISO) have adopted a reference standard extraterrestrial spectral distribution (ASTM E490-00a), and terrestrial reference spectral distributions for direct beam and total hemispherical (on a 37° tilted south facing surface) spectra at a prescribed air mass of 1.5 (ASTM G173-03).[2,10,11]

The extraterrestrial reference spectral distribution was assembled from a number of recent diverse space-based (satellite) measurement sources. The ETR reference spectrum is normalized to a total integrated irradiance of 1366.1 Wm^{-2}.

The terrestrial reference spectra required a more extensive set of criteria to be met. Specifically, a set of reasonable conditions that could occur rather commonly in nature should be used. The conditions for the terrestrial reference spectra were chosen to meet the following criteria:[12]

- Air mass 1.5 represents the condition where approximately 1/2 of the total available solar energy is available for air masses greater than and less than this condition, respectively.
- For solar thermal and photovoltaic systems using flat plate collectors, energy collection is optimized for collectors tilted south (in the northern hemisphere) at approximately the latitude of the site.
- The mean latitude of the contiguous 48 United States is approximately 37° North.
- Standard test conditions prescribed in photovoltaic standards require a total hemispherical irradiance on flat plate collectors of 1000 Wm^{-2} (a value that can obtained on a clear day around noon).
- Concentrating solar collector systems that utilize the direct beam radiation would be more likely deployed in areas with relatively low aerosol optical depth, to maximize direct beam utilization.[13]
- The terrestrial reference should be easily reproducible, preferably by a simple (but accurate) model calculation, and the model should be easily maintained and updated, and in the public domain.[14]
- The terrestrial reference spectrum should have uniform wavelength increments for ease of computation and comparison with measured spectral data.

These criteria resulted in the choice of a relatively simple, but accurate spectral model of Gueymard (SMARTS: Simple Model for Atmospheric Transmission of Sunshine) to compute the reference standard terrestrial spectra.[7]

The philosophy behind the SMARTS model is to parameterize the band model transmittance functions used by a very complex (50,000 line of FORTRAN code and about 200 subroutines) MODTRAN (MODerate resolution TRANSmittance) code[15] developed by the Air Force Geophysics Laboratory, for the most important at-

Table 1. Transmission expressions developed for SMARTS model.

Absorption Mechanism	Transmittance Expression
Rayleigh Scattering	$Tr(\lambda) = \exp\{(P/P_o)/[a_0(\lambda/\lambda_o)^4 + a_1(\lambda/\lambda_o) + a_2 + a_3(\lambda/\lambda_o)^{-2}]\}$
Ozone	$To(\lambda) = \exp[-m_o\, u_o\, Ao(\lambda)]$
Nitrogen Dioxide (NO$_2$)	$Tn(\lambda) = \exp[-m_n\, u_n\, An(\lambda)]$
Mixed and Trace Gases	$Tg(\lambda) = \exp[-(m_g\, u_g\, Ag(\lambda))]$
Water Vapor	$Tw(\lambda) = \exp[-(m_w\, u_w)B_w(\lambda)B_m(\lambda)B_p(\lambda)B_{aw}(\lambda)\, Aw(\lambda)]$
Aerosol	$Ta(\lambda) = \exp[-m_a\, \beta_i\, (\lambda/\lambda_1)^{-\alpha_i}]$

mospheric constituents, at a resolution of 0.5 nm in the ultraviolet less than 400 nm, 1 nm between 400 nm and 1700 nm, and 5 nm between 1700 nm and 4000 nm. These parameterized transmittance functions were developed to account for Rayleigh scattering (Tr), ozone (To), mixed gas (Tg), nitrogen dioxide (Tn), water vapor (Tw), and aerosol (Ta) transmission of the direct beam irradiance using Eq. 3:

$$E(\lambda) = E_o(\lambda)\, Tr(\lambda)\, To(\lambda)\, Tg(\lambda)\, Tn(\lambda)\, Tw(\lambda)\, Ta(\lambda) \tag{3}$$

at each wavelength (λ, in nm), where E is the terrestrial spectral irradiance, E_o is the extraterrestrial spectral irradiance, and the spectral transmittances are defined above. Table 1 summarizes the form of transmittance functions developed for the SMARTS model.

Table 1 expression parameters are; P = station pressure, P_o = standard pressure, a_i = fitting coefficients, m = air mass correction for path length, u = absorber abundances, A = absorption coefficients, a = absorber/wavelength dependent, Bs = water vapor band, airmass, and pressure scaling factors, $\alpha_i\, \beta_i$, = Ångstrom parameters, $i = 1$ for $\lambda < 500$ nm, $i = 2$ for $\lambda \geq 500$ nm,[*] λ_1: Reference wavelength (usually 1000 nm or 1 μm)

Figures 16 and 17 show percent difference between SMARTS MODTRAN results and one of many comparisons of measured and SMARTS spectral data. Agreement within the uncertainty limits of spectral irradiance measurements (1% in the visible, and 3% to 5% in the ultraviolet and infrared), is achieved.

Version 2.9.2 of the SMARTS spectral model used to generate the spectral reference standard, the users manual for the model, and a list of references, can be downloaded free of charge from the National Renewable Energy Laboratory Renewable Resource Data Center at the following URL: http://rredc.nrel.gov/solar/models/SMARTS/. A CD-ROM adjunct to the ASTM G-173-03 standard, with a copy of the model, manual, and reference material is available for purchase from ASTM.

The SMARTS input file is a straightforward assembly of fifteen to twenty parameters arranged as in a stack of input *cards*. Table 2 is an annotated input file used to generate the ASTM G173 standard spectra on a 37° tilted south facing plane. Note that we do not show all possible input combinations.

Figures 18 and 19 portray the ETR and terrestrial standard reference spectra.

---

[*] The common assumption the $\alpha_1 = \alpha_2 = \alpha$ can lead to errors for the urban, maritime, and rural aerosol profiles. The Angstrom exponents are determined as a function of aerosol type and relative humidity (cf. Appendix B of Gueymard.)[7]

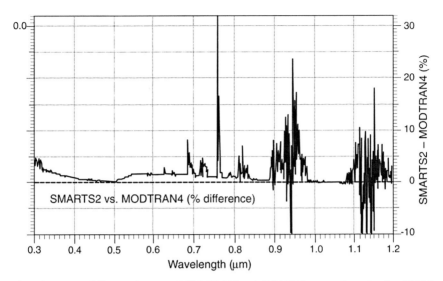

Fig. 16. Percent difference between MODTRAN and SMARTS spectral results, for ASTM reference spectra conditions. Largest differences due to SMARTS trace gases.

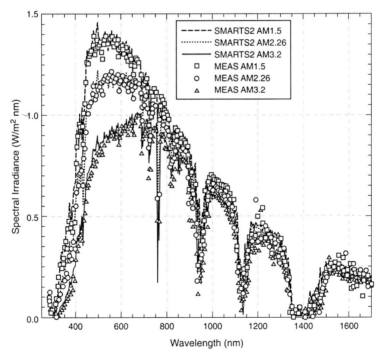

Fig. 17. SMARTS model results (lines) and measurements (symbols) at 5 nm resolution for 3 air masses at NREL, Sep 18, 2001.

Table 2. SMARTS version 2.9 input file for ASTM reference spectra G173-03.

Card ID	Value	Parameter/Description/Variable name
1	ASTM_G173_Std_Spectra	Comment line
2	1	Pressure input mode (1 = pressure and altitude): ISPR
2a	1013.25 0.	Station pressure (mb) and altitude (km): SPR, ALT
3	1	Standard Atmosphere Profile Selection (1 = use default atmosphere): IATM1
3a	USSA	Default Standard Atmosphere Profile: ATM (one of eleven choices, including user defined)
4	1	Water vapor input (1 = default from Atmospheric Profile): IH2O (may be user specified)
5	1	Ozone calculation (1 = default from Atmospheric Profile): IO3 (may be user specified)
6	1	Pollution level mode (1 = standard conditions/no pollution): IGAS (for 10 pollutant gases)
7	370	Carbon monoxide volume mixing ratio (ppm): qCO2
7a	1	Extraterrestrial spectrum (1 = SMARTS/Gueymard): ISPCTR (one of seven choices)
8	S&F_RURAL	Aerosol profile to use: AEROS (one of 10 choices, including user specified)
9	0	Specification for aerosol optical depth/turbidity input (0 = AOD at 500 nm): ITURB
9a	0.084	Aerosol optical depth @ 500 nm: TAU5
10	38	Far field spectral Albedo file to use (38 = Light Sandy Soil): IALBDX (on of 40 choices, including user defined)
10b	1	Specify tilt calculation (1 = yes): ITILT
10c	38 37 180	Albedo and Tilt variables—Albedo file to use for near field, Tilt, and Azimuth: IALBDG, TILT, WAZIM
11	280 4000 1.0 1367.0	Wavelength range—start, stop, mean radius vector correction, integrated solar spectrum irradiance: WLMN, WLMX, SUNCOR, SOLARC
12	2	Separate spectral output file print mode (2 = yes): IPRT: Spectral & broadband files
12a	280 4000 .5	Output file wavelength—Print limits, start, stop, minimum step size: WPMN, WPMX, INTVL
12b	2	Number of output variables to print: IOTOT (up to 32)
12c	8 9	Code relating output variables to print (8 = Hemispherical tilt, 9 = direct normal + circumsolar): OUT(8), OUT(9) [up to 32 spectral parameters available for output]
13	1	Circumsolar calculation mode (1 = yes): ICIRC
13a	0 2.9 0	Receiver geometry—Slope, View, Limit half angles: SLOPE, APERT, LIMIT
14	0	Smooth function mode (0 = none): ISCAN (Gaussian and triangle filter shapes can be specified)
15	0	Illuminance calculation mode (0 = none): ILLUM (Luminance and efficacy may be selected)
16	0	UV calculation mode (0 = none): IUV (UVA, UVB, action weighed dosages available)
17	2	Solar geometry mode (2 = Air Mass): IMASS (zenith and azimuth, date/time/lat/long available)
17a	1.5	Air mass value: AMASS

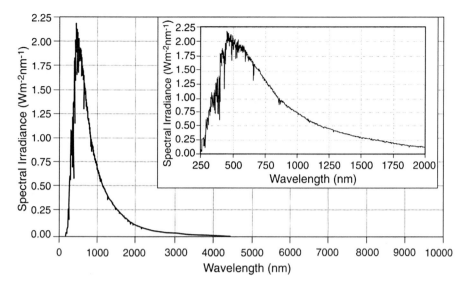

Fig. 18. ASTM E490-00a extraterrestrial reference spectrum. The actual spectral data go out to 100,000 nm (100 microns). Inset shows details in the 250 nm to 2000 nm region.

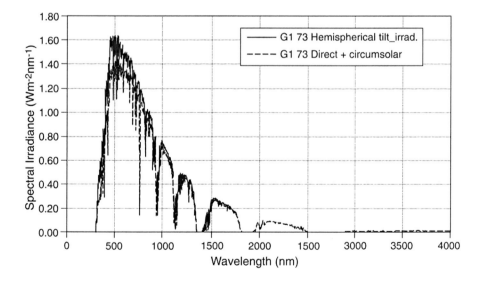

Fig. 19. ASTM G173-03 Terrestrial reference spectra for Air Mmass 1.5, conditions specified in Table 2.

6 Summary

The basic properties of the sun and the solar radiation received at the top of the Earth's atmosphere have been described in this chapter. The extraterrestrial solar spectral distribution is modified through interactions with the gases and particles in the atmosphere to produce terrestrial spectral distributions that vary with respect to both amplitude and wavelength over a very wide range. The solar resource to a specific solar renewable energy technology depends upon the spectral response of the systems and materials involved. This is true whether the technology addresses biomass (photosynthesis), daylighting, building heat loads, thermal energy conversion, or photovoltaic production of electricity. Each application utilizes one or mores specific regions of the terrestrial solar spectrum, either in isolation or in various combinations. The solar renewable energy community has developed a set of tools and standards to assist in the design and optimization of solar renewable systems, including hybrid systems that may combine solar and other renewable technologies, such as wind energy generation. The present set of extraterrestrial and terrestrial solar spectral standards have evolved over the past 30 years to keep abreast of the requirements that new, innovative renewable energy systems researchers, designers, and manufacturers require to meet their customers' needs.

References

1. Cox, A. N., ed. *Allen's Astrophysical Quantities*. 4th ed., AIP Press, Springer Verlag, New York, NY., 1999.
2. ASTM, Standard Solar Constant and Zero Air Mass Solar Spectral Irradiance Tables, Standard E490-00a, American Society for Testing and Materials, West Conshohocken, PA, 2000.
3. C. Frohlich and J. Lean, *Total Solar Irradiance Variations: The Construction of a Composite and it's Comparison with Models*, International Astronomical Union Symposium 185: New Eyes to See Inside the Sun and Stars, Dortrect, The Netherlands, Kluwer Academic, 1998.
4. C. A. Gueymard, The sun's total and spectral irradiance for solar energy applications and solar radiation models, *Solar Energy*,. **76**(4) 423 (2004).
5. R. L. Kurucz, *Synthetic Template Spectra. Highlights of Astronomy*, L. Appenzeller, ed., Vol. 10, The Hague, Netherlands, Aug 15-17, 1994, Kluwer Acad. (1995) pp. 407–409.
6. W. M. Farmer, *The Atmospheric Filter*, Vol. I., JCD Publishing, Winter Park, FL, 2001, p. 273.
7. C. Gueymard, Parameterized transmittance model for direct beam and circumsolar spectral irradiance, *Solar Energy* **71**(5) 325 (2001).
8. H. Field, *Solar Cell Spectral Response Measurements Related to Spectral Bandwidth and Chopped Light Waveform*, 26th IEEE Photovoltaic Specialists Conference, Institute of Electrical and Electronic Engineers, Anaheim, CA, 1997.
9. I. Reda, J. Hickey, C. Long, D. Myers, T. Stoffel, S. Wilcox, J. J. Michalsky, E. G. Dutton, and D. Nelson, Using a blackbody to calculate net-longwave responsivity of shortwave solar pyranometers to correct for their thermal offset error during outdoor calibration using the component sum method, *Journal of Atmospheric and Oceanic Technology* **22** 1531 (2005).

10. ASTM, Standard Tables for Reference Solar Spectral Irradiance at Air Mass 1.5: Direct Normal and Hemispherical for a 37° Tilted Surface, Standard G177-03. 2003 American Society for Testing and Materials, West Conshohocken, PA.
11. ISO, Solar energy—Reference solar spectral irradiance at the ground at different receiving conditions, pt. 1. International Standard 9845-1, International Organization for Standardization, 1992.
12. C. Gueymard, D. Myers, and K. Emery, Proposed reference irradiance spectra for solar energy systems testing, *Solar Energy,.* **73**(6) 443 (2002).
13. S. Kurtz, D. Myers, T. Townsend, C. Whitaker, A. Maish, R. Hulstrom, and K. Emery, Outdoor rating conditions for photovoltaic modules and systems. *Solar Energy Materials Solar Cells* **62** 379 (2000).
14. Myers, D., K. Emery, C. Gueymard, Revising and Validating Spectral Irradiance Reference Standards for Photovoltaic Performance Evaluation. ASME Journal of Solar Energy Engineering, 2004. **126**: p. 567-574.
15. G. P. Anderson, A. Berk, P. K. Acharya, M. W. Matthew, L. S. Bernstein, J. H. Chetwynd, Jr., H. Dothe, S. M. Adler-Golden, A. J. Ratkowski, G. W. Felde, J. A. Gardner, M. L. Hoke, S. C. Richtsmeier, B. Pukall, J. B. Mello, and L. S. Jeong, *MODTRAN4: Radiative Transfer Modeling for Remote Sensing*, in Optics in Atmospheric Propagation and Adaptive Systems III, Society of Photo-Optical Instrumentation Engineers Bellingham, WA., 1999.

3

Electrolysis of Water

Kevin Harrison and Johanna Ivy Levene

NREL, Golden, CO

1 Introduction

Hydrogen energy systems, based on renewable energy (RE) sources, are being proposed as a means to increase energy independence, improve domestic economies, and reduce greenhouse gas emissions from stationary and mobile fossil-fueled sources. In 2003, the United States consumed roughly 84.3 billion m^3 (7.6 billion kilograms) of hydrogen, the majority of which was produced via the widely established thermal process known as steam methane reforming (SMR).[1]. The electrolytic production of hydrogen, while not economically competitive today with SMR, is positioned to become the preferred method due to the inevitable price increase of natural gas and as environmental, social, and economic factors are weighed.

SMR constitutes roughly 50% of the 450–500 billion m^3 yr^{-1} (38–42 billion kg yr^{-1}) of global production of the gas.[1,2]. SMR, like hydrogen production from all fossil fuels, suffers from supply issues and climate-altering carbon-based pollution. The reforming process generates CO_2 as well as carbon monoxide (CO), which is poisonous to humans because the oxygen-transporting hemoglobin has 200 times the affinity to CO than O_2.[3] Electrolysis currently supplies roughly 4% of the world's hydrogen.

If hydrogen is to be used as a transportation fuel, the United State could conceivably replace the 140 billion gallons per year (gal yr^{-1}) of gasoline consumed in 2004 with domestically produced hydrogen. The energy equivalent of this much gasoline is 17.3×10^{15} BTU, assuming approximately 5.2 million BTU bbl^{-1} of motor gasoline.[4] The environmental gains hoped for by the transition to a hydrogen economy can only be achieved when renewable sources are ramped up to produce an increasing amount of the hydrogen gas.

From the early 1800s to the mid 1900s town gas was comprised of roughly 50% hydrogen that brought light and heat to much of America and Europe and can still be found in some parts of Europe, China and Asia. Due to hydrogen's thermal conductivity and low density the gas is being used to cool many large thermal electrical power generators.

Hydrogen is used in a wide variety of applications:[5]

- Chemicals
    - Ammonia and fertilizer manufacture
    - Synthesis of methanol
    - Sorbitol production
    - General pharmaceuticals and vitamins
- Electronics
    - Polysilicon production
    - Epitaxial deposition
    - Fiber optics
- Metals
    - Annealing/heat treating
    - Powder metallurgy
- Fuels
    - Petroleum refinement
    - Liquid rocket fuel
    - Some use in fuel cells
- Food and float glass
    - Fats/fatty acids
    - Blanketing

Renewable sources of electricity and off-peak hydroelectric can be used to produce a sustainable supply of hydrogen for transportation, peak-shaving applications and in some special cases to smooth the variability in the renewable source. Powering millions of hydrogen internal combustion engines and/or fuel cell vehicles with hydrogen generated with traditional fossil fuel sources (without carbon dioxide (CO_2) capture and storage or geological sequestration) is merely transferring the pollution from the tailpipe to the stack pipe. In the case of SMR, liquid natural gas imports would increase to replace today's 12.9 million bbl day^{-1} of oil imports here in the U.S.4.[4] As developing countries fall in love with motorized transportation, much like the developed countries already have, transportation's contribution to greenhouse gas emissions will grow from the 25% it holds today.[6]

Still today, the electrolytic production of hydrogen using renewable sources is the only way to produce large quantities of hydrogen without emitting the traditional by-products associated with fossil-fuels. The electrolysis of water is an electrochemical reaction requiring no moving parts and a direct electric current, making it one of the simplest ways to produce hydrogen. The electrochemical decomposition of water into its two constituent parts has been shown to be reliable, clean and with the removal of water vapor from the product capable of producing ultra-pure hydrogen (> 99.999%).

The primary disadvantage of electrolysis is the requirement of high-quality of electrical energy needed to disassociate the gas. Electricity is a convenient energy carrier as it can be transported to loads relatively easily. However, locating and constructing new transmission and distribution power lines is challenging and expensive. The cost of transporting electricity along power lines can constitute greater than 50% of the total cost at the point of end-use.[7] Historically, hydrogen production via electrolysis has only been viable where large amounts of inexpensive electricity have been available or the high purity product gas was necessary in a downstream process.

The potential environmental benefit of a hydrogen-based economy is hinged to a large degree on the ability to generate the gas from renewable resources in a cost-effective manner. An apparently ideal solution is to use wind-generated electricity to electrolyze water. Today, hydrogen production via electrolysis only meets the U.S. Department of Energy (DOE) goals of $2–$3 per kilogram (kg) in large installations where electrolyzer capital costs are low, less than $800 per kilowatt (kW), and those having access to inexpensive electricity, less than $0.04 per kilowatt-hour (kWh).[8]

Electricity from large-scale wind farms in Class 4 or better resource can be generated in the range of $0.05–$0.08 kWh^{-1}, not including today's $0.019 kWh^{-1} Federal production tax credit.[9,10] The out-of-pocket cost of fossil-fuels, whether for electricity production or as transportation fuels, has remained relatively low; limiting the expansion of renewable forms of energy. For example, if the external costs of production were taken into account the cost of coal-generated electricity would rise an additional $0.03–$0.06 kWh^{-1}.[11] Further limiting market penetration of renewable sources is that fossil fuels continue to receive the bulk of tax incentives here in the U.S.[12]

The term *renewable* defines these technologies as driven by natural and sustainable processes which are inherently variable, not intermittent. Natural processes vary over time but are not subject to the on-off switching that, for example, a light bulb connected to a switch is subjected to. Advocates may want to begin training themselves to describe RE as variable, not intermittent, to better describe their naturally occurring behavior. RE sources of energy can provide cost-effective, emission-free electricity with zero- or low-carbon impact making it one of the preferred methods for supplying energy to society. The large-scale wind energy facilities being installed throughout the world are a testament to the growing demand, environmentally preferred and cost-effectiveness of this RE technology.

2 Electrolysis of Water

"Personally, I think that 400 years hence the power question in England may be solved somewhat as follows. The country will be covered with rows of metallic windmills working electric motors, which in their turn supply current at a very high voltage to great electric mains. At suitable distances there will be great power stations where during windy weather the surplus power will be used for the electrolytic decomposition of water into hydrogen and oxygen…. In times of calm, the gases will be recombined in explosion motors working dynamos which produce electrical energy once more or more probably in oxidation cells."

Haldane, in his talk entitled, *Daedalus or Science and the Future*, Cambridge University, 1923.[13]

Hydrogen as an energy carrier and potentially widely-used fuel is attractive because it can be produced easily without emissions by splitting water. In addition, the readily available electrolyzer can be used in a home or business where off-peak or surplus electricity could be used to make the environmentally preferred gas. Electrolysis was first demonstrated in 1800 by William Nicholson and Sir Anthony Carlisle and has found a variety of niche markets ever since. Two electrolyzer technologies, alkaline and proton exchange membrane (PEM), exist at the commercial level with solid oxide electrolysis in the research phase.

Electrolysis is defined as splitting apart with an electric current. Decomposition of the water occurs when a direct current (DC) is passed between two electrodes immersed in water separated by a non-electrical conducting aqueous or solid electrolyte to transport ions and completing the circuit. The voltage applied to the cell must be greater than the free energy of formation of water plus the corresponding activation and ohmic losses before decomposition will proceed. Ion transport through the electrolyte is critical as the purest of water would only contain small amounts of ions making it a poor conductor.

Ideally, 39 kWh of electricity and 8.9 liters of water are required to produce 1 kg of hydrogen at 25 °C and 1 atmosphere pressure. Typical commercial electrolyzer system efficiencies are 56%–73% and this corresponds to 70.1–53.4 kWh/kg.[14] The U.S. consumes somewhere between 140–150 billion gallons of gasoline per year equating to the same number of kilograms if we were to use only hydrogen for transportation. This would result in needing 330 billion gallons of water to make that much hydrogen. If the hydrogen were used in a fuel cell that is two times as efficient as an internal combustion engine in a car the amount of water required would be half. For comparison, gasoline production uses 300 billion gallons per year, domestic water use tops 4800 billion gallons per year and thermal electric power generation 70 trillion gallons per year.[15]

When comparing literature from fuel cell (FC) models with water electrolysis work it is important to remember the differences between the system anode and cathode. This basic understanding may be trivial to most but is many times confused when switching between the two processes. The anode is always the electrode at which oxidation occurs, where electrons are lost. The cathode is defined as the electrode at which electrons enter, where reduction takes place. In electrolysis the cathode is the electrode where H_2 gas is created, in FC systems the anode is the electrode where H_2 is introduced.

2.1 Alkaline

The alkaline electrolyzer is a well-established technology that typically employs an aqueous solution of water and 25–30 wt.% potassium hydroxide (KOH). However, sodium hydroxide (NaOH), sodium chloride (NaCl) and other electrolytes have also been used. The liquid electrolyte enables the conduction of ions between the elec trodes and is not consumed in the reaction but does need to be replenished periodically due other system losses. Typically commercial alkaline electrolyzers are run with current densities in the range of 100–400 mA cm^{-2}. The reactions for the alkaline anode and cathode are shown in Eqs. 1 and 2 respectively, showing the hydroxyl (OH$^-$) ion transport.

$$4 \, OH^-_{(aq)} \rightarrow O_{2(g)} + 2 \, H_2O_{(l)} \tag{1}$$

$$2 \, H_2O_{(l)} + 2e^- \rightarrow H_{2(g)} + 2 \, OH^-_{(aq)} \tag{2}$$

The first water electrolyzers used the tank design and an alkaline electrolyte.[20] These electrolyzers can be configured as unipolar (tank) or bipolar (filter press) designs. In the unipolar design (see Figure 1), electrodes, anodes, and cathodes are alternatively suspended in a tank. In this design, each cell is connected in parallel and the entire system operated at 1.9–2.5 V_{dc}.

The advantage to the unipolar design is that it requires relatively few parts, is extremely simple to manufacture and repair because individual cells can be taken offline while the remaining cells remain productive. The disadvantage is that it usually operates at lower current densities and lower temperatures.[16] More recent unipolar designs include operation at high hydrogen pressure outputs (up to 6,000 psig).

The bipolar design (Fig. 2), often called the filter-press, has alternating layers of electrodes and separation diaphragms that are clamped together. The cells are connected in series and result in higher stack voltages. Since the cells are relatively thin, the overall stack can be considerably smaller than the unipolar design. The advantages to the bipolar design are the reduced stack footprints, higher current densities, and its ability to produce higher pressure gas. The disadvantage is that it cannot be repaired without servicing the entire stack.[16,17] Fortunately, it rarely needs servicing. Previously asbestos was used as a separation diaphragm, but manufacturers have replaced or are planning to replace this with new polymer materials such as Ryton®.[18]

2.2 Proton Exchange Membrane

A second commercially available electrolyzer technology is the solid polymer electrolyte membrane (PEM). PEM electrolysis (PEME) is also referred to as solid polymer electrolyte (SPE) or polymer electrolyte membrane (also, PEM), but all represent a system that incorporates a solid proton-conducting membrane which is not electrically conductive. The membrane serves a dual purpose, as the gas separation device and ion (proton) conductor. High-purity deionized (DI) water is required in PEM-based electrolysis, and PEM electrolyzer manufacturer regularly recommend a minimum of 1 MΩ-cm resistive water to extend stack life.

PEM technology was originally developed as part of the Gemini space program.[16] In a PEM electrolyzer, the electrolyte is contained in a thin, solid ion-conducting membrane rather than the aqueous solution in the alkaline electrolyzers. This allows the H$^+$ ion (proton) or hydrated water molecule (H$_3$O$^+$) to transfer from the anode side of the membrane to the cathode side, and separates the hydrogen and oxygen

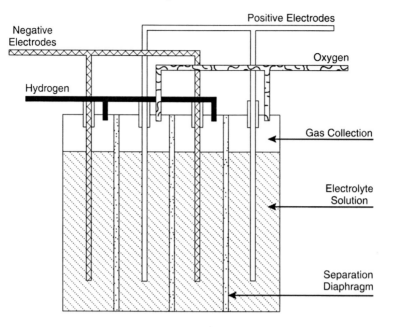

Fig. 1. Unipolar (tank) electrolyzer design.

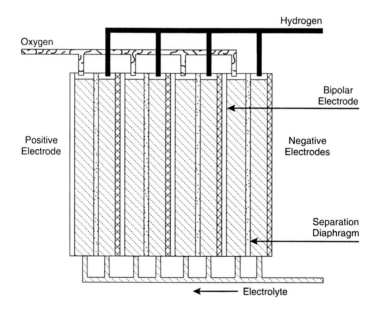

Fig. 2. Bipolar (filter-press) electrolyzer design.

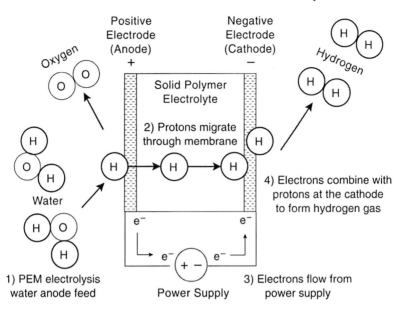

Fig. 3. PEM cell components and reaction showing the positive anode and negative cathode electrodes.

gases. Oxygen is produced at the anode side and hydrogen is produced on the cathode side. The most commonly used membrane material is Nafion® from DuPont. PEM electrolyzers use the bipolar design and can be made to operate at a high differential pressure across the membrane.

DI water is introduced at the anode of the cells, and a potential is applied across the cell to dissociate the water. The protons (H^+) are pulled through the membrane under the influence of an electric field and rejoin with electrons being supplied by the power source at the cathode to form hydrogen, H_2, gas. PEM electrolyzers are operated at higher current densities (> 1600 mA cm^{-2}) almost an order of magnitude higher than their alkaline counterparts. Stack efficiency decreases as current density increases but is necessary to increase hydrogen production to offset the higher capital costs of PEM cells. PEM advantages over alkaline include the ability to maintain a significant differential pressure across the anode and cathode avoiding the risk of high pressure oxygen. In addition, PEM electrolysis requires DI water but avoids the hazards surrounding KOH. The PEM anode and cathode reactions are described in Eqs. 3 and 4, respectively, and shown in Figure 3,

$$2\,H_2O \rightarrow 4\,H^+ + 4e^- + O_2 \tag{3}$$

$$4\,H^+ + 4e^- \rightarrow 2\,H_2 \tag{4}$$

Figure 4 shows the major water and hydrogen components inside Proton Energy Systems HOGEN 40RE® including the heart of the system: the PEM stack in front

Hydrogen
phase
separator

Combustible
gas detector

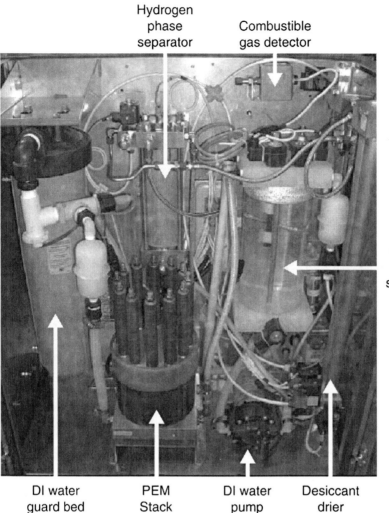

Oxygen
phase
separator

DI water   PEM   DI water   Desiccant
guard bed   Stack   pump   drier

Fig. 4. Internal components of HOGEN 40RE.

center. The compartment behind these systems (not viewable) contains the AC/DC power converter, ventilation fan, 24 Vdc power supply, system controller, radiator, and control relays. The RE version contains a DC/DC power converter and DC disconnects used to interconnect to a PV array. The combustible gas detector monitors hydrogen levels in this compartment and the oxygen phase separator.

Figure 5 shows the step currents from the power supplies and the resulting hydrogen flow in standard cubic feet per hour (scfh) from the system. Hydrogen production ripple is caused by the internal hydrogen phase separator pumping down the

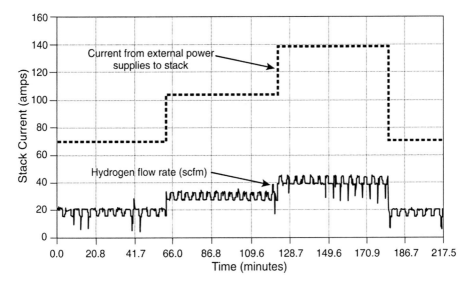

Fig. 5. Sample current step waveform from external power supplies and resulting hydrogen mass flow from HOGEN 40RE.

accumulated water and desiccant drying tube crossover. These system functions cause a drop in system pressure resulting in varying hydrogen production output.

The system efficiency (Eq. 5) is calculated using both the ancillary losses plus the stack energy. The system efficiency uses the higher heating value of hydrogen (39 kWh kg^{-1}), the energy consumed by the stack (kWh), efficiency of the DC power supplies, and the balance of plant ancillary loads like pumps, valves, sensors and controller (kWh). Stack efficiency (Eq. 6) is determined by calculating the ideal cell potential at the operating temperature and pressure multiplied by the number of cells in the stack and then divided by the measured stack voltage,

$$\text{System Efficiency} = \frac{\text{HHV}\left(\dfrac{\text{kWh}}{\text{kg}}\right)}{\dfrac{\left(\dfrac{\text{Stack Input Energy (kWh)}}{\text{Power Supply Efficiency}}\right) + \text{Ancillary Losses (kWh)}}{\text{Hydrogen Produced (kg)}}} \qquad (5)$$

$$\text{Stack Efficiency} = \frac{\text{Ideal Stack Potential}}{\text{Actual Stack Potential}} \qquad (6)$$

The HHV of hydrogen is 39 kWh kg^{-1} and the ideal stack potential is a function of temperature and pressure. All efficiencies are referenced to the HHV of hydrogen. The minimum amount of energy that must be consumed to split water into hydrogen

Table 1. Constants for heat capacities of gases in ideal state and liquid water.

	A	B	C	D
H_2	3.249	0.422×10^{-3}	0	0.083×10^5
O_2	3.639	0.506×10^{-3}	0	-0.227×10^5
H_2O	8.712	1.25×10^{-3}	-0.18×10^{-6}	0

and oxygen is known as the heat of formation (enthalpy) and corresponds to the HHV of hydrogen.

3 Fundamentals of Water Electrolysis

3.1 First Principles

The overall reaction of the electrolysis cell (Eq. 7) provides the required stoichiometric coefficients for the products and reactant used in Eq. 8. The sign convention is positive for products and negative for reactants with analogous definitions for ΔB, ΔC and ΔD. Data for the constants A, B, C and D are thermodynamic properties and are reproduced in Table 1 from,[19]

$$H_2O + 2e^- \rightarrow H_2 + \frac{1}{2}O_2 \tag{7}$$

$$\Delta A \rightarrow A_{H_2}\, \frac{1}{2} A_{O_2} - A_{H_2O} \tag{8}$$

The model uses the specific heat capacity of water and gases in the ideal state to determine the standard Gibbs free energy of reaction, ΔG°, i.e., Eq. 9,

$$\Delta G^\circ = RT \left[\frac{\Delta G_0^0 - \Delta H_0^0}{RT_0} + \frac{\Delta H_0^0}{RT} + \frac{1}{T}\int_{T_0}^T \frac{\Delta C_P^0}{R}dT - \int_{T_0}^T \frac{\Delta C_P^0}{R}\frac{dT}{T} \right] \tag{9}$$

where ΔH_0° and ΔG_0° are the standard enthalpy and Gibbs energy of formation of liquid water, respectively, at reference temperature T_0. The integrals of Eq. 9 take into account the temperature dependency of the heat capacities of the products and reactants and are reduced to Eqs. 10 and 11,

$$\int_{T_0}^T \Delta C_p^0 dT = (\Delta A)T_0(\tau - 1) + \frac{\Delta B}{2}T_0^2(\tau^2 - 1) + \frac{\Delta C}{3}T_0^3(\tau^3 - 1) + \frac{\Delta D}{T_0}\left(\frac{\tau - 1}{\tau}\right) \tag{10}$$

$$\int_{T_0}^T \frac{\Delta C_P^0}{R}\frac{dT}{T} = \Delta A \ln \tau + \left[\Delta B\, T_0 + \left(\Delta C\, T_0^2 + \frac{\Delta D}{\tau^2 T_0^2} \right)\left(\frac{\tau + 1}{2}\right) \right](\tau - 1) \tag{11}$$

where T is the reaction temperature (K), T_0 is the reference temperature (298 K), R is he universal gas constant (8.314 J mol-1 K-1) and tau (τ) is defined as $\tau \equiv T/T_0$.

Enthalpy is an intrinsic property of a substance and a function of temperature and pressure.[19] In practice the Gibbs free energy is the net internal energy available to do work, less work done by changes in pressure and temperature.[20] Exergy, on the other hand, is defined as the total amount of work that can be harnessed and becomes more relevant in high-temperature and high-pressure electrolysis.

At standard temperature and pressure (STP, 25 °C, 1atm) Gibbs free energy of formation is defined as the point of zero energy and is used to calculate the change in energy of a system. The reversible (i.e., the minimum) voltage required to electrolyze water into hydrogen and oxygen is determined by the change of Gibbs free energy of formation between the products and reactants. As described above, the Gibbs free energy of formation is not constant; it changes with temperature and state (liquid or gas),[20]

$$E_0 = -\frac{\Delta G^0}{zF} \qquad (12)$$

where E_0 is the theoretical minimum reversible voltage of a cell, z is the number of electrons (2) taking part in the reaction and F is Faraday's constant (96,485 Coulomb mol^{-1}). The actual voltage (V_{cell}) required to decompose water at any significant rate will require V_{cell} be greater than E_0. The difference between the voltages is known as overpotential, polarization or simply losses.

The Nernst potential (V_n) of Eq. 13 accounts for changes in the activity of the reaction and for nonstandard conditions,

$$V_n = E_0 + \frac{RT}{2F} \ln\left(\frac{P_{H_2} P_{O_2}^{1/2}}{P_{H_2O}}\right) \qquad (13)$$

where P_{H2}, P_{O2} and P_{H2O} represent the partial pressures of hydrogen, oxygen and water respectively.

The partial pressure of water is determined with the empirical formula from Ref. 21 and is shown in Eq. 14. This relationship enables the partial pressure of water to be calculated from experimental data as the temperature of the DI water into the stack anode varies,

$$P_{H_2O} = 610.78 \exp\left[\frac{T_c}{(T_c + 238.3)}(17.2694)\right] \qquad (14)$$

where T_c is the temperature into the stack in °C and should not be confused with T and T_0 from earlier that have units of Kelvin.

The partial pressures of hydrogen and oxygen are determined using measurements from the stack cathode and anode and Eqs. 15 and 16,

$$P_{H_2} = P_C - P_{H_2O} \qquad (15)$$

$$P_{O_2} = P_A - P_{H_2O} \tag{16}$$

where P_C and P_A are the experimental pressures (atm) of the cathode and anode respectively.

3.2 Overpotentials

Water electrolysis is an electrochemical reaction where water is split into hydrogen and oxygen in the presence of a catalyst and applied electric field. As current density increases the cell losses due to membrane, electrode, and interfacial resistances dominate and are referred to as ohmic overpotential.

At equilibrium (i.e., no current) there exist dynamic currents, measured in amps, at each electrode and are a fundamental characteristic of electrode behavior. The anode and cathode exchange current densities can be defined as the rate of oxidation and reduction respectively. The exchange current density is a measure of the electrode's ability to transfer electrons and occurs equally in both directions resulting in no net change in composition of the electrode.[22] A large exchange current density represents an electrode with fast kinetics where there is a lot of simultaneous electron transfer. A small exchange current density has slow kinetics and the electron transfer rate is less.

The anode and cathode exchange current density's can be fitted exponentially as a function of temperature. Experimentally is has been determined and intuition suggests that as temperatures increase the faster a chemical reaction will proceed. Arrhenius was the first to recognize that the higher kinetic energy due to higher temperature results in lowering the activation potential.[22] Lower activation losses reduces the amount of energy for the reaction to proceed, thus increasing stack efficiency. The conductivity coefficient was fitted linearly because it is primarily a function of current.

(i) Activation overpotential

Electrochemical reactions possess energy barriers which must be overcome by the reacting species. This energy barrier is called the 'activation energy' and results in activation overpotential, which are irreversible losses (heat) in the system. Activation energy is due to the transfer of charges between the electronic and the ionic conductors. The activation overpotential is the extra potential necessary to overcome the energy barrier of the rate-determining step of the reaction to a value such that the electrode reaction proceeds at a desired rate.[23] The anode (Eq. 17) and cathode (Eq. 18) activation overpotentials, η_A and η_C, represent irreversible losses of the PEM stack and dominate the overall overpotential at low current densities:

$$\eta_A = \frac{RT}{\alpha_a zF} \ln\left(\frac{i}{i_{a,o}}\right) \tag{17}$$

$$\eta_C = \frac{RT}{\alpha_c zF} \ln\left(\frac{i}{i_{c,0}}\right) \tag{18}$$

where α_a and α_c are the anode and cathode electron transfer coefficients respectively, $i_{a,o}$ and $i_{c,o}$ are the anode and cathode exchange current densities (A cm^{-2}) respectively and i is the current density of the stack (A cm^{-2}). The electron transfer coefficient is a measure of the symmetry of the activation energy barrier and can range from zero to unity.[22]

The higher the exchange current density the *easier* it is for reaction to continue when current is supplied to the stack. The cathode exchange current density is thus not the limiting parameter of the activation overpotential term and is often ignored. The current density (i) normalizes the stack current (I) to the active area of the cell.

(*ii*) Ohmic

Ohmic losses occur because of resistance to the flow of ions in the solid electrolyte and resistance to flow of electrons through the electrode materials. Because the ionic flow in the electrolyte obeys Ohm's law, the ohmic losses can be expressed by Ohm's law. The ohmic overpotential, η_o of Eq. 19 is a function of the stack current density (i), membrane thickness (φ), and the conductivity of the stack (σ),

$$\eta_o = \frac{\varphi}{\sigma} i \tag{19}$$

where σ (Siemen cm^{-1}) represents the sum of membrane resistance to ion transfer and bundled electrical resistances of electrodes and interconnections within the stack. As the stack ages, internal polarization losses increase and stack voltage will increase for a given current.

(*iii*) Anode exchange current density

Small exchange current densities exhibited by the anode give rise to slow charge transfer that is in turn an activation-controlled process. Unfortunately, for the case of both the anode and cathode exchange current densities, the range of values varies dramatically from author to author because of the various operating conditions, cell construction, and stack configurations. Typical values found in the literature for the anode exchange current density range from 10^{-7}–10^{-12} A cm^{-2}.[24,25]

(*iv*) Cathode exchange current density

The cathode exchange current density is typically four orders of magnitude greater than the anode exchange current density and supported by Choi and Berning in Ref. 24 and 25. The anode side is therefore limiting the reaction and dominates the activation overpotential.

(v) Conductivity

Specifically speaking, membrane conductivity represents only the membrane's resistance to flow of protons (H^+) and is highly dependant on its thickness (φ) and water content. Electrical resistance of electrodes, cell interconnects, and the formation of any insulating layer on the electrode surface are all bundled under the conductivity term. Voltage decreases for a given current as temperature increases and can be controlled to improve stack efficiency.

4   Commercial Electrolyzer Technologies

Electrolyzers produced in 2006 range in sizes from less than 0.1 kg day^{-1} to over 1000 kg day^{-1}. The smallest systems, in the under 0.5-kg day^{-1} range, are used for the production of hydrogen at the lab scale, and are designed as hazard-free alternatives to high pressure gas cylinders.[26] Electrolyzers are sized to meet the system requirements of the existing high purity hydrogen markets. However, only systems above the 0.55-kg day^{-1} production rate are viable for producing hydrogen as a transportation fuel, and the transportation fueling market could use systems larger than the existing 1000-kg day^{-1} unit. A 0.055-kg day^{-1} system would produce 200 kg year^{-1}, enough to fuel a single car. A 1000-kg day^{-1} system would produce 365,000 kg year^{-1}, enough to fuel 1,800 vehicles. Both of these values assume a 100% capacity factor on the electrolysis system. Typically, these units are capable of operations in the high 90% range.[27] The number of cars served by a system was determined by calculating that a car requires approximately 200 kg of hydrogen year^{-1}. This 200-kg requirement assumes that on average a car travels 12,000 miles year^{-1}, and that a vehicle will travel 60 miles kg^{-1} of hydrogen.[28]

Electrolysis systems that could be used for transportation fuel production can be categorized into five different size ranges: home, small neighborhood, neighborhood, small forecourt and forecourt.[28] The term forecourt refers to a refueling station. The number of cars served and hydrogen production rate for each size are as follows:

- The home size will serve the fuel needs of 1–5 cars with a hydrogen production rate of 200–1000-kg hydrogen year^{-1}.
- The small neighborhood size will serve the fuel needs of 5–50 cars with a hydrogen production rate of 1000–10,000-kg hydrogen year^{-1}.
- The neighborhood size will serve the fuel needs of 50–150 cars with a hydrogen production rate of 10,000–30,000-kg hydrogen year^{-1}.
- The small forecourt size, which could be a single hydrogen pump at an existing station, will serve 150–500 cars with a hydrogen production rate of 30,000–100,000-kg hydrogen year^{-1}.
- A full hydrogen forecourt will serve more then 500-cars per year with a hydrogen production rate of greater then 100,000-kg hydrogen year^{-1}.

A sampling of electrolyzer manufacturers in 2006 are presented in Table 2.

Table 2. Commercial electrolyzer manufacturers and selected performance data.

Manufacturer	Technology	Lower capacity (kg day^{-1})	Upper capacity (kg day^{-1})	Pressure	Location	Ref.
AccaGen SA	Alkaline, acid and PEM	0.043	215.7	up to 200 barg	Switzerland	29
Avalance	Alkaline	0.75	300.	up to 6500 psig	Connecticut, USA	6
ELT	Alkaline	6.47	1639.4	up to 30 barg	Germany	5
Gesellschaft fur Hochleistungselktrolyseure zur Wasserstofferzeugung	Unknown	25.9	1078.6	30 barg	Germany	30
Giner	PEM	11.8	11.8	3000 psig	Massachusetts, USA	31
Hamilton Sundstrand	PEM	129.3	129.3	up to 100 psig	Connecticut, USA	32
Hydrogenics	PEM and alkaline	10.8	129.4	up to 363 psig	Canada	33
Industrie Haute Technologie SA	Alkaline	1639.4	1639.4	up to 32 barg	Monthey, Switzerland	34
Linde	Unknown	10.8	539.3	25 barg	Germany	35
Norsk Hydro	Alkaline	129.4	1046.2	15 barg	Norway	36
Peak Scientific	Ion exchange membrane	0.01	0.01	0-100 psig	Scotland	30
PIEL division of ILT Technology s.r.l.	Alkaline	2.2	30.2	3 barg	Italy	31
Proton Energy System	PEM	1.1	12.9	up to 218 psig	Connecticut USA	32
Schmidlin-DBS AG	Membrane	0.005	0.026	1–155 psig	Neuheim, Switzerland and Padova, Italy	37
Siam Water Flame Co.	Alkaline	0.647	0.647	unknown	Bangkok, Thailand	38
Teledyne Energy Systems	Alkaline	6.5	129.4	up to 115 psig	Maryland, USA	39
Treadwell Corporation	PEM	2.6	22.0	up to 1100 psi	Connecticut, USA	40

5   Electrolysis System

The system used to produce hydrogen via electrolysis consists of more than just an electrolyzer stack. A typical electrolysis process diagram is shown in Fig. 6.[45] The primary feedstock for electrolysis is water. Water provided to the system may be stored before or after the water purification unit to ensure that the process has adequate feedstock in storage in case the water system is interrupted.

Water quality requirements differ between electrolyzers. Some units include water purification inside their hydrogen generation unit, while others require an external purification unit, such as a deionizer or reverse osmosis unit, before water is fed to the cell stacks. The high purity water will be mixed with KOH if the system is an alkaline system before being introduced to the hydrogen generation unit. Note that PEM units will not a KOH feed, as no electrolytic solution is needed. Each system has a hydrogen generation unit that integrates the electrolysis stack, gas purification

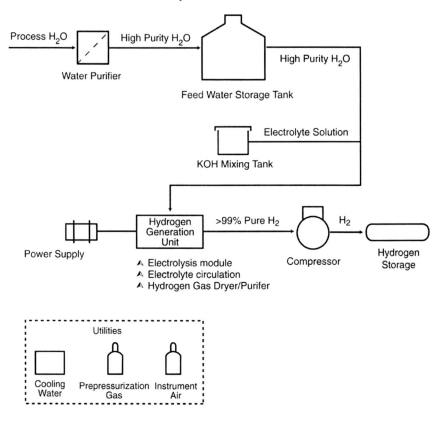

Fig. 6. Process flow diagram for a water electrolyzer system.

and dryer, and heat removal. Electrolyte circulation is also included in the hydrogen generation unit in alkaline systems. The hydrogen generation system is usually en closed in a container or is installed as a complete package. Oxygen and purified hydrogen are produced from the hydrogen generation unit. If desired, a compressor, hydrogen storage, and oxygen storage can be added to the system.

A second *feedstock* needed for electrolysis is electricity. Typically electricity is not considered a feedstock but a utility; however it is a critical component in the splitting of the water molecule into hydrogen and oxygen. An electrolyzer typically will convert supplied AC to DC, as the stack requires DC to split water.

Typical utilities that the electrolysis systems need include electricity for other pe ripheral equipment; cooling water for the hydrogen generation unit; pre pressurization gas; and instrumentation gas (Fig. 6).

5.1 Energy Efficiency

Energy efficiency is defined as the higher heating value (HHV) of hydrogen divided by the energy consumed by the electrolysis system per kilogram of hydrogen pro-

Table 3. Efficiency of selected electrolyzers in the market.

	Energy required system (kWh/kg)	System efficiency (%)	Production pressure	Ref.
AccaGen SA	74.5–52.4	52–74	Up to 200 bar	29
Hydrogenics	53.4	73	363 psi	37
IHT	46.7–51.2	84–76	Up to 32 bar	29
PIEL division of ILT Technology s.r.l.	77.9	50	3 bar	31

duced. See Eqs. 5 and 6 for details on this calculation. HHV is used as opposed to the lower heating value (LHV) because in commercial electrolyzers in 2005, the water electrolyzed is in a liquid state. To further clarify why HHV is used, the reaction of the formation of water is:

$$H_2 + \frac{1}{2} O_2 \rightarrow H_2O + energy \qquad (20)$$

At 25 °C and 1 atm, the heat of formation of liquid water, or the energy released when water is formed in the reaction above is 39 kWh kg^{-1} of hydrogen. This value is the higher heating value (HHV) of hydrogen. The heat of formation of steam is 33 kWh/kg of hydrogen, and is the lower heating value (LHV) of hydrogen. The electrolysis reaction is the opposite of the formation of water reaction:

$$H_2O + energy \rightarrow H_2 + \frac{1}{2} O_2 \qquad (21)$$

As a result, the amount of energy needed to create hydrogen from liquid water using electrolysis is 39 kWh kg^{-1}. The reason this distinction is important is because if using the lower heating value the efficiency of electrolyzers is misrepresented. If LHV is used to calculate electrolyzer efficiencies, the maximum hydrogen system efficiency is 33/39 or approximately 85%. Thus a 100% efficient electrolysis system on an LHV basis is actually thermodynamically impossible if you are electrolyzing liquid water. That is to say that an electrolyzer that converts every kWh of input energy into hydrogen energy will have only 85% efficiency, even though there are no losses.[41]

The energy efficiency of several electrolyzers is shown in Table 3. The energy efficiency ranges of commercial systems ranges from 47–77 kWh/kg (83–51%). An efficiency goal for electrolyzers in the future has been reported to be in the 46.9 kWh kg^{-1} range, or a system efficiency of 83%.[42] This 83% includes compression of the hydrogen gas to 6000 psig. Currently most electrolyzers reach a pressure ranging from 0–500 psig for the power requirements presented, with a few research stage electrolyzers reaching pressures in the 3000–6500-psig range. So most electrolyzers would need additional energy input beyond what is presented in the table below to compress to fueling pressures.

Note that the above values are energy requirement of the entire electrolysis system, excluding any additional compression beyond what the stack produces. This is an appropriate way to calculate system efficiency. As an example, the electrolyzer

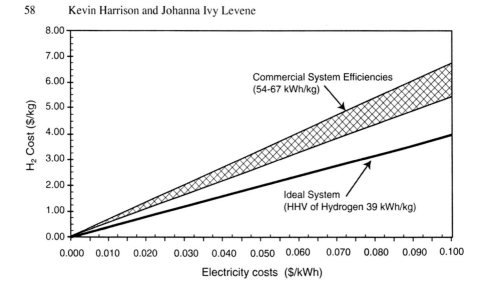

Fig. 7. Hydrogen costs via electrolysis with only electricity costs considered.

stack alone for a Hydrogenics system requires 46.8 kWh kg^{-1} (4.2 kWh Nm^{-3}), which corresponds to 83% efficiency when you divide the HHV of hydrogen by the electrolyzer power requirement. However, when you include the rectifier and auxiliaries the energy requirement becomes 53.4 kWh kg^{-1} (4.8 kWh Nm^{-3}) or 73% efficient.

5.2 Electricity Costs

Electricity costs are a key component when producing hydrogen via electrolysis. A boundary analysis was completed to determine the effects of electricity price on hydrogen costs, and the results are shown in Fig. 7.[28] For each electrolyzer, the specific system energy requirement is used to determine how much electricity is needed to produce hydrogen; no capital, operating or maintenance costs are included in the calculation. The system energy requirement used is the lowest energy requirement reported for each manufacturer. This graph shows that, at current electrolyzer efficiencies, in order to produce hydrogen at lower than $3.00 kg^{-1}, electricity costs must be between 4 and 5.5¢ kWh^{-1}. In order to produce hydrogen for less than $3.00 kg^{-1} with a system that is 100% efficient, electricity prices must be less than 7.5¢ kWh^{-1}.

The U.S. Department of Energy's Energy Information Administration (EIA) reports 2002 industrial, commercial, and residential electricity prices at 4.83, 7.89, and 8.45¢ kWh^{-1}, respectively.[28] Thus, if only electricity costs were incurred, current electrolyzers could produce hydrogen for $3.00 kg^{-1} at industrial electricity prices; an ideal system could produce hydrogen for $3.00 kg^{-1} at slightly lower then commercial prices. This analysis shows that regardless of any additional cost elements, electricity costs will be a major price contributor.

6 Opportunities for Renewable Energy

Integrating electrolyzers with renewable energy system can present challenges as well as unique benefits. Currently most renewable energy systems produce power and interconnect with thte electrical grid via some form of power electronics (PE). To use electrical grid power, today's commercial electrolyzers also have some type of power electronics interface that can represent a significant portion of the overall system cost.[43] The power electronics convert alternating current (AC) from the grid to direct current (DC) power required by the electrolysis cell stack. In addition to the DC requirements of the stack, the system also consumes additional AC power for the balance of plant or ancillary loads. At least one electrolyzer manufacturer offers a version of an electrolyzer that can accommodate a connection to photovoltaic (PV) panels in addition to having the standard AC to DC converter for utility operation.[44] The additional power electronics, incorporating maximum power point tracking (MPPT), converts all available DC power from the PV array to run the electrolysis stack. This system appears to be one of the firsts to incorporate dedicated PE to interface with a PV source.

In addition to using PV systems as electricity sources, wind energy can also be used. Today, the majority of wind to hydrogen demonstration projects are focused on installing commercially available electrolyzers and powering them from the AC power from wind turbines.[45–47] In these projects the AC from the wind turbines is sent out onto the grid and the electrolyzers tied into the grid achieving a loose coupling of source and load. Scheduling the power to the electrolyzer, based on an output signal from the wind turbines, is relatively straightforward in this case.

Capital costs of electrolysis equipment range from just under $1000 kW^{-1} for the largest alkaline systems to over $10,000 kw^{-1} for small proton exchange membrane (PEM) electrolyzers.[28,48] Merely taking an off-the-shelf wind turbine with its own PE and commercial electrolyzer with its own PE reduces overall energy transfer from the wind to hydrogen system. The potential exists to characterize electrolyzer performance under varying input power and design a single PE package and intelligent controller to achieve direct coupling between the stack and wind turbine output. This topology would not only eliminate the redundancy of power electronics that exists in the wind turbine and electrolyzer but also achieve gains in system cost and robustness. Characterizing the system demands of renewable energy sources and the requirements of the hydrogen-producing stack appears to have synergistic benefits. Ultimately, the detailed understanding of both systems and design of the directly coupled wind to electrolysis would reduce the cost of renewably generated hydrogen.

In renewable-based energy systems PEM electrolysis seems to have an advantage over alkaline in that the thin membrane and ion transport mechanism can react to nearly instantaneously with the rapidly changing energy output of renewable sources, especially wind. Stacks involving the circulation of a liquid electrolyte have inherently more inertia in the transport of ions in solution than the PEM systems.

On the turbine side, variable-speed wind turbines (which will soon be the norm as a result of enhanced energy capture relative to constant-speed machines) rely on power electronics to convert the variable frequency, variable voltage AC produced at the generator to DC. Small turbines used in battery-charging applications stop here;

however, larger turbines used to connect to the grid must then convert the DC back to AC at grid frequency: 60 Hertz (Hz). It is important to note that because of the economies of scale, it is the large wind turbines that are achieving highly competitive energy costs and will likely be the device of choice in large-scale wind-to-hydrogen operation.

The small wind-to-hydrogen systems (< 20 kW) being studied today are systems incorporating a common DC bus fixed with a battery bank to which the wind turbine and electrolyzer as well as fuel cells and PV panels are connected. Typically, the wind turbine is of the battery-charging type, which requires connection to a constant voltage DC bus (hence, the battery bank) and incorporates power electronics to convert wild AC to DC and to regulate power output. The electrolyzer stack accepts DC power input but the system would also include power electronics to regulate power input and possibly convert DC at one voltage level to another.

There are a number of weaknesses with this configuration, namely a redundancy of power electronics leading to increased cost and potential for failure. The inability to match wind turbine power output to electrolyzer power requirements because of separate power electronic controllers ultimately results in reduced energy capture.

An advanced topology would be the direct coupling of an electrolyzer with a wind turbine. This would allow hydrogen production that is proportional to the available wind energy and reduce electricity storage requirements.. The single point of control will allow the matching of wind turbine and electrolyzer electrical characteristics, thereby increasing the energy capture of the wind turbine. Finally, this solution will eliminate the need for a constant voltage DC bus and provide a true test of electrolyzer operation under fluctuating power-input conditions.

Renewable electrolysis can help overcome one of the key barriers to realizing a hydrogen-based economy by replacing the carbon-intensive one that exists today. There is an excellent opportunity for research in renewable hydrogen production both in terms of understanding the operation of the electrolyzer under variable sources and optimizing, in terms of efficiency, cost, and robustness, the link between a renewable source and electrolyzer stack.

7 Conclusions

There exists an opportunity to change the face of our energy consumption from one of polluting our air, water, and land to one more in harmony with the environment. The environment and the economy are often at odds for resources, but it does not have to be that way. Renewable hydrogen seems to possess the ability to transition the world's carbon-based economy into a near-carbon-free economy. Hydrogen can be extracted from all fossil fuels as well as split from water using the electricity from RE sources. However, without sequestering the climate-altering CO_2 produced using fossil fuels, the environmental benefits are completely lost and may be even worsened by the transition to hydrogen as an energy carrier.

If the environmental benefits of the long-term development of the hydrogen economy are to be realized, the production of hydrogen via electrolysis from RE sources will be a vital component. Today's commercially available electrolyzers are designed to use grid electricity, produce well regulated DC power to the electrolysis stack, and condition the output gas for applications different than that required by PEM fuel cells. The key element in hydrogen production from any electrical source is the electrolyzer stack that converts water and electricity into hydrogen, oxygen and heat. The electrolyzer stack is inherently a nearly constant, low-voltage, DC device requiring some form of control system and power electronics to connect it to a high-voltage, AC source of power.

The primary intent of this work is to design, build and verify a system capable of accurately varying important system variables that are normally strictly monitored and controlled by the commercial electrolyzers containing the same PEME stack. The goal of the experimental characterization of the stack, under varying conditions and power, is to enable an optimized interconnection between the stack and RE source. Such a coupled system specifically designed with the RE source in mind would reduce the overall cost of independent stand-alone systems and may eliminate the need for electrical storage components.

Electrical power provided to the electrolyzer in such a system would be controllable with excess power provided to the grid. Thus, a combined system would have more dispatch-ability than a wind-electric turbine or PV array alone. Such dispatch-ability might be used to provide the utility with a measure of control over the renewable energy systems total output that does not exist in current renewable based, grid-connect only systems. Using a variable RE source, like wind or PV, to generate the hydrogen gas will guarantee this energy carrier will be produced with nearly zero emissions.

References

1. B. Suresh, S. Schlag, and Y. Inogucji, *Chemical Economics Handbook Marketing Research Report*, SRI Consulting, 2004.
2. M. Momirlana and T.N.Veziroglub, The properties of hydrogen as fuel tomorrow in sustainable energy system for a cleaner planet, *International Journal of Hydrogen Energy* **30**, 795 (2005).
3. *Hemoglobin*, in Wikipedia, the Free Encyclopedia, Retrieved on June 22, 2006 from http://en.wikipedia.org/wiki/Hemoglobin.
4. *Annual Energy Review 2004*, EIA, http://www.eia.doe.gov/emeu/aer/pdf/aer.pdf, Report No. DOE/EIA-0384 (2004), August 2005.
5. *Safe Use of Hydrogen and Hydrogen Systems*, NASA Training Center, 2006.
6. J. B. Heywood, Fueling our transportation future, *Scientific American*. **295** 60 (2006).
7. W. E. Winshe, K. C. Hoffman, and F. J. Salzano, Hydrogen: Its future role in the nation's energy economy, *Science* **180** 1325 (1973).
8. J. Levene, B. Kroposki, and G. Sverdrup, *Wind Energy and Production of Hydrogen and Electricity - Opportunities for Renewable Hydrogen*, NREL Report No. CP-560-39534, 2006.
9. Comparative Cost of Wind and Other Energy Sources, American Wind Energy Association (AWEA), http://www.awea.org/pubs/factsheets/Cost2001.PDF, 2001.
10. *The Economics of Wind Energy*, American Wind Energy Association, http://www.awea.org/pubs/factsheets/EconomicsOfWind-Feb2005.pdf, February 2005.

11. R. L. Ottinger, D. Wooley, D. R. Hodas, N. A. Robinson, and S. E. Babb, *Pace University Center for Environmental Legal Studies; Environmental Costs of Electricity*, Oceana Publications, New York, 1990.

12. K. Silverstein, Clean tech goes mainstream, *EnergyBiz Insider*, http://www.energy central.com/centers/energybiz/ebi_detail.cfm?id=164, CyberTech Inc., 2006.

13. J. B. S. Haldane, *DAEDALUS or Science and the Future*, E. P. Kutton & Company, New York, 1923.

14. *Technology Brief: Analysis of Current-Day Commercial Electrolyzers*, NREL, Golden, CO NREL/FS-560-36705, September 2004.

15. J. A. Turner, Sustainable hydrogen production, *Science*. **305** 972 (2004).

16. A. Konopka and D. Gregory, *Hydrogen Production by Electrolysis: Present and Future*, in 10th Intersociety Energy Conversion Engineering Conference, IEEE Cat. No. 75CHO 983-7 TAB, 1975.

17. W. Kincaide, *Alkaline Electrolysis: Past, Present and Future*, in Hydrogen for Energy Distribution, Institute of Gas Technology, 1978.

18. *Ryton® PPS - Chevron Phillips Chemical Company LLC*, Retrieved on June 29, 2006, from http://www.cpchem.com/enu/ryton_pps.asp, 2006.

19. J. M. Smith, H. C. Ness, and M. M. Abbott, *Introduction to Chemical Engineering Thermodynamics*, 6th ed., Mc Graw Hill, New York, 2001.

20. J. Larminie and A. Dicks, *Fuel Cell Systems Explained*, 2nd ed., John Wiley and Sons, Ltd., West Sussex, England, 2002.

21. T. Padfield, *Moisture in air, Equations describing the physical properties of moist air*, retrieved on January 9, 2006, from http://www.natmus.dk/cons/tp/atmcalc/atmoclc1.htm, 1996.

22. A. J. Bard and L. R. Faulkner, *Electrochemical Methods, Fundamentals and Applications*, 2nd ed., John Wiley & Sons, Inc., New York, 2001.

23. S. H. Chan, K. A. Khor, and Z. T. Xia, A complete polarization model of a solid oxide fuel cell and its sensitivity to the change of cell component thickness, *Journal of Power Sources* **93** 130 (2001).

24. T. Berning and N. Djilali, "Three-Dimensional Computational Analysis of Transport Phenomena in a PEM Fuel Cell — A Parametric Study," *Journal of Power Sources*, vol. 106, pp. 284-292, 2003.

25. P. Choi, D. G. Bessarabov, and R. Datta, A simple model for solid polymer electrolyte (SPE) water electrolysis, *Solid State Ionics*, **175** 535 (2004).

26. *UHP Zero Air and Hydrogen Generators for Fuel Gas*, 2006 <http://www.chromtech. com/online_catalog/instruments/gas_gen/Hydrogen_ZeroAir2.pdf>, p. 3.

27. *Peak Scientific: The Future of Gas Generation*, 2007, Peak Scientific, 2006. <http://www. peakscientific.com/peak_products/product_detail.asp?GasID=2&ApplicationID=11& ProductID=25>.

28. J. Ivy, *Summary of Electrolytic Hydrogen Production: Milestone Completion Report*, NREL, Golden, CO, NREL/MP-560-35948, April 2004.

29. *AccaGen SA – Homepage*, Vol. 2006, AccaGen SA, 2006 <http://www.accagen.com/>.

30. *GHW - Gesellschaft für Hochleistungselektrolyseure zur Wasserstofferzeugung mbH*, 2006 <http://www.ghw-mbh.de/english/01_home/index.html>.

31. *Welcome to Giner Inc.*.Vol. 2006, Giner, Inc. and Giner Electrochemical Systems, LLC 2006 <http://www.ginerinc.com/>.

32. *Hamilton Sundstrand - System Solutions*, Vol. 2006, Hamilton Sundstrand, 2006 <http://www.snds.com/ssi/ssi/SystemSolutions/h2gen.html>.

33. Hydrogenics, *On-site hydrogen generation stations, hydrogen storage and compression*, Vol. 2006, 2006 <http://www.hydrogenics.com/onsite/products.asp>.

34. IHT, *Clean hydrogen solutions*. Vol. 2006, 2006 <http://www.iht.ch/>.

35. Linde, *Hydrogen Solutions - Supply > On-Site > Ecovar® | Linde Gas Division*, Vol. 2006, 2006 <http://www.linde-gas.com/international/web/lg/com/likelgcom30.nsf/>.
36. Hydro, *Hydrogen Technologies*, Vol. 2006, 2006 <http://www.hydro.com/electrolysers/en/>.
37. *On-site hydrogen generation stations, hydrogen storage and compression,* Vol. 2006: Hydrogenics Corporation, 2006 <http://www.hydrogenics.com/onsite/products.asp>.
38. *Clean hydrogen solutions,* Vol. 2006, IHT, 2006 <http://www.iht.ch/>.
39. *Hydrogen Solutions - Supply > On-Site > Ecovar® | Linde Gas Division*, Vol. 2006, Linde, 2006 <http://www.linde-gas.com/international/web/lg/com/likelgcom30.nsf/>.
40. *Hydrogen Technologies*, Vol. 2006, Norsk Hydro Electrolysers AS, 2006, <http://www.hydro.com/electrolysers/en/>.
41. R. Merer, *RE: H2A Update*, personal e-mail, 17 Mar. 2004.
42. *3.1 Hydrogen Production*, Multi-Year Research, Development and Demonstration Plan: Planned program activities for 2003-2010, Washington DC, US Department of Energy, Energy Efficiency and Renewable Energy, January 21, 2005, p. 51.
43. S. Hock, C. Elam, and D. Sandor, Can we get there? Technology advancements could make a hydrogen electric economy viable—and expand opportunities for all renewables, *Solar Today*, May-June 2004, p.24-28.
44. Proton Energy Manufactures Three Families of HOGEN® Hydrogen Generation Systems, *Proton Energy Systems - Products - HOGEN H Series, Hogen S Series, Hogen GC*, retrieved on July 6, 2006, from http://www.protonenergy.com/products.html, 2005.
45. *Hybrid Wind Energy System*, Retrieved on June 26, 2006, from http://energy.coafes.umn.edu/windenergy, University of Minnesota, Research and Demonstration Center, 2005.
46. Basin electric joins pilot project to marry wind, hydrogen, *Energy Services Bulletin*, retrieved on May 22, 2006, from http://www.wapa.gov/es/pubs/esb/2004/December/dec045.htm, 2006.
47. G. Schroeder, *Transition to the Hydrogen Age Transition to the Hydrogen Age: Myths and Realities*, retrieved on June 29, 2006 from http://fcgov.com/utilities/pdf/eps06-fuel-hydrogen.pdf, 2006.
48. A. F. G. Smith and M. Newborough, *Low-Cost Polymer Electrolysers and Electrolyser Implementation Scenarios for Carbon Abatement*, Heriot-Watt University, Edinburgh, Report to the Carbon Trust and ITM-Power PLC, November 200.

4

A Solar Concentrator Pathway to Low-Cost Electrolytic Hydrogen

Robert McConnell

NREL, Golden, CO

1 Direct Conversion of Concentrated Sunlight to Electricity

Concentrating sunlight through the use of mirrors or lenses is historically associated with the generation of heat. Legend has it that Archimedes used mirrors and the sun's energy to set attacking Roman ships on fire.[1] Children often discover that magnifying lenses can burn paper or tree leaves, sometimes after first burning their fingers. At the turn of the 19th century, several inventors and engineers used heat from solar concentrators to operate steam engines to pump water and later to generate electricity by means of rotating machinery.[2] Several solar concentrator technologies being developed today use heat and rotating machinery. Large systems based on this technology have been generating electricity successfully in California since the 1980s.

With the invention of the modern solar cell in 1955, scientists and engineers began developing a revolutionary new technology—photovoltaics (PV)—for converting sunlight directly into electricity. Photovoltaic technologies are based on high-technology semiconductors in which the sun's photons liberate an electric charge within the semiconductor and that charge is driven by an internal electric field to electrodes connected to an external load. In the 1960s and 1970s, these marvelous solar *batteries* were the only reliable power sources providing electricity for the first space satellites of the Cold War, as well as for later communications satellites.

In the 1970s, engineers demonstrated that concentrating sunlight and focusing the equivalent of hundreds of *suns* onto a solar cell could generate hundreds of times more electricity.[2] However, not all of the sunlight is converted to electricity, and engineers designed *heat sinks* to transfer heat away from the solar cells or actively cooled the solar cells using a cooling fluid because efficiency decreases when the solar cells heat up. So, for efficient electricity production, this solar heat was wasted. As we shall see, that ordinarily wasted heat can be used to augment the electrolytic production of hydrogen above the already dramatically high efficiencies of concentrator solar cells. We will describe how solar concentrator photovoltaic (CPV) systems can produce hydrogen from water at high efficiency. But before we do, we need to understand more about the characteristics of CPV systems and why they are just now entering into the world's energy markets.

Unlike flat-plate PV systems seen on roofs around the world today, solar concentrators need to track the sun. To focus sunlight onto a solar cell throughout the day, a tracking mechanism points the solar concentrator structure at the sun as it crosses the sky. Electrical output drops dramatically if the sun is not focused on the cell or if clouds block the sun. The resulting system consists of a solar concentrator using mirrors or lenses, a tracking mechanism, solar cells, and a heat sink. As shown in Figs. 1 and 2, these CPV systems are quite different from the flat-plate PV panels generating electricity throughout the world today. Sun-tracking also increases the daily energy production above that of non-tracking flat-plate PV panels. Utilities are interested in all solar tracking technologies because of this additional value in energy production.

2 The CPV Market

Sometimes technologies are developed that few people buy. For decades, CPV appeared to be one of those technologies with no market and no customers. Of some 1,500 megawatts (MW) of PV sold throughout the world in 2005, less than 1 MW were CPV systems. Although most of the world's PV installations in 2005 were on rooftops, CPV systems had not been developed for roofs. As the photos show, typical CPV systems are large and more suitable for a utility customer, although several

Fig. 1. Several 35-kilowatt (kW) CPV systems built by Amonix in Torrance, California, are installed at an Arizona Public Service power plant. The system uses Fresnel lenses to concentrate sunlight. The pickup truck in the shade gives an idea of size.

Fig. 2. Several 25-kW CPV systems built by Solar Systems in Hawthorn, Australia, and installed on aborigine lands. These systems use mirrors for concentration (see www.solarsystems.com.au). Note the people in the foreground for an idea of size.

companies are now developing smaller CPV products for rooftop markets. CPV systems generate little electricity in areas with cloud cover and, not surprisingly, CPV researchers often live in sunny areas such as the southwestern United States, Israel, Spain, and Australia. It is frequently stated that CPV will be competitive only in these sunny, cloudless regions; however, studies of CPV in less sunny locations suggest that the costs could still be competitive with those of other PV technologies if the solar cell efficiencies are high enough.[3,4] Nevertheless, CPV systems will certainly penetrate their first markets in these sunny areas—just as the first wind systems were installed in very windy locations before going into less windy sites as their costs declined.

In the 1980s, the U.S. Department of Energy (DOE) and the Electric Power Research Institute (EPRI), the research organization for electric utilities, funded CPV projects for utility applications. Both organizations curtailed their CPV studies in the early 1990s as rooftop PV markets started to become dominant. Recently, however, two companies—Amonix in California and Solar Systems in Australia—found customers for their systems shown in Figs. 1 and 2. Amonix now has a 10 MW/year production facility in a joint venture with the developer, Guascor, in Spain. And Solar Systems has been installing hundreds of kilowatts of CPV systems in Australian outback locations where electricity is expensive due to high transport cost of diesel fuel for diesel generators.[5] For almost two decades, these two companies have been persistent and innovative in developing several generations of CPV designs leading to their present products.

But more than technology development is needed for a new product to enter energy markets. Market incentives can be critical, especially for new technologies struggling to compete with deeply entrenched conventional energy technologies. The justification for society to provide market incentives can be the benefits of clean air, combating global climate change, providing local energy production and jobs, as well as avoiding the often-ignored problems of mining and waste removal associated with large-scale, conventional energy sources. Further, today's conventional energy sources have a long history of government incentives and support for justifiable reasons. As renewable technologies mature and energy needs increase, governments around the world are finding renewable energy market incentives both justifiable and effective in responding to society's energy concerns.

For PV systems, two main types of market incentives exist. Most government market support for PV in the United States is in the form of money refunded for the purchase and installation of a PV system. Therefore, many dollars per installed PV watt are returned to the customer, who, in turn, hands the money over to companies providing and installing the systems. These rebates were designed for companies and customers wanting to install small flat-plate PV systems for rooftops, which is the principal market for PV systems. Such rebates have been successful in developing PV markets for rooftops in Japan, as well as in the United States. Almost 20 states have some form of rebate for PV systems that can be combined with the new federal rebate approved by the U.S. Congress in 2005.

However, there is an issue with most rebates in that they are paid at or soon after the PV installation, with little or no requirements that the system perform well in 2, 5, or even 20 years from the time of sale. Addressing such a situation, Germany developed an effective *feed-in tariff* program that pays, at a declining rate, for the energy produced over 20 years. The State of Washington and Spain recently initiated their own programs for feed-in tariffs and California is beginning to move in this direction. These programs express a commitment by the governments to honor energy purchase agreements for as long as 15 or 20 years. The U.S. rebate for PV systems is presently planned to be available for only 2 years. Feed-in tariffs can be very important market incentives, especially ones designed to reward investors for energy production, to reduce their risk in recovering their investment and to promote long-term system reliability. Such tariffs have been instrumental in the market success of wind energy systems, presently totaling about 10 times more electricity generating capacity than the world's PV systems. In the case of CPV systems, feed-in tariffs open a market door for a technology that maximizes electricity production because CPV systems produce more kilowatt-hours (kWh) per kW than flat-plate PV systems. An attractive feed-in tariff provided the economic justification for the recent Amonix-Guascor CPV joint venture in Spain.

As Fig. 3 shows, an advantage exists today for CPV systems using high-efficiency solar cells in terms of energy produced for the same amount of capital invested in different PV systems.[6] This is a very simple comparison between total project cost and annual energy produced for the different systems. It avoids the many assumptions required in other techno-economic analyses, such as the levelized cost of electricity. The comparison is made between a typical non-tracking flat-plate PV system, a single-axis-tracking flat-plate system, a CPV system using standard CPV silicon technology, and a CPV system with today's new high-efficiency solar cells. A $1000 investment in a technology using today's high-efficiency CPV cells could yield 450 kWh per year—almost 2-1/2 times more electricity than that generated by $1000 paid for fixed flat-plate PV systems. The increased *bang for the buck* is huge for investment in CPV technologies using new high-efficiency cells.

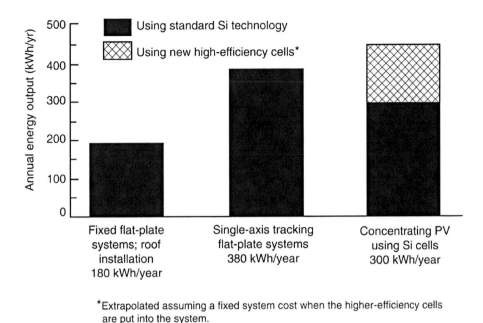

*Extrapolated assuming a fixed system cost when the higher-efficiency cells are put into the system.

Fig. 3. CPV systems using new high-efficiency solar cells generate considerably more electricity for the same amount of money than do the alternatives.[6]

3 Higher and Higher Conversion Efficiencies

With the advent of funding from the DOE in the late 1970s and early 1980s came plans and goals to develop PV technologies through improving performance (efficiency), reducing cost, and assuring reliability of operation. CPV systems offered the possibility of lower cost because expensive solar cells are replaced with less costly structural steel holding mirrors or lenses. However, early CPV systems showed the importance of optical efficiencies as optical losses typically reduced the CPV system efficiency by 15% to 20%. To compensate for optical losses, CPV systems needed the highest-quality, highest-performing solar cells to compete with flat-plate PV systems.

Early PV researchers, principally Martin Green in Australia and Richard Swanson and Vahan Garboushian in the United States, developed innovative designs for crystalline silicon solar cells, leading to the record efficiencies of the 1980s and 1990s. Today's CPV systems using high-efficiency crystalline silicon solar cells have system efficiencies approaching 20%. Installed CPV system costs are comparable today with those of utility-scale flat-plate PV systems at about $6/watt.[5] But the

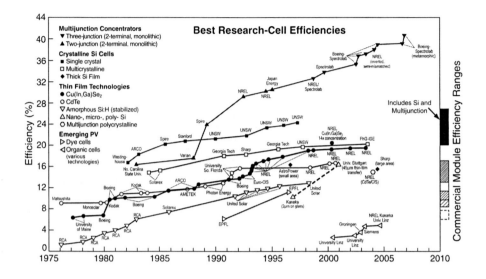

Fig. 4. The highest-efficiency solar cells, both crystalline silicon and multijunction concentrator devices, have been most suitable for solar concentrator systems. Replace with graph having the new 40.7 % result in 2006.

dramatically higher efficiency solar cells—above 40% now—as shown in Fig. 4, are creating considerable excitement about CPV systems.[5]

Research on multijunction solar cells began in the 1980s as part of a DOE effort to explore new solar cell materials and new solar conversion processes to improve cell efficiency. A single-junction solar cell is tuned to just one wavelength of the solar spectrum so that maximum efficiency occurs only at that color. (A semiconductor junction refers to an interface between a p-type semiconductor material and an n-type material. P and n refer to semiconductor charge carriers and are a reminder that solar cells behave like batteries in that they have positive terminals, negative terminals, and generate direct current.) Early solar cell researchers calculated that an infinite number of junctions would be the most effective means to harvest energy from each and every color in the solar spectrum and that such a stacked set of junctions could theoretically convert more than 80% of the sunlight into electricity. Yet, the first monolithic two-junction solar cell, made almost three decades after the discovery of the modern solar cell, demonstrated efficiencies less than that of a single-junction cell. The materials and chemical science difficulties encountered in making the first monolithic two-junction solar cells were significant. These multijunction PV technologies are based on elements in columns III and V of the Periodic Table, and they are often referred to as III-V solar cells. Soon thereafter, two-junction III-V solar cells were developed with efficiencies higher than those of the best silicon solar cells.

Table 1. Benchmark (10-MW) System parameters and impact of multijunction (III-V) solar cell efficiency on a CPV utility reference system.[7] High-efficiency solar cells are installed in essentially identical solar concentrator structures, and the cost per watt drops from about $6/watt to well under $2/watt while electricity costs fall below 10 cents per kWh. Higher production levels can lead to even lower levelized costs of energy (LCOE).[7]

System size	MW	10	12.5	16
Module price	$/W_dc	4.13	3	1.56
Cell efficiency	%	26 (Si)	32 (III-V)	40 (III-V)
Module size	kW_pdc	40	50	64
Module efficiency	%	20	25	32
Installed system price	$/W_dc	5.95	4.3	2.52
LCOE	$/kWh_ac	0.15–0.27	0.10–0.15	0.06–0.11

The U.S. Department of Defense recognized the potential of these new solar cells for powering satellites and supported the development of their manufacturing processes. Again, a new PV technology found a commercial niche in space power markets. Today, almost every commercial and defense satellite—as well as the Mars Rover instrumentation packages—use multijunction III-V solar cells for their electrical power sources. Just before the turn of the century, collaborative research and development by the National Renewable Energy Laboratory and Spectrolab, a division of Boeing, demonstrated a three-junction solar cell with a higher efficiency than that of two-junction cells. Efficiencies are now over 40% in the laboratory, with reasonable quantities of 35% cells available from suppliers. This technology may be coming back to earth more quickly than the early silicon cell technology did as today's governments and investors respond to the world's demands for more and cleaner energy sources. As shown in Table 1, performance pays.[7] Capturing those economic benefits involves replacing the crystalline silicon solar cells in essentially identical solar concentrator structures with new high-efficiency III-V multijunction cells.

The pioneering companies of Amonix and Solar Systems developed their CPV structures around crystalline silicon solar cells, but both are rapidly incorporating the new high-efficiency multijunction cells into CPV products that they expect to have available in the near future—within 2 to 5 years. Can the companies making multijunction III-V solar cells (e.g., Spectrolab and Emcore in the United States) meet this new and imminent market demand? Today's annual manufacturing capacity for multijunction solar cells is about 1 MW under 1-sun illumination. Remember that these multijunction cells are used in non-concentrator versions in space. And as the market for satellites undergoes its own demand cycles, there are periods in which substantial portions of the production facilities are available for other markets, such as the terrestrial CPV market. Concentration provides a huge lever to this production capacity. A solar concentration ratio of 1000 suns means that a manufacturing capacity of 1 MW of flat-plate space PV panels could be the solar power sources for 1000 MW of CPV systems. The total production capacity throughout the world for III-V multijunction solar cells is already about 1 MW/year. The potential capacity therefore exists for CPV technology to make a dramatic leap from megawatts to gigawatts in the market in the very near future.

However, companies want to be sure that these new multijunction solar cells will operate reliably in their CPV systems. After all, the new solar cells typically operate at higher voltages, generate higher current, and behave differently under environmental conditions of temperature cycles and humidity than do crystalline silicon solar cells. And there is a long history for crystalline silicon operation on earth, whereas very few multijunction III-V solar cells have been deployed in field installations. However, early demonstrations are promising. One CPV company, Concentrating Technologies, has operated Spectrolab's triple-junction solar cells for more than one year at an Arizona Public Service test site. Nevertheless, companies integrating these new solar cells into their solar concentrator structures expect it will take 2 to 5 years to assure reliable products for the marketplace.[5]

4 CPV Reliability

Today, flat-plate crystalline silicon technologies are renowned for their reliability in generating electricity for decades. What is often forgotten is that before flat-plate PV test standards were established in the early 1980s, large projects of flat-plate PV systems sometimes failed catastrophically. Standards organizations provide an important service for all technology development activities by providing a forum for companies, customers, and independent engineers to create a set of agreed-upon tests for identifying weaknesses in products before they go to market. Test standards, especially military standards, were critical to the success of space solar cells developed for defense satellites in the 1960s and 1970s. However, with the first efforts to bring space PV down to earth, project leaders discovered that the test standards for space solar cells were inadequate for terrestrial PV systems. Programs begun in the early 1980s at the Jet Propulsion Laboratory led to the successful development of qualification standards for crystalline silicon flat-plate PV technologies. Today, crystalline silicon solar cells are renowned for their long-term reliability, and few people are aware of the early disasters.

Many early CPV systems suffered the same fate in that reliability was a serious issue. Professor Charles Backus, a CPV pioneer and mechanical engineer, noted that electrical or electronic engineers developing PV systems and PV standards were unaccustomed to solving mechanical engineering problems or developing standards for large mechanical structures.[2] In the late 1980s, Sandia National Laboratories developed a set of stress tests (accelerated environmental testing) for CPV systems based on their early CPV field tests funded by the DOE. This work served as the basis for the first CPV qualification standard developed in the late 1990s and finally published by the International Electrical and Electronics Engineers (IEEE) standards organization in 2001. This first IEEE standard was most suitable for CPV systems using Fresnel lenses, typical of many U.S. CPV designs.

The International Electrotechnical Commission (IEC), based in Geneva, Switzerland, is voting on the first international CPV draft standard suitable for testing all of the CPV geometries and technologies. The IEC standard builds on the concept of testing representative sample assemblies for key elements of different designs under conditions of high temperature, low temperature, temperature cycling, humidity, electrical performance under wet and humid conditions, and outdoor performance. This new IEC standard is expected to play an important role as companies work rapidly to integrate III-V multijunction solar cells into their solar concentrator structures.[8]

5 Following in Wind Energy's Footsteps

Wind energy is one renewable energy technology developed successfully by mechanical engineers. In the 1970s, wind systems and PV systems started out on nearly the same footing: only a few experimental systems for each were installed around the world. Today, there are roughly 10 times more wind energy systems than PV systems installed—50,000 MW of wind systems versus 5,000 MW of PV systems. Why were wind energy technologies able to surpass PV systems? One reason was that wind developers were able to quickly demonstrate economies of production just as a market opportunity appeared. The state of California offered long-term standard-offer contracts from 1985 to 1989 to purchase the electricity over 20 years from large-scale renewable energy projects. These long-term contracts were similar in many ways to the successful European feed-in tariffs. Solar concentrators producing heat to drive electric generators—called concentrating solar power (CSP) systems—also took advantage of the California opportunity; almost 400 MW of CSP systems were installed in the 1980s, and they have been generating solar electricity ever since then. Fabrication facilities are relatively inexpensive for both wind and CSP systems when compared with PV manufacturing facilities. Wind production facilities resemble automobile assembly lines.[9]

PV production facilities, although not as complex or costly as those of the integrated-circuit industry processing semiconductor silicon, still cost roughly 10 times more than wind production facilities. For early investors, this is an important issue. Consider the investment choice. Crystalline silicon and amorphous thin-film flat-plate PV production facilities both cost about $100 million or more for 100 MW per year manufacturing plants.[10,11] A wind production facility of the same size might cost $10 million for the same annual production. Investment in production facilities for new technologies entails significant risk, and the lower risk for investing in wind facilities was one reason investors provided funds for large 1,000 MW wind projects in the 1980s. PV was not able to demonstrate the economies of production quickly enough to take advantage of the small window of opportunity provided by California's standard-offer contracts. In the 1980s, the very-high-efficiency solar cells needed by CPV systems were still in the research laboratory.

The California market incentives helped wind and CSP developers and investors move their technologies forward, reducing cost and acquiring valuable operational experience that improved reliability. Wind engineers developed their qualifications standards during this same period; like early PV technologies, wind systems often suffered from poor reliability until their certification standards were established and required in the marketplace.

Fig. 5. An Amonix production facility in Los Angeles is strikingly different from a flat-plate PV manufacturing facility. This difference results in much lower capital costs for the facility.

Some noteworthy similarities exist between wind energy systems and CPV systems.[9] They both employ relatively common materials, particularly steel. Wind system costs are typically less than $1 per watt; they depend mainly on the cost of steel, whereas flat-plate PV is linked to the availability and cost of expensive semiconductor silicon. But solar concentrator structures are also amenable to an auto-assembly type of production (see Fig. 5), and CPV developers estimate CPV production facility costs are much closer to those of wind systems than to those of flat-plate PV production facilities. In early EPRI cost studies, CPV production facility costs were estimated (on the same costing basis as the crystalline and amorphous silicon facilities) to be about $28 million for a 100 MW per year installation—about one-quarter the cost of the conventional silicon PV facilities.[10] These lower investment costs can lead to a faster scale-up of manufacturing facilities because investor risk is relatively smaller than the risks entailed in investing in conventional PV production facilities.

Further, cost studies in Spain and Israel estimate CPV installed system costs will, like wind systems costs, finish below $1 per watt when gigawatt levels of CPV production are reached.[7,12] Both CPV and wind energy technologies are modular, like flat-plate PV modules, but the sizes are different. Wind units are now megawatts in size whereas CPV units range from kilowatts to tens of kilowatts. Flat-plate PV modules are usually less than 100 watts. And, obviously, both wind and CPV systems have moving parts, yet moving parts have not limited the success of wind systems. Finally, wind systems first penetrated the energy marketplace in sites with very high and steady winds, whereas CPV systems will almost certainly enter markets in locations with considerable sunlight and almost no clouds, similar to the climates of the southwestern United States, Spain, Australia, and Israel.

6 Low-Cost Hydrogen from Hybrid CPV Systems

We are now in a position to understand why this new high-efficiency solar electric technology provokes a fresh look at the challenge of generating hydrogen from water using sunlight. In addition to generating solar electricity at low cost, CPV systems have the potential to produce hydrogen through an electrolysis process. The generation of electrolytic hydrogen from solar energy is critically important to the world's long-term energy needs for several reasons. The feedstock (water) and supplied energy (solar) are inherently carbon free so that on a life cycle basis the total carbon emissions will be significantly less than from fossil-based options for generating hydrogen. And there is the potential to generate hydrogen near its markets, thus minimizing transportation costs. In the past, the principal criticism of photovoltaics for generating hydrogen has been the high cost of PV electricity and the inefficiencies of the conversion processes, particularly the PV process.

As we have seen, CPV systems have the potential for generating lower-cost electricity, primarily due to developing high-efficiency multijunction III-V solar cells with efficiencies above 40%. But it is the heat boost from CPV systems that can dramatically improve and enhance the electrolysis efficiency of water in a high-temperature solid-oxide electrolyzer. This heat boost—40% was measured in the 1990s by the company Solar Systems in Australia above 1100°C[13,14]—has been substantiated in recent theoretical analyses.[15] This new pathway provides significant engineering and economic benefits for generating electrolytic hydrogen from solar energy, thereby creating opportunities for PV to contribute to future transportation markets directly with low-cost hydrogen or by producing liquid hydrogen-carrier fuels such as methanol.[16]

Solar-to-hydrogen conversion efficiencies of 40%, including optical losses, are attainable in the near-term (within the next few years) using high-efficiency III-V multijunction solar cells, whereas efficiencies of 50% and higher are realistic targets within 5 to 10 years. These efficiencies are dramatically higher, by roughly a factor of 3 or 4, than those of any of the other methods previously considered for generating electrolytic hydrogen from solar electricity.[16] These results, based on the long-term potential for CPV systems to be mass produced at costs of less than $1/W, lead to hydrogen production costs comparable with the energy costs of gasoline—recognizing that 1 kg of hydrogen has the energy equivalent of one U.S. gallon of gasoline.[17,18]

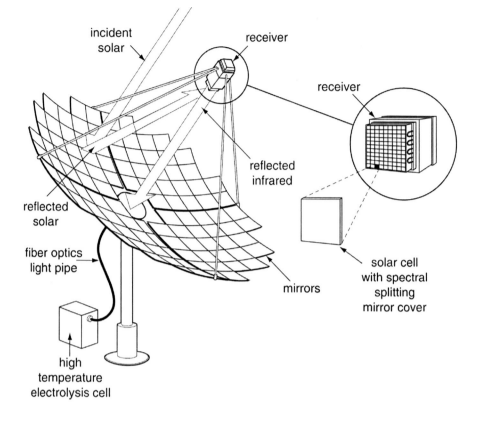

Fig. 6. Schematic of system shows sunlight reflected and focused on the receiver, with reflected infrared directed to a fiber-optics light pipe for transport to a high-temperature solid-oxide electrolysis cell. Solar electricity is sent to the same electrolysis cell, which is able to use both heat and electricity to split water.

7 Describing the Hybrid CPV System

This approach first proposed by Solar Systems in Australia employs a dish concentrator that reflects sunlight onto a focal point (see Fig. 6). At the focal point is a spectral splitter (heat mirror) that reflects infrared solar radiation and transmits the visible sunlight to high-efficiency solar cells behind the spectral splitter. Figure 7 schematically shows the transmission across the solar spectrum wavelengths.

The reflected infrared radiation is gathered by a fiber-optics "light pipe" and conducted to the high-temperature solid-oxide electrolysis cell. The electrical output of the solar cells also powers the electrolysis cells. About 120 megajoules are needed—whether in electrical or thermal form, or both—to electrolyze water and generate 1 kg of hydrogen. The result is that more of the solar energy is used for

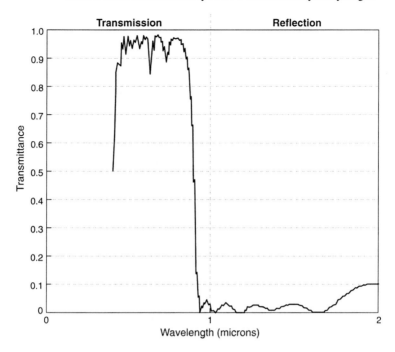

Fig. 7. Transmittance (and reflectance) of a spectral splitter mirror as a function of solar wavelength in microns. This response depicts that of a "hot mirror" in which light is transmitted in the visible region and reflected in the infrared.

hydrogen production. And we shall see that the additional costs for the hybrid solar concentrator components—the spectral splitter and fiber-optics light pipe—are relatively small compared with the boost in hydrogen production.

The testing of components shown in Fig. 6 occurred in the mid-1990s and has been described previously[13,14,17] on a scale considerably smaller than that of Fig. 2. The solar concentrator was a paraboloidal dish 1.5 m in diameter, with two-axis tracking, and is capable of more than 1000-suns concentration (Fig. 8). The full dish was not needed and most of it was shaded appropriately for use with the small electrolysis cell. At that time, the solar cell was a GaAs cell with an output voltage of 1 to 1.1 V at maximum power point, with a measured efficiency of about 19%. The voltage was an excellent match for direct connection to the electrolysis cell when operating at 1000 °C. The tubular solid-oxide electrolysis cell was fabricated from yttria-stabilized zirconia; the cell had platinum electrodes because the test temperature was higher than that of typical solid oxide cells. Figure 9 shows a schematic of the solid-oxide electrolysis cell operation.

A metal tube surrounded the cell to uniformly distribute the solar flux over the cell's surface. The test occurred during a 2-hour period of operation, with an excess of steam applied to the electrolysis cell. The output stream of unreacted steam and generated hydrogen was bubbled through water and the hydrogen was collected

Fig. 8. This photo, taken in the 1990s, shows John Lasich demonstrating how cool his concept is for conducting infrared energy through a fiber-optics light pipe. The dish reflects sunlight to a "heat mirror" that reflects long-wavelength solar radiation to the fiber-optic bundle along the axis of the parabolic dish. The visible light seen at the end of the light pipe in Lasich's hand is a result of partial reflection of visible light by the heat mirror.

and measured. During a definitive 17 minutes of system operation in steady state, 80 mL of hydrogen were collected. The ratio of the thermoneutral voltage of 1.47 V to the measured electrolysis cell voltage of 1.03 V was 1.43, corresponding to a boost of more than 40% in hydrogen production due to the input of thermal energy. This was also confirmed by energy balance. Combining the optical efficiencies of the concentrator dish (85%), solar cell efficiency, and thermal-energy boost, the total

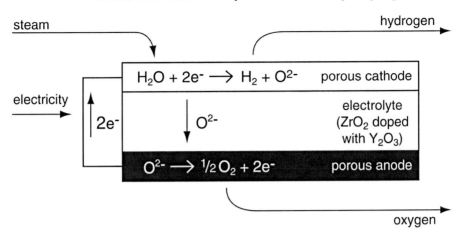

Fig. 9. Schematic of high-temperature electrolysis in a solid-oxide cell. The geometry can be planar or tubular as in the case of the first demonstration of the hybrid solar concentrator PV system. Operating the electrolysis cell in reverse corresponds to electricity and heat production in solid-oxide fuel cell operation.

system efficiency was 22% for conversion of solar energy to hydrogen. At the time of these measurements in the mid-1990s, the efficiency was almost three times better than that recorded for any other technology converting solar energy to hydrogen.

These early tests were not conducted with the most efficient solar cells available at that time. The record efficiency then was about 30% for a laboratory cell (see Fig. 4) and those cells were not easily obtainable. Today's record efficiency is 40.7%, and 35%-efficient cells are commercially available.[18] Therefore, 40% solar-to-hydrogen efficiency is expected in the near term assuming a heat boost of 40%, a multijunction solar cell efficiency of 35%, and an optical efficiency of 85%. A 40% multijunction solar cell would yield a solar-to-hydrogen conversion efficiency of almost 50%. Nevertheless, electrochemical theoretical results calculated by Licht, shown in Figure 10, are consistent with these predictions based on Solar Systems' early experiments.[15]

Two cost analyses have been reported for this concept.[17,19] Although the resulting hydrogen costs agreed within their costing uncertainties, the hydrogen generation plants were quite different in nature, as were the financial assumptions in their cost analyses. The first analysis was conducted in 2004 and reported in 2005, and it used a set of financial and plant assumptions developed by the DOE Hydrogen Program.[19,20] Because of the complexity of the DOE H2A plant assumptions, a *back of the envelope* calculation with simplified financial assumptions was made to highlight key cost elements.[17] This analysis is presented below and compared with the H2A analysis. The principal difference between the two analyses is the additional costs of operation, transportation, storage and distribution in the H2A analysis.

The largest cost for the hybrid solar concentrator system will be for the dish concentrator and PV receiver, shown in Fig. 6. Algora recently completed an extensive cost analysis based on previously collected data for CPV systems.[7] Many of the

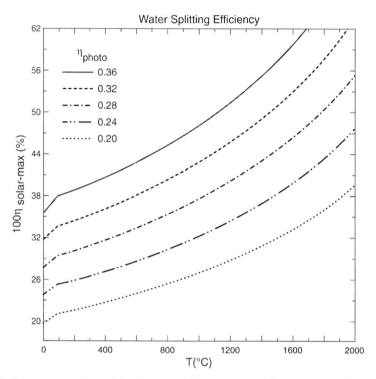

Fig. 10. Energy conversion efficiency of solar-driven water splitting to generate H2 as a function of temperature and photovoltaic conversion efficiency at AM1.5 insolation, at pH2O = 1 bar. Reprinted with permission from *J. Phys. Chem. B* **2003**, 107, 4253-4260. Copyright 2003 American Chemical Society."

project costs came from installed costs for the 480-kW reflective CPV system in Tenerife, Spain. The analysis included a wide range of parameters, including cumulative production of 10 MW for present-day systems to cumulative production of 1000 MW for the mid-term systems where *learning* cost reductions are incorporated. Concentrations ranged from 400 to 1000 suns, with solar cell efficiencies ranging from 32% to 40%. Module efficiencies ranged from 24.8% to 32.2%, and the plant's AC annual efficiency ranged conservatively from 18.2% to 23.6%. Present-day base costs were 2.34 euro/W (almost $3/W with today's exchange rate). The lowest projected system costs ranged from 0.5 to 1 euro/W for efficiencies of 40%, 1000-suns concentration, and cumulative production of 1000 MW.

We wanted to compare the results of our simplified engineering cost analysis of hydrogen generated by this hybrid solar concentrator system with those of more extensive cost analyses:

1. the electrolytic generation of hydrogen by wind systems, where cumulative production of this highly developed technology is approximately 50 gigawatts (GW); and,

Table 2. Component and system costs for 10-MW hybrid CPV project for solid-oxide electrolytic production of hydrogen.

	Component costs assuming 1000-MW technology ($/kW)
Concentrator PV	800
Spectral splitter	15
Fiber optics	25
Electrolysis cell	400
Total System Cost	1240

2. the conventional production of hydrogen by reforming natural gas. So, for our analysis, we used cost estimates for mature CPV technology.

Cost studies for conceptual high-temperature nuclear reactors (projected for mature 600-MW designs) suitable for high-temperature electrolysis cells face similar problems because both the hybrid solar concentrator and high-temperature nuclear reactor are in early stages of exploratory research and development for hydrogen generation. Further, high-temperature solid-oxide electrolysis cells will be required in large sizes (500 kW to 500 MW) for integration with nuclear reactors.[21] Unit sizes ranging from 20 to 50 kW could be used with hybrid solar concentrators. Although solid-oxide fuel cells are commercially available, solid-oxide electrolysis cells are beginning development. It is important to note that solid-oxide electrolysis cells have been demonstrated to date in small sizes equivalent to hundreds of watts. The modular character and size of the hybrid CPV system is commensurate with the development of solid-oxide electrolysis cell technology, also in early stages of development.

Using a set of assumptions for a well-developed technology, we acquired costs in $/kW for solid-oxide electrolysis cells from a developer of solid-oxide electrolysis cells.[22] Table 2 summarizes the cost data for a well-developed technology (1000-MW cumulative production) for the hybrid CPV system and high-temperature solid-oxide electrolysis cell. Table 3 summarizes the hydrogen production costs for a 10-MW project built with the well-developed technology assuming a 20% rate of return per year not including operating, storage, transmission or distribution costs.[19] It also contain the estimated costs from the H2A analysis that includes the additional operating, storage, transmission and distribution costs expected for a distant, centralized hydrogen generation plant.[17]

Table 4 compares these production costs with those of other hydrogen production technologies.

8 Discussion

The literature contains many cost analyses for hydrogen production, but the assumptions behind the analyses vary dramatically. The DOE, through its Hydrogen Program, is establishing a cost-analysis structure for comparing different hydrogen and

Table 3. Hydrogen production data for mature 10-MW plant.[17] The estimated hydrogen cost of $2.48/kg has considerable uncertainty related to technology immaturity and simplistic assumptions. The H2A costs include operating, storage, transmission and distribution costs.[19]

	Hydrogen cost data for mature technology
Plant size (MW)	10
Plant cost ($ million)	12.4
H_2 produced (kg/yr)	10^6
Hydrogen cost ($/kg)[17]	2.48
H2A Hydrogen cost ($/kg)[19]	3.18

fuel cell technologies within a common set of assumptions. The analysis in Table 3 is a preliminary study needing additional work to fit within that framework. CPV systems are just beginning to enter the energy market, so cost uncertainties are significant compared with those of highly developed wind systems with a worldwide installed capacity approaching 50 GW. Nevertheless, these preliminary hydrogen costs are comparable with hydrogen costs from wind electrolysis, so additional cost studies are warranted. Today, wind system costs are in the $800/kW range—as are the estimated costs for highly developed CPV systems—whereas wind electrolysis does not have an opportunity for a heating boost in electrolysis efficiency. Assuming these cost analyses continue to be positive, it will be worth demonstrating this hybrid solar concentrator technology on a larger scale.

The uncertainties in this cost analysis arise principally from the early stage of technology development for solar concentrators, high-efficiency solar cells, and solid-oxide electrolysis cells. There are many positive indications that these technologies can progress and achieve their performance and cost potentials, but additional work will be needed.

9 Hydrogen Vision Using Hybrid Solar Concentrators

The U.S. National Research Council and National Academy of Engineering believes that one of the four most fundamental technological and economic challenges for the hydrogen economy is:

"To reduce sharply the costs of hydrogen production from renewable energy sources over a time frame of decades"[23]

Table 4. Cost comparison for the hybrid CPV production of electrolytic hydrogen. Note that 1 kg of hydrogen has the energy equivalent of one U.S. gallon of gasoline.

Process	Hydrogen Production Cost ($/kg)
Gas reformation[20]	1.15
Wind electrolysis[20]	3.10
Hybrid CPV electrolysis[17] (approximating distributed generation)	2.48
Hybrid CPV electrolysis[19] (assuming centralized generation of H2A analysis)	3.18

Wind electrolysis is a strong renewable energy option, and this study indicates hybrid CPV electrolysis could be another. Also, the solar energy resource is considered larger and more widely distributed than that of wind energy. And totally new system configurations may be possible with hybrid solar concentrator electrolysis. Small 50-kW systems could be part of hydrogen filling stations, reducing hydrogen distribution costs. Systems could incorporate backup-heating sources, probably natural gas in the near term, to improve the electrolysis system capacity factor.[22]

Electricity providers throughout the world are considering large-scale CPV projects, some in the range of 100s of MW. The hybrid CPV system could generate both electricity and hydrogen for future electric utilities. With low-cost tank storage on utility land, the solid-oxide electrolysis cell could be designed to operate in a regenerative mode, producing electricity from hydrogen during non-solar periods. This design would greatly increase the value of solar electricity to utilities.

Probably the most dramatic impact of this study has been the realization that the hybrid CPV system is a PV option that could provide transportation fuel on a large scale. In a scenario where hydrogen is used in fuel cell vehicles—which can have double the efficiency of standard internal combustion cars—the *effective cost* of solar hydrogen would be half, i.e., $1.24/kg. For customers paying $3 per U.S. gallon for gasoline, which is more than twice the effective hydrogen cost, the potential for a very large market clearly exists. To determine the final price of solar hydrogen to the customer, we would need to factor in the additional costs of operation, distribution, retailing, and taxes, as well as consider the society cost benefits due to the *clean and renewable* value of solar hydrogen.

The International Energy Agency's Photovoltaic Power Systems Program recently published a study on the feasibility of very large-scale PV systems.[24] Entitled *Energy from the Desert*, the study explores the concept of using the world's deserts to provide electricity at terawatt (10^{12} watt) levels of production. While it is evident that CPV systems could be a major contributor to such large projects, the hybrid solar concentrator opens the possibility for a new vision, one where the world's oceans could be harvested for hydrogen. Such a vision would again follow in the footsteps of wind technology as wind projects continue to appear off of the world's coastlines.

The hybrid solar concentrator is a potential *leap frog* technology that may rapidly lower the cost of clean hydrogen in light of the following: the imminent market entry of CPV systems for electricity production; solar cell efficiencies above 40%, with clearer ideas for 50%-efficient solar cells; and the opportunity to use wasted solar heat for augmenting solar electrolysis.

10 Conclusions

An innovative hybrid CPV electrolysis technology has been described that offers a potential cost of hydrogen lower than that from wind electrolysis and in the same range as gasoline for much of the world's population. The analysis is preliminary, but additional cost analysis and technology demonstrations are warranted.

This study has described several reasons supporting the argument that this is a technology on the horizon that could be significant throughout the world. The reasons include the following:

- The existing world's annual production capacity for manufacturing high-efficiency III-V multijunction solar cells is enough today for 1000 MW annually of CPV systems. This level provides a good jumpstart into the market.
- Society's recent concerns about energy security, global climate change, clean air, and high-technology economic opportunities are leading to the appearance of market openings based on feed-in tariffs in regions with solar resources suitable for CPV systems.
- The CPV community has completed the first CPV qualification standards in time to respond to market opportunities, although additional safety, performance, and tracker standards are still needed.
- The early success of wind energy technologies, similar in several respects to CPV technologies, augurs rapid and dramatic success for CPV systems.
- The possibility of efficiently producing hydrogen by splitting water with low-cost solar electricity opens a new pathway for production of transportation fuels.

This innovative renewable energy technology could *leap frog* other renewable energy technologies for electrolytic production of hydrogen—a potentially important transportation fuel for our future.

Acknowledgements

The author acknowledges the valuable work of many CPV pioneers who were responsible for the research progress described in this Chapter. At the risk of overlooking others, the author has particularly benefited from numerous articles by and discussions with Charles Backus, Andreas Bett, Vahan Garboushian, Martin Green, Richard King, Sarah Kurtz, John Lasich, Antonio Luque, Jerry Olson, Gabriel Sala, Richard Schwartz, Richard Swanson, and Masafumi Yamaguchi. Tomorrow's CPV companies will be building on the successes of these early leaders in developing high-efficiency solar cells and solar CPV technologies. In the field of hydrogen and electrolysis, the author acknowledges valuable discussions with John Turner, Krishnan Rajeshwar, Joe Hartvigsen, S. Srinivansan, and again, John Lasich, the originator of the hybrid solar CPV concept for electrolytic hydrogen production.

References

1. http://web.mit.edu/2.009/www/experiments/deathray/10_ArchimedesResult.html, October 2005.
2. C. E. Backus, *A Historical Perspective on Concentrator Photovoltaics*, Proceedings of the International Solar Concentrator Conference for the Generation of Electricity or Hydrogen, Alice Springs, Australia, November 2003.

3. R. M. Swanson, *Straight Talk About Concentrators*, Future Generation Photovoltaic Technologies: First NREL Conference, Denver, Colorado, American Institute of Physics Conference Proceedings #404, October 1997, p. 277-284.
4. R. M. Swanson, The promise of concentratiors, *Prog. Photovolt. Res. Appl.* **8**, John Wiley and Sons, Ltd., Hoboken, NJ (2000) pp. 93-111.
5. R. McConnell, S. Kurtz, and M. Symko-Davies, Concentrating PV technologies: Review and market prospects, *ReFOCUS*, Elsevier Ltd, , July/August 2005, p. 35.
6. NREL Frequently Asked Questions (FAQs): *What's New in Concentrating PV?*, Report No. FS-520-36542; DOE/GO-102005-2027, February 2005.
7. C. Algora, *Next Generation Photovoltaics,* Chapter 6, Ed. by A. Marti and A. Luque, Institute of Physics Publishing, Bristol and Philadelphia, 2004.
8. L. Ji and R. McConnell, *New Qualification Test Procedures for Concentrator Photovoltaic Modules and Assemblies*, Proceedings of the 2006 IEEE 4th World Conference on Photovoltaic Energy Conversion, Waikoloa, Hawaii, May 2006.
9. R. McConnell, *Large-Scale Deployment of Concentrating PV: Important Manufacturing and Reliability Issues*, Proceedings of the First International Conference on Solar Electric Concentrators, New Orleans, Louisiana, NREL/EL-590-32461, May 2002.
10. R. Whisnant, S. Wright, P. Champagne, and K. Brookshire, *Photovoltaic Manufacturing Cost Analysis: A Required-Price Approach*, Vols. 1 and 2, EPRI AP-4369, Electric Power Research Institute, Palo Alto, CA, 1986.
11. R. Whisnant, S. Johnston, and J. Hutchby, Economic analysis and environmental aspects of photovoltaic systems, Ch. 21, *Handbook of Photovoltaic Science and Engineering*, Ed. by A. Luque and S. Hegedus, John Wiley and Sons, Ltd., 2003.
12. D. Faiman, D. Raviv, and R. Rosenstreich, *The Triple Sustainability of CPV with the Framework of the Raviv Model*, Proceedings of the 20th European Photovoltaic Solar Energy Conference and Exhibition, Barcelona, Spain, June 2005.
13. J. Lasich, U.S. Patent No. 5658448, August 19, 1997.
14. J. Lasich, U.S. Patent No. 5973825, October 26, 1999.
15. S. Licht, *J. Phys. Chem. B* **107**, 4253–4260, 2003 (also see Chapter 5 in this book).
16. N.Lewis, http://www7.nationalacademies.org/bpa/SSSC_Presentations_Oct05_Lewis.pdf, August 2006.
17. R. D. McConnell, J.B. Lasich, and C. Elam, *A Hybrid Solar Concentrator PV System for the Electrolytic Production of Hydrogen*, Proceedings of the 20th European Photovoltaic Solar Energy Conference and Exhibition, Barcelona, Spain, June 2005.
18. R. McConnell, M. Symko-Davies, and D. Friedman, *Multijunction Photovoltaic Technologies for High Performance Concentrators*, Proceedings of the 2006 IEEE 4th World Conference on Photovoltaic Energy Conversion, Waikoloa, Hawaii, May 2006.
19. J. Thompson, R. McConnell, and M. Mosleh, *Cost Analysis of a Concentrator Photovoltaic Hydrogen Production System*, NREL/CD-520-38172, Proceedings of the International Conference on Solar Concentrators for the Generation of Electricity or Hydrogen, Scottsdale, Arizona, May 2005.
20. D. Mears, M. Mann, J. Ivy, and M. Rutkowski, *Overview of Central H2A Results*, U.S. Hydrogen Conference Proceedings, April 2004.
21. R. Anderson, S. Herring, J. O'Brien, C. Stoots, P. Lessing, J. Hartvigsen, and S. Elangovan, Proceedings of the National Hydrogen Association Conference, 2004.
22. J. Hartvigsen, private communication; also see www.ceramatec.com 2006.
23. National Research Council and National Academy of Engineering, in *The Hydrogen Economy*, National Academies Press, Washington DC, 2004.

24. Photovoltaic Power Systems Executive Committee of the International Energy Agency, *Energy from the Desert: Feasibility of Very Large Scale Photovoltaic Power Generation Systems*, Ed. by K. Kurokawa, James and James, London 2003; also see http://www.iea-pvps.org/products/rep8_01s.htm.

5

Thermochemical and Thermal/Photo Hybrid Solar Water Splitting

Stuart Licht

University of Massachusetts, Boston, MA

1 Introduction to Solar Thermal Formation of Hydrogen

1.1 Comparison of Solar Electrochemical, Thermal & Hybrid Water Splitting

Solar electrochemical, solar thermal,[1,2] and solar thermal/electrochemical hybrid[3] hydrogen generation are introduced in this Section. Water electrolysis and electrolysis using solar concentrator technology were discussed in Chapters 3 and 4. The thermal and the hybrid processes will be discussed in depth in subsequent Sections of this chapter. At high temperatures (> 2000 °C), water chemically disproportionates to hydrogen and oxygen. Hence, in principle, by using solar energy to directly heat water to very high temperatures, hydrogen and oxygen gases can be generated. This is the basis for all direct thermochemical solar water splitting processes.[1] However, catalysis, gas recombination, and containment materials limitations above 2000 °C have led to very low solar efficiencies for direct solar thermal hydrogen generation. In another approach, the utilization of a multi-step, indirect, solar thermal reaction processes to generate hydrogen at lower temperatures has been extensively studied, and a variety of pertinent reaction processes considered.[2] These reactions are conducted in a cycle to regenerate and reuse the original reactions, ideally, with the only net reactant water, and the only net products hydrogen and oxygen. However, such cycles suffer from challenges often encountered in multi-step reactions.[4] While these cycles can operate at lower temperatures than the direct thermal chemical generation of hydrogen, efficiency loses can occur at each of the steps in the multi-step sequence, resulting in low overall solar to hydrogen energy conversion efficiencies.

Electrochemical water splitting, generating H_2 and O_2 at separate electrodes, largely circumvents the gas recombination and high temperature limitations occurring in thermal hydrogen processes. Thus a hybrid of thermal dissociation and electrolysis provides a pathway for efficient solar energy utilization. The hybrid method expands on existing solar electrochemical processes, which are therefore discussed briefly here. There has been significant, ongoing experimental[5-15] and theoretical.[5,16,17] interest in utilizing solar generated electrical charge to drive electrochemical water splitting (electrolysis) to generate hydrogen. In each of the above referenced studies, water electrolysis occurs at, or near, room temperature. Photoelectrochemical models predict a maximum ~30% solar water splitting conversion efficiency by eliminating

- the linkage of photo to electrolysis surface area,
- non-ideal matching of photo and electrolysis potentials, and incorporating the effectiveness of contemporary
- electrolysis catalysts, and
- efficient multiple bandgap photoabsorbers (semiconductors).[18]

However, these models did not incorporate solar heat effects on the electrolysis energetics as elaborated below.

The UV and visible energy rich portion of the solar spectrum is transmitted through water (Chapters 1 and 2). Therefore a mediator for light absorption, such as a semiconductor, is required to drive the electrical charge for the water-splitting process. The PV (photovoltaic) process refers to a solar panel connected ex situ to electrochemically drive water splitting, e.g., an illuminated semiconductor-based photovoltaic device wired to an electrolyzer. On the other hand, the photoelectrochemical process refers to in situ immersion of the illuminated semiconductor in a chemical solution (electrolyte) to electrochemically drive water splitting, as described in Chapter 7. The significant fundamental components of PV and photoelectrochemical hydrogen generation are identical, but from a pragmatic viewpoint the PV process seems preferred, as it isolates the semiconductor from contact with and corrosion in the electrolyte. The UV and visible energy rich portion of the solar spectrum is transmitted through H_2O. Semiconductors, such as TiO_2, can split water, but their wide bandgap limits the photoresponse to a small fraction of the incident solar energy. Solar photoelectrochemical attempts to split water have utilized TiO_2,[20] InP,[21] and also multiple bandgap semiconductors.[19,22,23] Photoelectrochemical water splitting studies have generally focused on diminishing the high bandgap apparently required for solar water splitting, by tuning (decreasing) the bandgap of the semiconductor, E_g, to better match the water splitting potential, E_{H_2O}. Multiples of electrolyzers and photovoltaics can be combined to produce an efficient match of the generated and consumed power, as shown in Fig. 1. Also multiple bandgap semiconductors can be combined to generate a single photovoltage well-matched to the electrolysis cell, and over 18% conversion energy efficiency of solar to hydrogen was demonstrated, albeit at room temperature (still without the potential benefits of hybrid thermal hydrogen generation.)[23]

Unlike room temperature solar PV and photoelectrochemical electrolysis, the hybrid approach utilizes energy of the full solar spectrum, leading to substantially higher solar energy efficiencies. The IR radiation is energetically insufficient to drive conventional solar cells, and this solar radiation is normally discarded (by reflectance or as re-radiated heat.) On the other hand, in the hybrid approach, as seen in Fig. 2

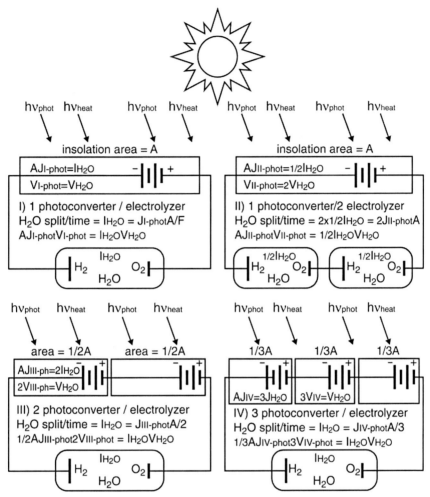

Fig. 1. Alternate configurations varying the number of photo harvesting units and electrolysis units for solar water splitting.[3] The photoconverter in the first system generates the requisite water electrolysis voltage and in the second system generates twice that voltage, while the photoconverter in the third and fourth units generate respectively only half or a third this voltage.

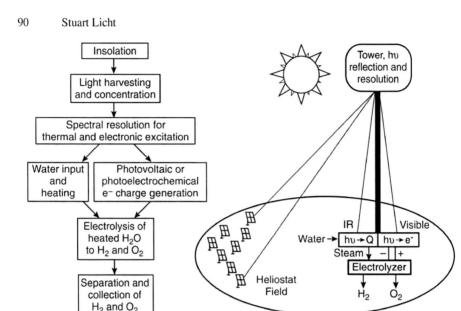

Fig. 2. Schematic representations of solar water electrolysis improvement through excess solar heat utilization.[3]

and as described in a latter Section of this chapter, the IR wavelengths are not discarded, but instead utilized to heat water. This in turn substantially decreases the necessary electrochemical potential to split the water, and substantially increases the solar hydrogen energy conversion efficiencies.

2 Direct Solar Thermal Water Splitting to Generate Hydrogen Fuel

2.1 Development of Direct Solar Thermal Hydrogen

The direct thermochemical process, to generate hydrogen by splitting water involves heating water to a high temperature and separating the hydrogen from the equilibrium mixture. Although conceptually simple, the single-step thermal dissociation of water has been impeded by the need for a high-temperature heat source to achieve a reasonable degree of dissociation, and—to avoid ending up with an explosive mixture—by the need for an effective technique to separate H_2 and O_2. Unfortunately, the decomposition of water does not proceed substantially until the temperature is very high. Generally temperatures of 2500 K have been considered necessary for direct thermal water splitting. The Gibbs function (ΔG, or free energy) of the gas reaction $H_2O \leftrightarrow H_2 + \frac{1}{2} O_2$, does not become zero until the temperature is increased

Table 1. The pressure equilibrium constants of the water dissociation reaction.[24]

	Temperature (K)		
	2500	3000	3500
K_1	1.34×10^{-4}	8.56×10^{-3}	1.68×10^{-1}
K_2	4.22×10^{-4}	1.57×10^{-2}	2.10×10^{-1}
K_3	1.52×10^{3}	3.79×10^{1}	2.67×10^{0}
K_4	4.72×10^{3}	7.68×10^{1}	4.01×10^{0}

to 4310 K at 1 bar pressure of H_2O, H_2 and O_2.[3] At a water pressure of 0.1 bar or greater, significant mole fractions of hydrogen are not spontaneously formed at temperatures below 2200 K. The entropy (ΔS), driving the negative of the temperature derivative of the Gibbs function change, is simply too small to make direct decomposition feasible at this time.[4]

2.2 Theory of Direct Solar Thermal Hydrogen Generation

In the high temperature gas phase equilibrium of water, in addition to H_2O, H_2 and O_2, the atomic components H and O must be considered. These components are relatively insignificant at temperatures below 2500 K, as the pressure equilibrium constants for either diatomic hydrogen or oxygen formation from their atoms are each greater than 10^3 at T $\leq$ 2500 K. However, the atomic components become increasingly significant at higher temperatures. The pressure equilibrium constants of the water dissociation reaction are summarized in Table 1 for water splitting at temperatures at which significant, spontaneous formation of H_2 occurs:

$$H_2O \leftrightarrow HO + H \qquad K_1 \qquad (1)$$

$$HO \leftrightarrow H + O \qquad K_2 \qquad (2)$$

$$2\,H \leftrightarrow H_2 \qquad K_3 \qquad (3)$$

$$2\,O \leftrightarrow O_2 \qquad K_4 \qquad (4)$$

Kogan has calculated that, at a pressure of 0.05 bar, water dissociation is barely discernible at 2000 K.[1] By increasing the temperature to 2500 K, 25% of water vapor dissociates at the same pressure. A further increase in temperature to 2800 K under constant pressure causes 55% of the vapor to dissociate.[1] These basic facts indicate the difficulties that must be overcome in the development of a practical hydrogen production by solar thermal water splitting:

(a) attainment of very high solar reactor temperatures,

(b) solution of the materials problems connected with the construction of a reactor that can contain the water spitting products at the reaction temperature, and

(c) development of an effective method for in situ separation of hydrogen from the mixture of water splitting products.

2.3 Direct Solar Thermal Hydrogen Processes

The problems with materials and separations at such a high temperature make direct decomposition not attractive at this time. The production of hydrogen by direct thermal splitting of water generated a considerable amount of research during the period 1975–1985. Fletcher and co-workers stressed the thermodynamic advantages of a one-step process with heat input at as high a temperature as possible.[25-27] The theoretical and practical aspects were examined by Olalde,[28] Lede,[29-31] Ounalli,[32] Bilgen[33,34] and by Ihara.[35,36] However, no adequate solution to the crucial problem of separation of the products of water splitting has been worked out so far. Effort was spent to demonstrate the possibility of product separation at low temperature after quenching the hot gas mixtures by heat exchange cooling,[8] by immersion of the irradiated, heated target in a reactor of water liquid,[9,10] by rapid turbulent gas jets,[29,30] or rapid quench by injecting a cold gas.[31] Based on a theoretical evaluation, Lapique[18] concluded that by quenching under optimal conditions it should be possible to recover up to 90% of the hydrogen formed by thermal water splitting. However, the quench introduces a significant drop in the efficiency and produces an explosive gas mixture.[2]

To attain efficient collection of solar radiation in a solar reactor operating at the requisite 2500 K, it is necessary to reach a radiation concentration of the order of 10000 suns. This is a rather stringent requirement. By way of example, a 3-MW solar tower facility, consisting of a field of 64 slightly curved heliostats, has each heliostat capable of concentrating solar radiation approximately by a factor of 50. Even by directing all the heliostats to reflect the sun rays towards a common target, a concentration ratio of only 3000 may be obtained.[37] It is possible, however, to enhance the concentration ratio of an individual heliostat by the use of a secondary concentration optical system, and such systems have been explored.[37,38]

Ordinary steels cannot resist temperatures above a few hundred degrees centigrade, while the various stainless steels, including the more exotic ones, fail at less than 1300 K. In the range 3000–1800 C alumina, mullite or fused silica may be used. A temperature range of about 2500 K requires use of special materials for the solar reactor. However, higher melting point materials can have additional challenges; carbide or nitride composites are likely to react with water splitting products at the high temperatures needed for the reaction. A list of candidate materials of high temperature oxide, carbide and nitride ceramics, is presented in Table 2.

Separation of the generated hydrogen from the mixture of the water splitting products, to prevent explosive recombination, is another challenge for thermochemically generated water splitting processes. Separation of the thermochemically generated hydrogen from the mixture of the water splitting products by gas diffusion through a porous ceramic membrane can be relatively effective. Membranes that have been considered include commercial and specially prepared porous zironias, although sintering was observed to occur under thermal water splitting conditions,[1,39] and ZrO_2-TiO_2-Y_2O_3 oxides.[40] In such membranes, it is necessary to maintain a Knudsen flow regime across the porous wall.[24] The molecular mean free path λ in

Table 2. Melting points of refractory materials, modified from Ref. 1.

Material	Type	Melting Point (°C)
SiO$_2$	oxide	1720
Quartz	oxide	1610
TiO$_2$	oxide	1840
Cr$_2$O$_3$	oxide	1990–2200
Al$_2$O$_3$	oxide	2050
UO$_2$	oxide	2280
Y$_2$O$_3$	oxide	2410
BeO	oxide	2550
CeO$_2$	oxide	2660–2800
ZrO$_2$	oxide	2715
MgO	oxide	2800
HfO$_2$	oxide	2810
ThO$_2$	oxide	3050
SiC	carbide	2200(decomp)
B$_4$C	carbide	2450
WC	carbide	2600(decomp)
TiC	carbide	3400–3500
Electrolytic graphite	carbide	3650(subl)
HfC	carbide	4160
Si$_3$N$_4$	nitride	1900
BN	nitride	3000(decomp)

the gas must be greater than the average pore diameter, ϕ. By kinetic theory, $\lambda = \sqrt{\pi}vd^2$; where d = molecular diameter (cm), v = molecular density = $273.15pv_o/T$, p = pressure (bar), T = temperature (K), and $v_o = 2.685\times10^{19}$ molecules/cm^3.[1] A double-membrane configuration has been suggested as superior to a single-membrane reactor.[41]

In recent times, there have been relatively few studies on the direct thermochemical generation of hydrogen by water splitting[1,24,39-45] due to continuing high temperature material limitations. Recent experimental work has been performed by Kogan and associates[1,37,40-41] and the cross section of one of their solar reactors is shown in Fig. 3. This reactor consists of a cylindrical zirconia housing of 10-cm inside diameter and 20-cm length, and insulated by 2-in thickness of Zircar felt and board. One end of the housing is closed by a circular disc with a central aperture 3 cm in diameter. A zirconia crucible having a porous wall is installed at the opposite end of the housing, and sintering of the crucible considerably limited performance. In 2004, Bayara reiterated that conversion rates in direct thermochemical processes are still quite low, and new reactor designs, operation schemes and materials are needed for new breakthroughs in this field.[45]

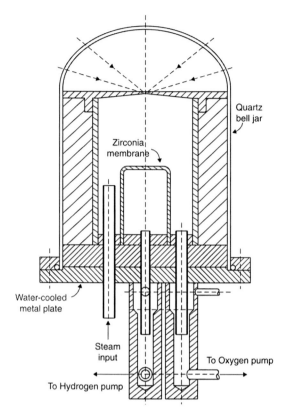

Fig. 3. Example of a solar reactor for direct thermochemical water splitting and solar hydrogen generation. Reprinted with permission from Ref. 1. Copyright (1998) International Journal of Hydrogen Energy.

3 Indirect (Multi-step) Solar Thermal Water Splitting to Generate Hydrogen Fuel

3.1 Historical Development of Multi-Step Thermal Processes for Water Electrolysis

As mentioned earlier, direct thermal dissociation of water requires temperatures above approximately 2500 K. Since there are not yet technical solutions to the materials problems, the possibility of splitting water instead, by various reaction sequences, has been probed. Historically, the reaction of reactive metals and reactive metal hydrides with water or acid was the standard way of producing pure hydrogen in small quantities. These reactions involved sodium metal with water to form hydrogen or zinc metal with hydrochloric acid or calcium hydride with water. All these

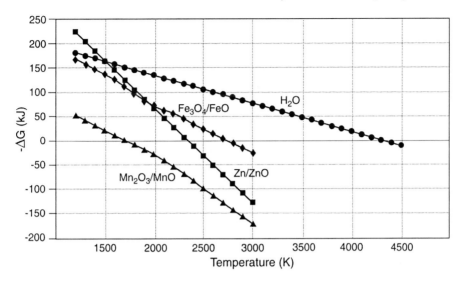

Fig. 4. Temperature variation of the free energy for several decomposition reactions pertinent to hydrogen generation. Reprinted with permission from Ref. 72. Copyright (2004) International Journal of Hydrogen Energy.

methods are quite outdated and expensive, including the reaction of metallic iron or ferrous oxide with steam at elevated temperatures to produce hydrogen.

The possibility to produce hydrogen in multi-reaction processes from water, with a higher thermal efficiency has been extensively studied. As summarized by Perkins and Weimer, and as shown in Fig. 4, a variety of pertinent, spontaneous processes can be considered that have a negative reaction free energy at temperatures considerably below that for water. These reactions are conducted in a cycle to regenerate and reuse the original reactions, ideally, with the only net reactant water, and the only net products hydrogen and oxygen. However, such cycles suffer from challenges often encountered in multi-step reactions. While these cycles operate at much lower temperatures than the direct thermal chemical generation of hydrogen, conversion efficiencies are insufficient and interest in these cycles has waned. Efficiency losses occur at each of the steps in the multiple step sequence, resulting in low overall solar to hydrogen energy conversion efficiencies. Interest in indirect thermal chemical generation of hydrogen started approximately 40 years ago with an average of less than a handful of publications per year in the decade starting 1964. A dramatic upsurge in interest occurred in the subsequent years with an average of over 70 papers per year from 1975 through 1985. Following that time, and because of the lack of clear successes in the field, interest waned in subsequent years and has averaged only ca.10 publications per year.[4]

3.2 Comparison of Multi-step Indirect Solar Thermal Hydrogen Processes

Early studies performed on H_2O-splitting thermochemical cycles were mostly characterized by the use of process heat at temperatures below about 1200 K, available from nuclear and other thermal sources. These cycles required multiple steps (more than two) and had inherent inefficiencies associated with heat transfer and product separation at each step. An overview of indirect thermochemical processes for hydrogen generation using more than two steps has been presented by Funk,[4] and several of these cycles are summarized in Table 3. An example includes cycle No. 2 in Table 3, which utilizes the following reaction steps:

$$CaBr_2 + 2\ H_2O \rightarrow Ca(OH)_2 + 2\ HBr \qquad\qquad T = 1050\ K \qquad (5)$$

$$2\ HBr + Hg \rightarrow HgBr_2 + H_2 \qquad\qquad T = 450\ K \qquad (6)$$

$$HgBr_2 + Ca(OH)_2 \rightarrow CaBr_2 + HgO + H_2O \qquad\qquad T = 450\ K \qquad (7)$$

$$HgO \rightarrow CaBr_2 + Hg + \tfrac{1}{2}\ O_2 \qquad\qquad T = 900\ K \qquad (8)$$

Status reviews on multiple-step cycles have been presented,[46,47] and include the leading candidates GA's 3-step cycle based on the thermal decomposition of H_2SO_4 at 1130 K,[48] and the UT3's 4-step cycle based on the hydrolysis of $CaBr_2$ and $FeBr_2$ at 1020 and 870 K:[49] This process involving two Ca and two Fe compounds has received some attention.[4] The process is operated in a cyclic manner in which the solids remain in their reaction vessels and the flow of gases is switched when the desired reaction extent is reached:

$$CaBr_2(s) + H_2O(l) \rightarrow CaO(s) + 2\ HBr(g) \qquad\qquad T = 973–1050\ K \quad (9)$$

$$CaO(s) + Br_2(g) \rightarrow CaBr_2(s) + O_2(g) \qquad\qquad T = 773–8\ K \qquad (10)$$

$$Fe_3O_4(s) + 8\ HBr(g) \rightarrow 3FeBr_2(s) + 4\ H_2O(g) + 2\ Br_2(g) \quad T = 473–573\ K \quad (11)$$

$$3\ FeBr_2(s) + 4\ H_2O(g) \rightarrow Fe_3O_4(s) + 6\ HBr(g) + H_2\ (g) \quad T = 823–873\ K \quad (12)$$

3.3 High-Temperature, Indirect-Solar Thermal Hydrogen Processes

More recently, higher temperature processes have been considered (at T > 2000 K), such as two-step thermal chemical cycles using metal oxide reactions.[2] The first step is solar: the endothermic dissociation of the metal oxide to the metal or the lower-valence metal oxide. The second step is non-solar, and is the exothermic hydrolysis of the metal to form H_2 and the corresponding metal oxide. The net reaction is $H_2O = H_2 + 0.5\ O_2$, but since H_2 and O_2 are formed in different steps, the need for high-temperature gas separation is thereby eliminated:

Table 3. Summary of multi-step chemical cycles for indirect thermochemical hydrogen generation, from Ref. 4.

No.	Elements in cycle	Maximum temperature, K	Total reaction steps in cycle
1	Hg,Ca,Br	1050	4
2	Hg,Ca,Br	1050	4
3	Cu,Ca,Br	1070	4
4	Hg,Sr,Br	1070	3
5	Mn,Na,(K)	1070	3
6	Mn,Na,(K),C	1120	4
7	V,Cl,O	1070	4
8	Fc,Cl,S	1070	4
9	Hg,Ca,Br,C	1120	5
10	Cr,Cl,Fe,(V)	1070	4
11	Cr,Cl,Fe,(V)Cu	1070	5
12	Fe,Cl	1070	5
13	Fe,Cl	1070	5
14	Fe,Cl	1120	5
15	Mn,Cl	1120	3
16	Fe,Cl	920	3
17	I,S,N	1120	6
18	S (hybrid)	1120	2
19	I,S,N,Zn	1120	4
20	Br,S (hybrid)	1120	3
21	Fe,Cl	920	5
22	Fe,Cl	920	4
23	S,I	1120	3
24	S,I	1120	3

$$\text{1st step (solar):} \qquad M_xO_y \rightarrow xM + y/2O_2 \qquad (13)$$

$$\text{2nd step (non-solar):} \quad xM + yH_2O \rightarrow M_xO_y + yH_2 \qquad (14)$$

where M is a metal and M_xO_y is the corresponding metal oxide. Such a two-step cycle was originally proposed[50] using the redox pair Fe_3O_4/FeO. The solar step, i.e., the thermal dissociation of magnetite to wustite at above 2300 K, has been thermodynamically examined[51] and experimentally studied in a solar furnace.[52,53] It was found necessary to quench the products in order to avoid re-oxidation, but quenching introduces an energy penalty of up to 80% of the solar energy input. The redox pair TiO_2/TiO_x (with $x < 2$) has been considered.[54,55] Solar experiments on the thermal reduction of TiO_2, conducted in an Ar atmosphere up to 2700 K, experienced losses due to the chemical conversion limited by the interfacial rate at which O_2 diffuses.

Other redox pairs, such as Mn_3O_4/MnO and Co_3O_4/CoO have also been considered, but the yield of H_2 in the reaction has been too low to be of any practical interest.[53] H_2 may be produced instead by reacting MnO with NaOH at above 900 K in a 3-step cycle.[56] Steinfeld further suggests[2] that partial substitution of iron in Fe_3O_4 by other metals (e.g., Mn and Ni) forms mixed metal oxides of the type $(Fe_{1-x}M_x)_3O_4$ that may be reducible at lower temperatures than those required for the reduction of Fe_3O_4, while the reduced phase $(Fe_{1-x}M_x)_{1-y}O$ remains capable of splitting water.[57-59]

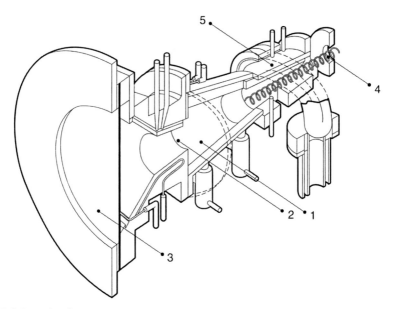

Fig. 5. Schematic of a *rotating-cavity* solar reactor concept for the thermal dissociation of ZnO to Zn and O_2 at 2300 K, modifed from Ref. 2. It consists of a rotating conical cavity-receiver (#1) that contains an aperture (#2) for access of concentrated solar radiation through a quartz window (#3). ZnO particles are continuously fed by means of a screw powder feeder located at the rear of the reactor (#4). The gaseous products Zn and O_2 continuously exit via an outlet port (#5) and are quenched.

One of the most actively studied candidate metal oxide redox pair for the 2-step cycle, is ZnO/Zn. As reviewed by Steinfeld,[2] several chemical aspects of the thermal dissociation of ZnO have been investigated.[55,60,61] The reaction rate law and Arrhenius parameters for directly irradiated ZnO pellets has been derived.[62] The condensation of zinc vapor in the presence of O_2 by fractional crystallization in a temperature-gradient tube furnace was studied.[63] Alternatively, electro-thermal methods for in situ separation of Zn(g) and O_2 at high temperatures have been experimentally demonstrated to work in small-scale solar furnace reactors.[64-67] High-temperature separation further enables recovery of the latent heat of the products (e.g., 116 kJ/mol during Zn condensation). Figure 5 shows the schematic configuration of a solar chemical reactor concept that features a windowed rotating cavity-receiver lined with ZnO particles that are held by centrifugal force.[68] In this arrangement, ZnO is directly exposed to high-flux solar irradiation and serves simultaneously the functions of radiant absorber, thermal insulator, and chemical reactant. Solar tests carried out with a 10 kW prototype subjected to a peak solar concentration of 4000 suns proved the low thermal inertia of the reactor system. The ZnO surface temperature reached 2000 K in 2 s, and was resistant to thermal shocks.[2] Cycles incorporating ZnO continue to be of active research interest.[69-73]

4 Hybrid Solar Thermal/Electrochemical/Photo (STEP) Water Splitting

4.1 Historical Development of Hybrid Thermal Processes

The solar driven room temperature electrolysis of water, discussed in the introduction to this chapter, can be substantially enhanced by heating the water with excess solar thermal energy. Nicholson and Carlisle first generated hydrogen by water electrolysis in 1800. Modifications, such as high temperature electrolysis of steam[74] or water electrolysis by photo-illuminated semiconductors[75] had been reported by the 1970s. With increasing temperature, the quantitative decrease in the electrochemical potential necessary to split water to hydrogen and oxygen had been well known by the 1950s[76] and as early as in 1980, Bockris had noted, that solar thermal energy could decrease the necessary energy for the electrolytic generation of hydrogen.[77] Over the ensuing two decades, designs were intermittently introduced to utilize this principle.[78–87] However, the process combines elements of solid state physics and electrochemical theory, complicating rigorous theoretical support of the process. The thermodynamic feasibility of the solar thermal electrochemical generation of hydrogen was initially shown in 2002.[88,89] The theory combined photodriven charge transfer, with excess sub-bandgap insolation to lower the water potential and demonstrated water splitting efficiencies in excess of 50%. In 2004, experimental support, which is described in the latter Sections, was provided in support of this theory.[90]

4.2 Theory of Hybrid Solar Hydrogen Generation

Thermally assisted solar electrolysis consists of (i) light harvesting, (ii) spectral resolution of thermal (sub-bandgap) and electronic (super-bandgap) radiation, the latter of which (iiia) drives photovoltaic or photoelectrochemical charge transfer $V(i_{H_2O})$, while the former (iiib) elevates water to temperature T, and pressure, p; finally (iv) $V(i_{H_2O})$ driven electrolysis of $H_2O(T,p)$. A schematic representation for this solar thermal water electrolysis (photothermal electrochemical water splitting) is presented in Figure 2, and rather than a field of concentrators, systems may use individual solar concentrators. This hybrid process provides a pathway for efficient solar energy utilization. Electrochemical water splitting, generating H_2 and O_2 at separate electrodes, largely circumvents the gas recombination limitations of direct solar thermochemical hydrogen formation and the multiple-step Carnot losses of indirect thermochemical processes.

Photodriven charge transfer through a semiconductor junction does not utilize photons which have energy below the semiconductor bandgap. Hence a silicon photovoltaic device does not utilize radiation below its bandgap of ~1.1 eV, while an AlGaAs/GaAs multiple bandgap photovoltaic device does not utilize radiation of energy less than the 1.43-eV bandgap of GaAs. As will be shown, this unutilized, available long wavelength insolation represents a significant fraction of the solar spectrum. This long wavelength insolation can be filtered and used to heat water prior to electrolysis. The thermodynamics of heated water dissociation are more favorable than that room temperature. This is expressed by a free-energy chemical shift and a decrease in the requisite water electrolysis potential, which can considerably enhance solar water splitting efficiencies.

The spontaneity of the H_2 generating water splitting reaction is given by the free energy of formation, ΔG°_f, of water and with the Faraday constant, F, the potential for water electrolysis:

$$H_2O \rightarrow H_2 + \tfrac{1}{2} O_2 \tag{15}$$

$$-\Delta G^0_{split} = \Delta G^0_{f,H_2O} \tag{16}$$

where $\Delta G^0_{f,H_2O}$ (25 °C, 1 bar, H_2O_{liq}) = –237.1 kJ mol^{-1}, and

$$E^0_{H_2O} = \frac{\Delta G^0_{f,H_2O}}{2F} \tag{17}$$

where $E^0_{H_2O}$ (25°C, 1 bar, H_2O_{liq}) = 1.229 V

Reaction 15 is endothermic and the electrolyzed water will undergo self-cooling unless external heat is supplied. The enthalpy balance and its related thermoneutral potential, E_{tneut}, are given by:

$$-\Delta H_{split} = \Delta H_{f,H_2O_{liq}} \tag{18}$$

where $\Delta H^0_{f,H_2O_{liq}}$ (25 °C, 1 bar, H_2O_{liq}) = –285.8 kJ mol^{-1}, and

$$E_{tneut} = \frac{-\Delta H_{f,H_2O}}{2F} \tag{19}$$

where E^0_{tneut} (25 °C, 1 bar, H_2O_{liq}) = 1.481 V.

The water electrolysis rest potential is determined from extrapolation to ideal conditions. Variations of the concentration, c, and pressure, p, from ideality are respectively expressed by the activity (or fugacity for a gas), as $a = \gamma c$ (or γp for a gas), with the ideal state defined at 1 atmosphere for a pure liquid (or solid), and extrapolated from $p = 0$ or for a gas or infinite dilution for a dissolved species. The formal potential, measured under real conditions of c and p can deviate significantly from the (ideal thermodynamic) rest potential, as for example the activity of water, a_w, at, or near, ambient conditions generally ranges from approximately 1 for dilute solutions to less than 0.1 for concentrated alkaline and acidic electrolytes.[91–93] The potential for the dissociation of water decreases from 1.229 V at 25 °C in the liquid phase to 1.167 V at 100 °C in the gas phase. Above the boiling the point, pressure is used to express the variation of water activity. The variation of the electrochemical potential for water in the liquid and gas phases are given by:

$$E_{H_2O_{liq}} = E^0_{H_2O_{liq}} + \frac{RT}{2F} \ln \frac{\gamma_{H_2} p_{H_2} (\gamma_{O_2} p_{O_2})^{1/2}}{a_w} \tag{20}$$

$$E_{H_2O_{gas}} = E^0_{H_2O_{gas}} + \frac{RT}{2F} \ln \frac{\gamma_{H_2} p_{H_2} (\gamma_{O_2} p_{O_2})^{1/2}}{\gamma_{H_2O} p_{H_2O}} \tag{21}$$

The critical point of water is 374 °C and 221 bar. Below the boiling point, E^0_{H2O} is similar for 1 bar and high water pressure, but diverges sharply above these conditions. Values of E^0_{H2O} include at $p_{H_2O} = 1$ bar: 1.229 V (25 °C), 1.167 V (100 °C), 1.116 V (300 °C), 1.034 V (600 °C), 0.919 V (1000 °C), 0.771 V (1500 °C), and at $p_{H_2O} = 500$ bar: 1.224 V (25 °C), 1.163 V (100 °C); 1.007 V (300 °C); 0.809 V (600 °C); 0.580 V (1000 °C). Due to overpotential losses, ζ, the necessary applied electrolysis potential is:

$$V_{H_2O}(T) = E^0_{H_2O}(T) + \zeta_{anode} + \zeta_{cathode} \equiv (1 + \zeta)E^0_{H_2O}(T) \tag{22}$$

The water electrolysis potential energy conversion efficiency occurring at temperature, T, is $\eta_{echem}(T) \equiv E^0_{H2O}(T)/V_{H2O}(T)$. Solar water splitting processes utilize ambient temperature water as a reactant. An interesting case occurs if heat is introduced to the system; that is when electrolysis occurs at an elevated temperature, T, using water heated from 25 °C. The ratio of the standard potential of water at 25° C and T, is $r = E^0_{H2O}(25\ °C) / E^0_{H2O}(T)$. As shown in Fig. 6, $E^0_{H2O}(T)$ diminishes with increasing temperature, as calculated using contemporary thermodynamic values summarized in Table 4.[94,95] In this case, an effective water splitting energy conversion efficiency of $\eta'_{echem} > 1$ can occur, to convert 25 °C water to H₂ by electrolysis at T:

$$\eta'_{echem} = r \cdot \eta_{echem}(T) = r \cdot \frac{E^0_{H_2O}(T)}{V_{H_2O}(T)} = \frac{E^0_{H_2O}(25\ °C)}{V_{H_2O}(T)} \tag{23}$$

For low overpotential electrolysis, $V_{H2O}(T > 25\ °C)$ can be less than E^0_{H2O} (25 °C), resulting in $\eta'_{echem} > 1$ from Eq. 23. Whether formed with pn or Schottky type junctions, the constraints on photovoltaic (solid state) driven electrolysis are identical to those for photoelectrochemical water splitting, although the latter poses additional challenges of semiconductor/electrolyte interfacial instability, area limitations, catalyst restrictions, and electrolyte light blockage. The overall solar energy conversion efficiency of water splitting is constrained by the product of the available solar energy electronic conversion efficiency, η_{phot}, with the water electrolysis energy conversion efficiency.[5] For solar *photothermal* water electrolysis, a portion of the solar spectrum will be used to drive charge transfer, and an unused, separate portion of the insolation will be used as a thermal source to raise ambient water to a temperature T:

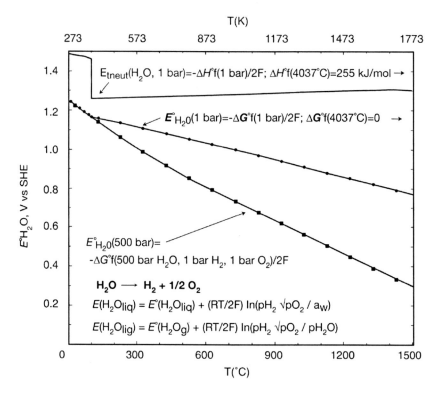

Fig. 6. Thermodynamic and electrochemical values for water dissociation to H_2 and O_2 as a function of temperature.[3] The curves without squares are calculated at one bar, for liquid water through 100 °C and for steam at higher temperatures. The high pressure utilized in this additional curve ($pH_2O = 500$ bar; $pH_2 = pO_2 = 1$ bar) is of general interest as (i) the electrolysis potential is diminished compared to that of water at 1 bar, (ii) the density of the high pressure fluid is similar to that the liquid and (iii) may be generated in a confined space by heating or electrolyzing liquid water.

$$\eta_{solar} = \eta_{phot} \cdot r \cdot \eta_{echem} = \eta_{phot} \cdot \frac{E^0_{H_2O}\left(25\ ^\circ C\right)}{E^0_{H_2O}(T)} \cdot \frac{E^0_{H_2O}(T)}{V_{H_2O}(T)} = \eta_{phot} \cdot \frac{1.229}{V_{H_2O}(T)} \quad (24)$$

Conditions of $\eta_{solar} > \eta_{phot}$ can be shown to place specific restrictions on the photoabsorber. When $V_{H2O} < E_{tneut}$, heat must flow to compensate for the self-cooling which occurs at the electrolysis rate. That is, for an enthalpy balanced system any additional required heat must flow in a flux equivalent to $i_{heat} = i_{H2O}$, and at an average power P_{heat}, such that:

$$E_{tneut} = \frac{V_{H_2O} + P_{heat}}{i_{H_2O}} \quad (25)$$

Table 4. Thermodynamic free energy and enthalpy of water formation for (a) all constituents at 1 bar, and (b) 500 bar water and 1 bar H_2 and O_2.

T(K)	P of $H_{2,gas}$ = 1 bar P of $O_{2,gas}$ = 1 bar P of H_2O = 1 bar				P of $H_{2,gas}$ = 1 bar P of $O_{2,gas}$ = 1 bar P of H_2O = 500 bar			
	P_{H2O} (bar)	H_2O state	ΔG°_f kJ/mol	ΔH°_f kJ/mol	P_{H2O} (bar)	H_2O state	ΔG°_f kJ/mol	ΔH°_f kJ/mol
298	1	liquid	237.1	285.8	500	liquid	236.2	285.0
300	1	liquid	236.8	285.8	500	liquid	235.9	285.0
373	1	liquid	225.2	280.2				
373	1	gas	225.2	239.5				
400	1	gas	223.9	243.0	500	liquid	220.1	282.0
500	1	gas	219.1	243.8	500	liquid	204.9	278.8
600	1	gas	214.0	244.8	500	liquid	190.5	274.9
647				critical point P = 221 bar				
700	1	gas	208.8	245.7	500	super critical	176.9	268.3
800	1	gas	203.5	246.5	500	super critical	164.6	258.1
900	1	gas	198.1	247.2	500	super critical	153.2	254.6
1000	1	gas	192.6	247.9	500	super critical	142.0	253.2
1100	1	gas	187.0	248.4	500	super critical	130.9	252.5
1200	1	gas	181.4	248.9	500	super critical	119.9	252.1
1300	1	gas	175.7	249.4	500	super critical	108.8	252.0
1400	1	gas	170.1	249.9	500	super critical	97.8	251.9
1500	1	gas	164.4	250.2	500	super critical	86.8	251.9
1600	1	gas	158.6	250.5	500	super critical	75.8	251.9
1700	1	gas	152.9	250.8	500	super critical	64.8	252.0
1800	1	gas	147.1	251.1	500	super critical	53.8	252.0
2300		gas	118.0	252.1				
2800		gas	88.9	252.8				
3300		gas	59.6	253.5				
3800		gas	30.3	254.2				
4310		gas	0.0	255.1				

A photoelectrolysis system can contain multiple photo-harvesting units and electrolysis units, where the ratio of electrolysis to photovoltaic units is defined as R. Efficient water splitting occurs with the system configured to match the water electrolysis and photopower maximum power point. This is illustrated in Fig. 1 representing the photosensitizers as power supplies driving electrolysis with a photo-driven charge from a photon flux to generate a current density (electrons per unit area) to provide the two stoichiometric electrons per split water molecule. For example, due to a low photopotential, a photodriven charge from three serial arranged Si energy gap devices may be required to dissociate a single room temperature water molecule, as described in the lower right portion of the figure. Alternately, as in a multiple bandgap device such as AlGaAs/GaAs, the high potential of a single photo-driven charge may be sufficient to dissocate two room temperature water molecules, as described in the upper right portion of the figure. In the figure, consider, four different photoelectrolysis systems, each functioning at the same efficiency for solar conversion of electronic power, η_{phot}, and the same efficiency for solar conversion of thermal power, η_{heat}. Whereas the photoconverter in the first system generates the requisite water electrolysis potential, that in the second system generates twice that potential (albeit at one half the photocurrent), while the photoconverter in the third and fourth units generate respectively only half or a third this potential (albeit at twofold or threefold the photocurrent to retain the same efficiency). The harvested photon power for electronic energy per unit insolation area will be the same in each of the four cases. Furthermore, the number of harvested photons for thermal energy, and the total thermal power available to heat water, will be the same of the four cases. For example in Case II, although twice the number of electrolysis units are utilized, each operates at only half the hydrolysis current compared to Cases I, III and IV, splitting the same equivalents of water.

For solar driven charge transfer, this maximum power is described by the product of the insolation power, P_{sun}, with η_{phot}, which is then applied to electrolysis, $\eta_{phot}P_{sun}$ = P_{echem} = $i_{H2O}V_{H2O}$. Rearranging for i_{H2O}, and substitution into Eq. 25, yields for heat balanced solar electrolysis at conditions of T and p, initiating with 25 °C, 1 bar water:

$$E_{tneut} = 1.481\ V = V_{H_2O}(T, p)\frac{1 + P_{heat}}{\eta_{phot} P_{sun}} \qquad (26)$$

As also elaborated in Chapter 2, Fig. 7 presents the available insolation power, $P_{\lambda max}$ (mW cm^{-2}) of the integrated solar spectrum up to a minimum electronic excitation frequency, v_{min} (eV), determined by integrating the solar spectral irradiance, S(mWcm^{-2}nm^{-1}), as a function of a maximum insolation wavelength, λ_{max} (nm). This $P_{\lambda max}$ is calculated for the conventional terrestial insolation spectrum either above the atmosphere, AM0, or through a 1.5 atmosphere pathway, AM1.5. Relative to the total power, P_{sun}, of either the AM0 or AM1.5 insolation, the fraction of this power available through the insolation edge is designated P_{rel} = $P_{\lambda max}$ / P_{sun}. In solar energy balanced electrolysis, excess heat is available primarily as photons without sufficient energy for electronic excitation. The fraction these sub-bandgap photons in insolation is α_{heat} = $1 - P_{rel}$, and comprises an incident power of $\alpha_{heat}P_{rel}$.

Figure 8 presents the variation of the minimum electronic excitation frequency, v_{min} with α_{heat}, determined from P_{rel} using the values of $P_{\lambda max}$ summarized in Fig. 7. A semiconductor sensitizer is constrained not to utilize incident energy below the bandgap. As seen in Figure 8 by the intersection of the solid line with v_{min}, over one third of insolation power occurs at v_{min} < 1.43 eV (867 nm), equivalent to the IR not absorbed by GaAs or wider bandgap materials. The calculations include both the AM0 and AM1.5 spectra. In the relevant visible and IR range from 0.5 to 3.1 eV (±0.03 eV) for both the AM0 and AM1.insolation spectra, $v_{min}(\alpha_{heat})$ in the figure are well represented ($R^2 \geq 0.999$) by polynomial fits.

When captured at a thermal efficiency of η_{heat}, the sub-bandgap insolation power is $\eta_{heat}\alpha_{heat}P_{sun}$. Other available system heating sources include absorbed super-bandgap photons which do not effectuate charge separation, P_{recomb}, and non-insolation sources, P_{amb}, such as heat available from the ambient environment heat

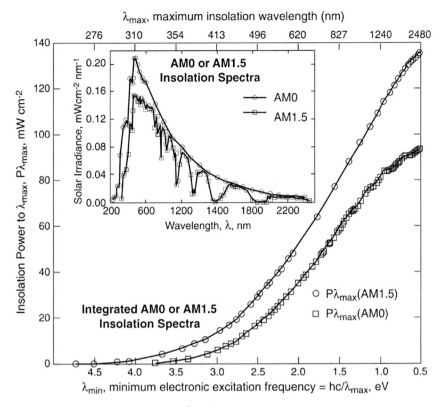

Fig. 7. The solar irradiance (mW cm^{-2} nm^{-1}) in the figure inset, and the total insolation power (mW cm^{-2}) in the main figure of the solar spectrum;[3] see also Chapter 2.

sink, and P_{recov}, such as heat recovered from process cycling or subsequent H$_2$ fuel utilization. The power equivalent for losses, such as the low power consumed in delivering the heated water to electrolysis, P_{pump}, can also be incorporated.

In Figs. 9 to12, determinations of the solar water splitting energy conversion are summarized, calculated using the $E^0_{H2O}(T,p)$ data in Fig. 8, and for various solar water splitting system's minimum allowed bandgap, $E_{g-min}(T,p)$ for a wide temperature range. Figures 9 and 10, or 8 and 11, are respectively modeled based on AM1.5, or AM0, insolation. Figures 9 and 11 are calculated, at various temperatures for p_{H2O} = 1 bar; while Figures 10 and 12 repeat these calculations for p_{H2O} = 500 bar. The $\eta_{solar-max}$ value is significantly greater for higher pressure photoelectrolysis (p_{H2O} = 500 bar). However as seen comparing the minimum bandgap in these figures (or in Figures 11 and 12 in the analogous AM0 models), at these higher pressures, this higher rate of efficiency increase with temperature is offset by lower accessible temperatures (for a given bandgap). Larger ζ in V_{H2O} will diminish η_{solar}, but will extend the usable small bandgap range. Together, Figs. 9–12 show the constraints on η_{solar} from various values of η_{phot}.

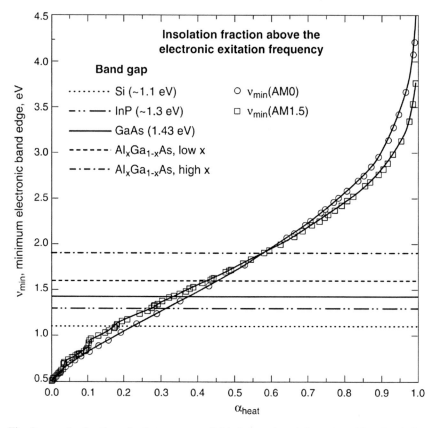

Fig. 8. α_{heat}, the fraction of solar energy available below the minimum sensitizer insolation frequency used to drive charge transfer, ν_{min}.[3] The α_{heat} is determined as $1 - P_{rel}$, with $P_{rel} = P_{\lambda max}/P_{sun}$, and using values of $P_{\lambda max}$ summarized in Fig. 7. The available incident power below ν_{min} will be $\alpha_{heat}P_{sun}$. The bandgaps of various semiconductors are superimposed as vertical lines in the figure.

The high end of contemporary experimental high solar conversion efficiencies ranges from 100% $\eta_{phot} = 19.8\%$ for multicrystalline single junction photovoltaics to 27.6% and 32.6% for single junction and multiple junction photovoltaics.[97] The efficiency of solar thermal conversion tends to be higher than solar electrical conversion, η_{phot}, particularly in the case of the restricted spectral range absorption used here with values of $\eta_{heat} = 0.5, 0.7$ or 1. While a small bandgap, $E_g < 1.23$ eV, is insufficient for water cleavage at 25 °C, its inclusion in Figs. 9–12 is of relevance in two cases,

1. where high temperature decreases $V_{H2O}(T)$ below E_g, and
2. where this E_g is part of a multiple bandgap sensitization contributing a portion of a larger overall photopotential.

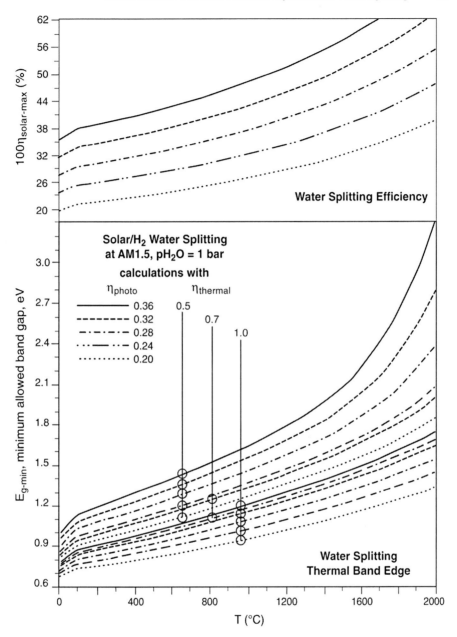

Fig. 9. Energy conversion efficiency of solar driven water splitting to generate H_2 as a function of temperature for AM1.5 insolation, with the system minimum bandgap determined at p_{H2O} = 1 bar.[3] The maximum photoelectrolysis efficiency is shown for various indicated values of η_{phot}.

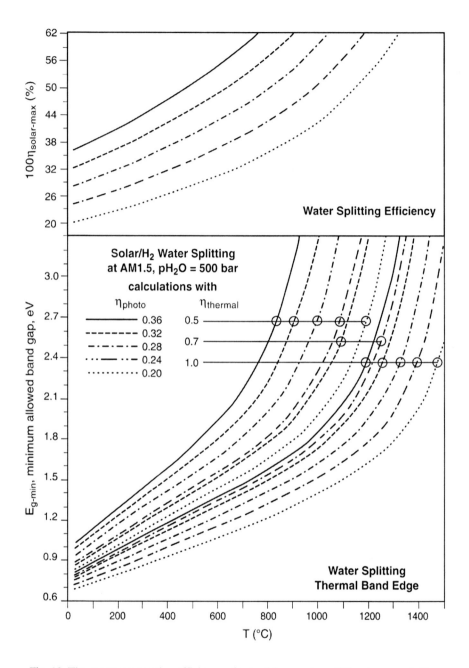

Fig. 10. The energy conversion efficiency of solar driven water splitting to generate H2 as a function of temperature at AM1.5 insolation, with the system minimum bandgap determined at p_{H2O} = 500 bar.[3] The maximum photoelectrolysis efficiency is shown for various indicated values of η_{phot}.

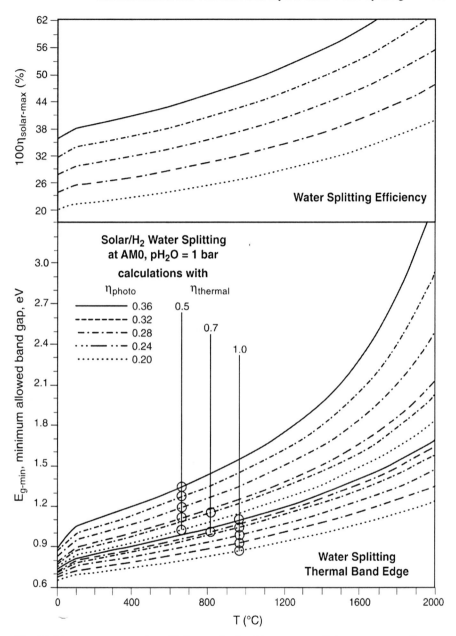

Fig. 11. The energy conversion efficiency of solar driven water splitting to generate H_2 as a function of temperature at AM0 insolation, with the system minimum bandgap determined at $p_{H2O} = 1$ bar.[3] The maximum photoelectrolysis efficiency is shown for various indicated values of η_{phot}.

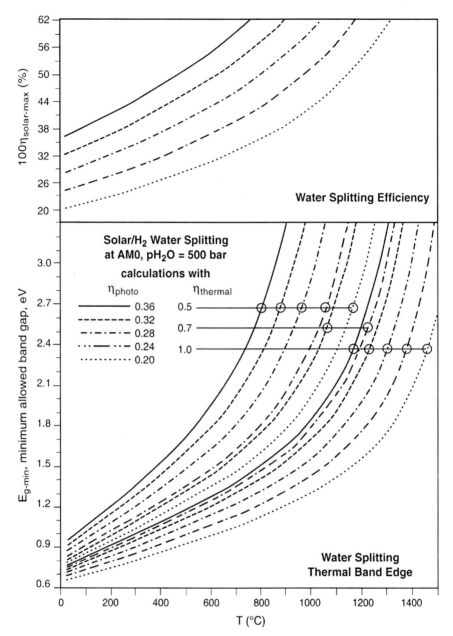

Fig. 12. The energy conversion efficiency of solar driven water splitting to generate H2 as a function of temperature at AM0 insolation, with the system minimum bandgap determined at $p_{H2O} = 500$ bar.[3] The maximum photoelectrolysis efficiency is shown for various indicated values of η_{phot}.

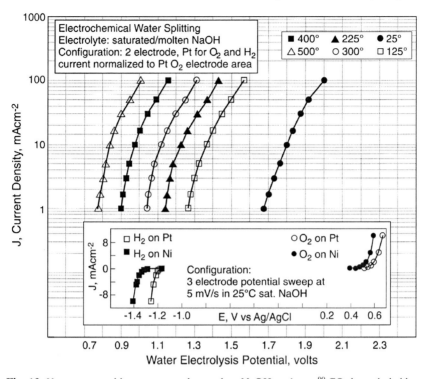

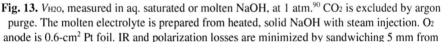

Fig. 13. V_{H2O}, measured in aq. saturated or molten NaOH, at 1 atm.[90] CO_2 is excluded by argon purge. The molten electrolyte is prepared from heated, solid NaOH with steam injection. O_2 anode is 0.6-cm^2 Pt foil. IR and polarization losses are minimized by sandwiching 5 mm from each side of the anode, two interconnected Pt gauze (200 mesh, 50 cm^2 = 5 cm x 5 cm x 2 sides) cathodes. Inset: At 25 °C, 3 electrode values at 5 mV/s versus Ag/AgCl, with either 0.6-cm^2 Pt or Ni foil, and again separated 5 mm from two 50-cm^2 Pt gauze acting as counter electrodes.

It is also noted that $E_g > 3.0$ eV is in adequate for efficient use of the solar spectrum. Representative results from Fig. 12 for solar water splitting to H_2 systems from AM1.5 insolation include a 50% solar energy conversion for a photoelectrolysis system at 638 °C with $p_{H_2O} = 500$; $p_{H_2} = 1$ bar and $\eta_{phot} = 0.32$.

4.3 Elevated Temperature Solar Hydrogen Processes and Components

Fletcher, repeating the fascinating suggestion of Brown that saturated aqueous NaOH will never boil, hypothesized that a useful medium for water electrolysis might be saturated, aqueous solutions of NaOH at very high temperatures.[98] These do not reach boiling point at 1 atm due to the high salt solubility, binding solvent, and changing saturation vapor pressure, as reflected in their phase diagram.[98] This domain is considered here, and also electrolysis in an even higher temperature domain

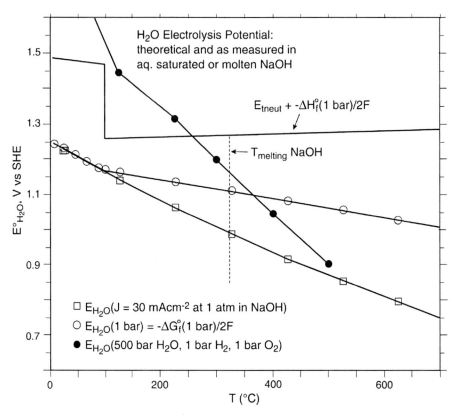

Fig. 14. Measured V_{H2O} (30 mA cm^{-2}) in aq. saturated or molten NaOH compared to thermo-dynamic E_{H2O} values.[90]

above which NaOH melts (318 °C) creating a molten electrolyte with dissolved water, resulting in unexpected V_{H2O}.

Figure 13 summarizes measured $V_{H2O}(T)$ in aqueous saturated and molten NaOH electrolytes. As seen in the inset, Pt exhibits low over potentials to H$_2$ evolution, and is used as a convenient quasi-reference electrode in the measurements which follow. As also seen in the inset, Pt exhibits a known large overpotential to O$_2$ evolution as compared to a Ni electrode or to E^0_{H2O} (25 °C) = 1.23 V. This overpotential loss diminishes at moderately elevated temperatures, and as seen in the main portion of the figure, at 125 °C there is a 0.4 V decrease in the O$_2$ activation potential at a Pt surface. Through 300 °C in Fig. 13, measured V_{H2O} remains greater than the calculated thermodynamic rest potential. Unexpectedly, V_{H2O} at 400 °C and 500 °C in molten NaOH occurs at values substantially smaller than that predicted. These measured values include voltage increases due to *IR* and hydrogen overpotentials, and hence provide an upper bound to the unusually small electrochemical potential.

This phenomenon is summarized in Fig. 14, in which even at relatively large rates of water splitting (30 mA cm^{-2}) at 1 atm, a measured V_{H2O} below that predicted by theory is observed at temperatures above the NaOH melting point. Theoretical calculations over an expanded temperature range are presented in Fig. 8, with calculations described in that Section. As seen in Fig. 14, the observed value at high temperature of V_{H2O} approaches that calculated for a thermodynamic system of 500 bar, rather than 1 bar, H_2O.

A source of this anomaly is described in Fig. 15. Shown on the left hand is the single compartment cell utilized here. Cathodically generated H_2 is in close proximity to the anode, while anodic O_2 is generated near the cathode. Their presence will facilitate the water forming back reaction, and at the electrodes this recombination will diminish the potential. In addition to the observed low potentials, two observations support this recombination effect. The generated H_2 and O_2 is collected, but is consistent with a Coulombic efficiency of $\approx 50\%$ (varying with T, j, and interelectrode separation.) Consistent with the right hand side of Fig. 15, when conducted in separated anode / cathode compartments, this observed efficiency is 98%–100%. Here however, all cell open circuit potentials increase to beyond the thermodynamic potential, and at $j = 100$ mA cm^{-2} yields measured V_{H2O} values of 1.45V, 1.60V, 1.78V at 500°, 400°, and 300°, which are approximately 450 mV higher than the equivalent Fig. 13 values for the single configuration cell.

The recombination phenomenon offers advantages (low V_{H2O}), but also disadvantages (H_2 losses), requiring study to balance these competing effects to optimize energy efficiency. In molten NaOH, the effects of temperature variation of ΔG^0_f (H_2O) and the recombination of the water splitting products can have a pronounced effect on solar driven electrolysis. As compared to 25 °C, in Fig. 13 only half the potential is required to split water at 500 °C over a wide range of current densities.

The unused thermal photons which are not required in semiconductor photodriven charge generation, can contribute to heating water to facilitate electrolysis at an elevated temperature. The characteristics of one, two, or three series interconnected

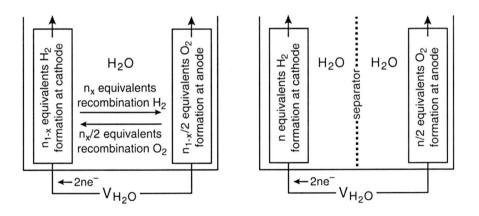

Fig. 15. Interelectrode recombination can diminish V_{H2O} and occurs in open (left) but not in isolated (right) configurations; such as those examined with or without a Zr_2O mix fiber separator (ZYK-5H, from Zircar Zirconia, FL, NY) situated between the Pt anode and cathodes.[90]

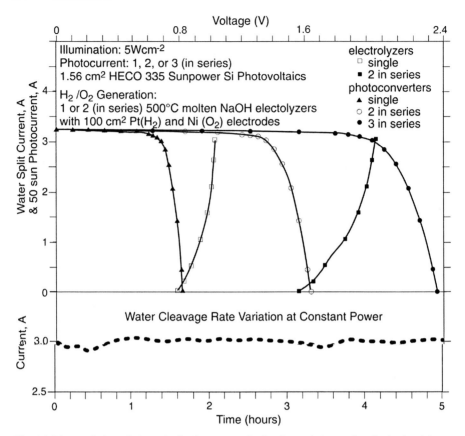

Fig. 16. Photovoltaic and electrolysis charge transfer for thermal electrochemical solar driven water splitting.[90] Photocurrent is shown for one, two or three 1.561 cm² HECO 335 Sunpower Si photovoltaics in series at 50 suns. Photovoltaics drive 500-°C molten NaOH steam electrolysis using Pt gauze anode and cathodes. Inset: electrolysis current stability.

solar visible efficient photosensitizers, in accord with the manufacturer's calibrated standards, are presented in Fig. 16. These silicon photovoltaics are designed for efficient photoconversion under concentrated insolation (η_{solar} = 26.3% at 50 sun). Superimposed on the photovoltaic response curves in the figure are the water electrolysis current densities for one, or two series interconnected, 500 °C molten NaOH single compartment cell configuration electrolyzers.

Constant illumination generates for the three series-connected cells, a constant photopotential for stability measurements at sufficient power to drive two series molten NaOH electrolyzers. At this constant power, and as presented in the lower portion of Fig. 16, the rate of water splitting appears fully stable over an extended period. In addition, as measured and summarized in the upper portion of the figure, for the overlapping region between the solid triangle and open square curves, a single Si photovoltaic can drive 500 °C water splitting, albeit at an energy beyond the maximum power point voltage, and therefore at diminished efficiency. This appears to be the first case in which an external, single, small bandgap photosentizer can cleave water, and is accomplished by tuning the water splitting electrochemical potential to decrease to a point below the Si open circuit photovoltage. $V_{H2O-tunned}$ is accomplished by two phenomena:

1. the thermodynamic decrease of E_{H2O} with increasing temperature, and
2. a partial recombination of the water splitting products.

$V_{H2O-tunned}$ can drive system efficiency advances, e.g., AlGaAs/GaAs, transmits more insolation, $E_{IR} < 1.4$ eV, than Si to heat water, and with η_{photo} over 30%, prior to system engineering losses, calculates to over 50% η_{solar} to H_2.

Without inclusion of high temperature effects, we had already experimentally achieved $\eta_{solar} > 0.18$, using an $\eta_{phot} = 0.20$ AlGaAs/Si system.[5] Our use of more efficient, $\eta_{solar} = 26.3\%$ at 50 suns, and inclusion of heat effects and the elevated temperature decrease of the water electrolysis potential, substantially enhances η_{solar}.[90] Existing, higher η_{phot} (= 0.28 to 0.33) systems should achieve proportionally higher results. Experimental components, for example as described in Fig. 2, for solar driven generation of H_2 fuel at 40-50% conversion efficiencies appear to be technologically available. In the high efficiency range, photoelectrochemical cells tend to be unstable, which is likely to be exacerbated at elevated temperatures, and the model system will be particularly conducive to photovoltaic, rather than photoelectrochemical, driven electrolysis.

As has already been elaborated in Chapter 4, the photovoltaic component is used for photodriven charge into the electrolysis component and does not contact the heated electrolyte. Stable photovoltaics are commonly driven with concentrated insolation[97] and specific to the system model here, heat will be purposely filtered from the insolation prior to incidence on the photovoltaic component. Dielectric filters used in laser optics split insolation without absorption losses (Chapter 4). For example, in a system based on a parabolic concentrator, a casegrain configuration may be used, with a mirror made from fused silica glass with a dielectric coating acting as band pass filter. The system will form two focal spots with different spectral configuration, one at the focus of the parabola and the other at focus of the casegrain.[99] The thermodynamic limit of concentration is 46000 suns, the brightness of the surface of the sun. In a medium with refractive index greater than one, the upper limit is increased by two times the refractive index, although this value is reduced by reflective losses and surface errors of the reflective surfaces, the tracking errors of the mirrors and dilution of the mirror field. Specifically designed optical absorbers, such as parabolic concentrators or solar towers, can efficiently generate a solar flux with concentrations of ~2000 suns, generating temperatures in excess of 1000 °C.[100,101]

Commercial alkaline electrolysis occurs at temperatures up to 150 °C and pressures to 30 bar,[96] and super-critical electrolysis to 350 °C and 250 bar.[102] Although less developed than their fuel cell counterparts which have 100 kW systems in operation and developed from the same oxides,[103] zirconia and related solid oxide based electrolytes for high temperature steam electrolysis can operate efficiently at 1000 °C,[104,105] and approach the operational parameters necessary for efficient solar driven water splitting. Efficient multiple bandgap solar cells absorb light up to the bandgap of the smallest bandgap component. Thermal radiation is assumed to be split off (removed and utilized for water heating) prior to incidence on the semiconductor and hence will not substantially effect the bandgap. Highly efficient photovoltaics have been demonstrated at a solar flux with a concentration of several hundred suns. AlGaAs/GaAs has yielded a η_{phot} efficiency of 27.6% and a GaInP/GaAs cell 30.3% at 180 suns concentration, while GaAs/Si has reached 29.6% at 350 suns, InP/GaInAs 31.8%, and GaAs/GaSb 32.6% with concentrated insolation.[97]

5 Future Outlook and Concluding Remarks

A hybrid solar thermal/electrochemical process combines efficient photovoltaic devices and concentrated excess sub-bandgap heat into highly efficient elevated temperature solar electrolysis of water to generate H_2 fuel. Efficiency is further enhanced by excess super-bandgap and non-solar sources of heat but diminished by losses in polarization and photo-electrolysis power matching. As also shown earlier in Chapter 4 and elaborated further here, solar concentration can provide the high temperature and diminish the requisite surface area of efficient electrical energy conversion components. High temperature electrolysis components are commercially available, suggesting that highly efficient solar generation of H_2 will be ultimately attainable.

References

1. A. Kogan, Direct solar thermal splitting of water and on-site separation of the products. II. Experimental feasibility study, *Int. J. Hydrogen Energy*, **23** 89-98 (1998).
2. A. Steinfeld, Solar thermochemical production of hydrogen-a review, *Solar Energy*, **78** 603–615 (2005).
3. S. Licht, Solar water splitting to generate hydrogen fuel–A photothermal electrochemical analysis, *Int. J. Hydrogen Energy*, **30** 459–470 (2005).
4. J. E. Funk, Thermochemical hydrogen production: past and present, *Int. J. Hydrogen Energy*, **26** 185–190 (2001).
5. S. Licht, B. Wang, S. Mukerji, T. Soga, M. Umeno, and H. Tributsh, Over 18% solar energy conversion to generation of hydrogen fuel; theory and experiment for sefficient solar water splitting, *Int. J. Hydrogen Energy*, **26** 653–659 (2001).
6. K. Agabossu, R. Chahine, J. Hamelin, F. Laurencelle, A. Anouar, J.-M. St-Arnaud, and T. K. Bose, Renewable energy systems based on hydrogen for remote applications, *J. of Power Sources* **96** 168–172 (2001).
7. T. Ohmori, H. Go, N. Yamaguchi, A. Nakayama, H. Mametisuka, and E. Suzuki, Photovoltaic water electrolysis using the sputter-deposited a-Si/c-Si solar cells, *Int. J. of Hydrogen Energy* **26** 661–664 (2001).
8. T. Tani, N. Sekiguchi, M. Sakai, and D. Otha, Optimization of solar hydrogen systems based on hydrogen production cost, *Solar Energy* **68** 143–149 (2000).

9. P. Hollmuller, J.-M. Jouibert, B. Lachal, and K. Yvon, Evaluation of a 5-kW$_p$ photovoltaic hydrogen production and storage installation for a residential home in Switzerland, *Int. J. of Hydrogen Energy* **25** 97–109 (2000).
10. S. Schulien, G. Sandstede, and H. W. Hahn, Hydrogen and carbon dioxide as raw materials for ecological energy – technology, *Int. J. of Hydrogen Energy* **24** 299–303 (1999).
11. C. Meurer, H. Barthels, W. A. Brocke, B. Emonts, and H. G. Groehn, Phoebus – and autonomous supply system with renewable energy: six years of operational experience and advanced concepts, *Solar Energy* **67** 131–138 (1999).
12. A. Szyszka, Ten year of solar hydrogen demonstration project at Neunburg vorm Wald, Germany, *Int. J. of Hydrogen Energy* **23** 849–860 (1998).
13. P. A. Lehman, C. E. Chamberlin, G. Pauletto, and M. A. Rocheleau, Operating experience with photovoltaic-hydrogen energy system, *Int. J. of Hydrogen Energy* **22** 465–470 (1997).
14. S. Galli and M. Stefanoni, Development of a solar-hydrogen cycle in Italy, *Int. J. of Hydrogen Energy* **22** 453–458 (1997).
15. J. W. Hollenberg, E. N. Chen, K. Lakeram, and D. Modroukas, Development of a photovoltaic energy conversion system with hydrogen energy storage, *Int. J. of Hydrogen Energy* **20** 239–243 (1995).
16. E. Bilgen, Solar hydrogen from photovoltaic-electrolyzer systems, *Energy Conversion & Management* **42** 1047–1057 (2001).
17. M. P. Rzayeva, O. M. Salamov, and M. K. Kerimov, Modeling to get hydrogen and oxygen by solar water electroclysis, *Int. J. of Hydrogen Energy*, **26** 195–201 (2001).
18. S. Licht, Multiple bandgap semiconductor/electrolyte solar energy conversion, *J. Phys., Chem. B,* **105** 6281–6294 (2001).
19. Semiconductor Electrodes and Photoelectrochemistry, Edited by S. Licht, Wiley-VCH, Weinheim, 2002.
20. A. Fujishima and K. Honda, *Nature* **37** 238 (1972).
21. A. Heller, E. Asharon-Shalom, and W. A. Bonner, B. Miller, Hydrogen-evolving semiconductor photocathodes: nature of the junction and function of the platinum group metal catalyst, *J. Am. Chem. Soc.* **104** 6942–6948 (1982).
22. O. Khaselev and J. A. Turner, A monolithic photovoltaic-photoelectrochemical device for hydrogen production via waer splitting, *Science,* **280** 425–427 (1998).
23. S. Licht, B. Wang, S. Mukerji, T. Soga, M. Umeno, and H. Tributsch, Efficient Solar Water Splitting, Conversion, *J. Phys., Chem., B,* **104** 8920–8924 (2000).
24. H. Ohya, M. Yatabe, M. Aihara, Y. Negishi, and T. Takeuchi, Feasibility of hydrogen production above 2500 K by direct thermal decomposition reaction in membrane reactor using solar energy, *Int. J. Hydrogen Energy,* **27** 369–376 (2002).
25. E. A.Fletcher and R. L.Moen, Hydrogen and oxygen from water, *Science,* **197** 105 (1977).
26. J. E. Noring, R. B. Diver and E. A. Fletcher, Hydrogen and oxygen water V. The ROC system, *Energy,* **6** 109 (1981).
27. R. B. Diver, S. Pederson, T. Kappauf, and E. A. Fletcher, Hydrogen and oxygen from water: VI. Quenching the effluent from a solar furnace, *Energy* **8** 947 (1983).
28. G. Olalde, D. Gauthier, and A. Vialaron, Film boiling around a zirconia target. Application to water thermolysis, *Adv. Ceramics,* **24** 879–883 (1988).
29. J. Lede, F. Lapigque, J. Villermaux, B. Cales, A. Ounalli, J. F. Baumard, and A. M. Anthony, Production of hydrogen by direct thermal decomposition of water: Preliminary investigations, *Int. J. Hydrogen Energy,* **7** 939–950 (1982).
30. F. Lapigque, J. Lede, L. Villermaux, A. Cales, J. Baumard, A. M. Anthony, E. Abdul Aziz, D. Peuchberty, and M. Ledoux, *Entropie,* **110** 42 (1983).

31. J. Lede, J. Villermaux, R. Ouzane, M. A. Hossain, and R. Ouahes, Production of hydrogen by simple impingement of a turbulent jet of steam upon a high temperature zirconia surface, *Int. J. Hydrogen Energy*, **12** 3–11 (1987).

32. A. Ounalli, B. Cales, K. Dembrinski, and J. F. Baumard, *C. R. Acad. Sci. Paris*, **292**(11) 1185 (1981).

33. E. Bilgen, Solar hydrogen production by direct water decomposition process: a preliminary engineering assessment, *Int. J. Hydrogen Energy*, **9** 53–48 (1984).

34. E. Bilgen, M. Duccarroir, M. Foex, F. Silieude, and F. Trombe, Use of solar energy for direct and two-step water decomposition cycles, *Int. J. Hydrogen Energy*, **2** 251–257 (1977).

35. S. Ihara, Feasibility of hydrogen production by direct water splitting at high temperature, *Int. J. Hydrogen Energy*, **3** 287–296 (1978).

36. S. Ihara, On the study of hydrogen production from water using solar thermal energy, *Int. J. Hydrogen Energy*, **5** 527–534 (1980).

37. A. Yogev, A. Kribus, M. Epstein and A. Kogan, Solar "Thermal Reflector" systems: A new approach for high-temperature solar plants, *Int. J. Hydrogen Energy* **26** 239–245 (1998).

38. A. Kribus, P. Doron, R. Rubin, J. Karni, R. Reuven, S. Duchan, and E. Taragan, A multistage solar receiver: the route to high temperature, *Solar Energy*, **67** 2–11 (2000).

39. A. Kogan, Direct solar thermal splitting of water and on-site separation of the products. IV. Development of porous ceramic membranes for a solar thermal water-splitting reactor, *Int. J. Hydrogen Energy*, **25** 1043–1050 (2000).

40. H. Naito and H. Arashi, Hydrogen production from direct water splitting at high temperatures using a ZrO_2-TiO_2-Y_2O_3 membrane, *Solid State Ionics*, **79** 366–370 (1995).

41. R. P. Omorjan, R. N. Paunovic, M. N. Tekic, and M. G. Antov, Maximal extent of an isothermal reversible gas-phase reaction in single- and double-membrane reaction; direct thermal splitting of water, *Int. J. Hydrogen Energy*, **26** 203–212 (2001).

42. A. Kogan, Direct solar thermal splitting of water and on-site separation of the products. I. Theoretical evaluation of hydrogen yield, *Int. J. Hydrogen Energy*, **22** 481–486 (1997).

43. A. Kogan, E. Spiegler, and M. Wolfshtein, Direct solar thermal splitting of water and on-site separation of the products. III. Improvement of reactor efficiency by steam entrainment, *Int. J. Hydrogen Energy*, **25** 739–745 (2000).

44. S. Z. Baykara, Hydrogen production by direct solar thermal decomposition of water, possibilities for improvement of process efficiency, *Int. J. Hydrogen Energy*, **29** 1451–1458 (2004).

45. S. Z. Baykara, Experimental solar water thermolysis, *Int. J. Hydrogen Energy*, **29** 1459–1469 (2004).

46. N. Serpone, D. Lawless, and R. Terzian, Solar fuels: status and perspectives, *Solar Energy*, **49** 221–234 (1992).

47. J. Funk, Thermochemical hydrogen production past and present, *Int. J. Hydrogen Energy*, **26** 185–190 (2001).

48. D. OKeefe, C. Allen, G. Besenbruch, L. Brown, J. Norman, R. Sharp, and K. McCorkle, Preliminary results from bench-scale testing of sulfur-iodine thermochemical water-splitting cycle, *Int. J. Hydrogen Energy*, **7** 381–392 (1982).

49. M. Sakurai, E. Bilgen, A. Tsutsumi, and K. Yoshida, Solar UT-3 thermochemical cycle for hydrogen production, *Solar Energy*, **57** 51–58 (1996).

50. T. Nakamura, Hydrogen production from water utilizing solar heat at high temperatures, *Solar Energy*, **19** 467–475 (1977).

51. A. Steinfeld, S. Sanders, and R. Palumbo, Design aspects of solar thermochemical engineering, *Solar Energy*, **65** 43–53 (1999).

52. A. Tofighi, Ph.D. Thesis, L'Institut National Polytechnique de Toulouse, France, 1982.

53. F. Sibieude, M. Ducarroir, A. Tofighi, and J. Ambriz, High-temperature experiments with a solar furnace: the decomposition of Fe_3O_4,Mn_3O_4, CdO, *Int. J. Hydrogen Energy*, **7** 79–88 (1982).
54. R. D. Palumbo, A. Rouanet, and G. Pichelin, The solar thermal decomposition of TiO_2 above 2200 K and its use in the production of Zn from ZnO, *Energy - Int. J.*, **20** 857– 868 (1995).
55. R. Palumbo, J. Lede, O. Boutin, E. Elorza Ricart, A. Steinfeld, S. Moeller, A. Weidenkaff, E.A. Fletcher, and J. Bielicki, The production of Zn from ZnO in a single step high temperature solar decomposition process, *Chem. Eng. Sci.*, **53** 2503–2518 (1998).
56. M. Sturzenegger and P. Nuesch, Efficiency analysis for a manganese-oxide-based thermochemical cycle, *Energy*, **24** 959–970 (1999).
57. K. Ehrensberger, A. Frei, P. Kuhn, H.R. Oswald, and P. Hug, Comparative experimental investigations on the water-splitting reaction with iron oxide $Fe_{1-y}O$ and iron manganese oxides $(Fe_{1-x}Mn_x)_{1-y}O$, *Solid State Ionics*, **78** 151–160 (1995).
58. Y. Tamaura, A. Steinfeld, P. Kuhn, and K. Ehrensberger, Production of solar hydrogen by a novel, 2-step, watersplitting thermochemical cycle, *Energy*, **20** 325–330 (1995).
59. Y. Tamaura, M. Kojima, Y. Sano, Y. Ueda, N. Hasegawa, and M. Tsuji, Thermodynamic evaluation of watersplitting by a cation-excessive (Ni, Mn) ferrite, *Int. J. Hydrogen Energy*, **23** 1185–1191 (1998).
60. A. Weidenkaff, A. Reller, A. Wokaun, and A. Steinfeld, Thermogravimetric analysis of the ZnO/Zn water splitting cycle, *Thermochim. Acta*, **359** 69–75 (2000).
61. A. Weidenkaff, A. Reller, F. Sibieude, A. Wokaun, and A. Steinfeld, Experimental investigations on the crystallization of zinc by direct irradiation of zinc oxide in a solar furnace, *Chem. Mater.*, **12** 2175–2181 (2000).
62. S. Moeller and R. Palumbo, Solar thermal decomposition kinetics of ZnO in the temperature range 1950–2400 K, *Chem. Eng. Sci.*, **56** 4505–4515 (2001).
63. A. Weidenkaff, A. Wuillemin, A. Steinfeld, A. Wokaun, B. Eichler, and A. Reller, The direct solar thermal dissociation of ZnO: condensation and crystallization of Zn in the presence of oxygen, *Solar Energy*, **65** 59–69 (1999).
64. E. A. Fletcher, Solar thermal and solar quasi-electrolytic processing and separations: zinc from zinc oxide as an example, *Ind. Eng. Chem. Res.*, **38** 2275–2282 (1999).
65 E. A. Fletcher, F. Macdonald, and D. Kunnerth, High temperature solar electrothermal processing II. Zinc from zinc oxide, *Energy*, **10** 1255–1272 (1985).
66. D. J. Parks, K.L. Scholl, and E.A. Fletcher, A study of the use of Y_2O_3 doped ZrO_2 membranes for solar electro-thermal and solar thermal separations, *Energy*, **13** 121 –136 (1988).
67. R. D. Palumbo and E. A. Fletcher, High temperature solar electro-thermal processing. III. Zinc from zinc oxide at 1200–1675 K using a non-consumable anode, *Energy*, **13** 319–332 (1988).
68. P. Haueter, S. Moeller, R. Palumbo, and A. Steinfeld, The production of zinc by thermal dissociation of zinc oxide - solar chemical reactor design, *Solar Energy* **67** 161–167 (1999).
69. H. Aoki, H. Kaneko, N. Hasegawa, H. Ishihara, A. Suzuki, and Y. Tamaura, The $ZnFe_2O_4/(ZnO+Fe_3O_4)$ system for H_2 production using concentrated solar energy, *Solid State Ionics*, **172**, 113-116, 2004
70. M. Inoue, N. Hasewaga, R. Uehara, N. Gokon, H. Kaneko, and Y. Tamaura, Solar hydrogen generation with $H_2O/ZNO/MnFe_2O_4$ system, *Solar Energy*, **76** 309–315 (2004).
72. C. Perkins and A. W. Weimer, Likely near-term solar-thermal water splitting technologies, *Int. J. of Hydrogen Energy*, **29** 1587–1599 (2004).
73. H. Kaneko, N. Gokon, N. Hasewaga, and Y. Tamaura, Solar thermochemcial process for hydrogen production using ferrites, *Energy*, **30** 2171–2178 (2005).

74. P. Blum, Cell for electrolysis of steam at high temperture, U.S. Patent 3, 993,653, Dec. 9, 1975.
75. D. I. Tcherev, Device for solar energy Conversion by photo-electrolytic decomposition of water, U.S. Patent 3, 925,212, Nov. 23, 1976.
76. A. J. DeBethune, T. S. Licht, and N. S. Swendemna, The temperature coefficient of Electrode Potentials, *J. Electrochem. Soc.*, **106** 618–625 (1959).
77. J. O'M. Bockris, *Energy Options*, Halsted Press, NY, 1980.
78. D. E. Monahan, Process and apparatus for generating hydrogen and oxygen using solar energy, U.S. Patent 4,233,127, Nov. 11, 1980.
79. L. E. Crackel, Spectral converter, U.S. Patent 4,313,425, Feb. 2, 1982.
80. C. Alkan, M. Sekerci, and S. Kung, Production of hydrogen using Fresnel lens-solar electrochemical cell, *Int. J. of Hydrogen Energy*, **20** 17–20 (1995).
81. C. W. Neefe, Passive hydrogel fuel generator, U.S. Patent 4,511,450, April 16, 1985.
82. D. E. Soule, Hybrid solar energy generating system, U.S. Patent 4,700,013, Oct. 13, 1987.
83. G. Tindell, Electrical energy production apparatus, U.S. Patent 4,841,731, June 27, 1989.
84. J. B. Lasich, Production of hydrogen from solar radiation at high efficiency, U.S. Patent 5,973,825, Oct. 26, 1999.
85. S. R. Vosen and J. O. Keller, Hybrid energy storage systems for stand-alone electric power systems: optimization of system performance and cost through control strategies, *Int. J. of Hydrogen Energy*, **24** 1139–1156 (1999).
86. J. Padin, T. N. Veziroglu, and A. Shahin, Hybrid solar high-temperature hydrogen production system, *Int. J. of Hydrogen Energy*, **25** 295–317 (2000).
87. H. Izumi, Hybrid solar collector for generating electricity and heat by separating solar rays into long wavelength and short wavelength, U.S. Patent 6,057,504, May 2, 2000.
88. S. Licht, Efficient solar generation of hydrogen fuel - a fundamental analysis, *Electrochemical Communications*, **4** 790–795 (2002).
89. S. Licht, Solar water splitting to generate hydrogen fuel: Photothermal electrochemical analysis, *J. Phys. Chem. B*, **107** 4253–4260 (2002).
90. S. Licht, L. Halperin, M. Kalina, M. Zidman, and N. Halperin, Electrochemical Potential Tuned Solar Water Splitting, *Chemical Communications*, 3006-3007 (2003).
91. S. Licht, pH measurement in conentrated alkaline solutions, *Anal. Chem.*, **57** 514–519 (1987).
92. T. S. Light, T. S, Licht, A. C. Bevilacqua, and Kenneth R. Morash, Conductivity and resistivity of ultrapure water, *Electrochem. Solid State Lett.*, **8** E16–E19 (2005)
93. S. Licht, Analysis in highly concentrated solutions: Potentiometric, conductance, evanescent, densometric, and spectroscopic methodolgies, in *Electroanalytical Chemistry*, Vol. 20, Edited by A. Bard and I. Rubinstein, Marcel Dekker, NY, 1998, pp. 87–140.
94. M. W. Chase, *J. Phys. Chem. Ref. Data* 14, Monograph 9 (JANF Thermochemical Tables, 4th edition), 1998.
95. M. W. Chase, *J. Phys. Chem. Ref. Data* Supplement No. 1 to 14, (JANF Thermochemical Tables, 3rd edition), 1986.
96. W. Kreuter and H. Hofmann, Electrolysis: The important energy transformer in a world of sustainable energy, *Int. J. Hydrogen Energy*, **23** 661–669 (1998).
97. M. A. Green, K. Emery, D. L. King, S. Igari, and W. Warta, Solar Efficiency Tables (Version 17), *Progr. Photovolt*, **9** 49–56 (2001).
98. E. Fletcher, *J. Solar Energy Eng*, **123** 143 (2001).
99. A. Yogev, *Quantum Processes for Solar Energy Conversion*, Weizmann Sun Symp. Proc., Rehovot, Israel, 1996.
100. R. Kribus, J. Doron, P. Rubin, J. Karni, S. Reuven, S. Duchan, and T. Tragan, A multistage solar receiver, *Solar Energy*, **67** 3–11 (1999).

101. E. Segal and M. Epstein, The optics of the solar tower reflector, *Solar Energy*, **69** 229–241 (2001).
102. B. Misch, A. Firus, and G. Brunner, An alternative method of oxidizing aqueous waste in supercritical water: oxygen supply by means of electrolysis, *J. Supercritical Fluids*, **17** 227–237 (2000).
103. O. Yamamoto, Solid oxide fuel cells: fundamental aspects and prospects, *Electrochimica Acta*, **45** 2423–35 (2000).
104. D. Kusunoki, Y. Kikuoka, V. Yanagi, K. Kugimiya, M. Yoshino, M. Tokura, K. Watanabe H. Miyamoto, S. Ueda, M. Sumi, and S. Tokunaga, Development of Mitsubishi - planar reversible cell - Fundamental test on hydrogen-utilized electric power storage system, *Int. J. Hydrogen Energy*, **20** 831–834 (1995).
105. K. Eguchi , T. Hatagishi, and H. Arai, Power generation and steam electrolysis characteristics of an electrochemical cell with a zirconia- or ceria-based electrolyte, *Solid State Ionics*, **86**-8 1245–1249 (1996).

6

Molecular Approaches to Photochemical Splitting of Water

Frederick M. MacDonnell

University of Texas at Arlington, Arlington, TX

1 Scope

This chapter focuses on the progress and challenges in the field of photocatalysis as applied towards the water splitting reaction (Eq. 1.). More specifically, homogeneous molecule-based systems that mimic the natural photosynthetic system are examined for their potential to drive reaction 1. A number of molecules, including porphyrins, metalloporphyrins and phthalocyanines,[1-7] transition metal complexes of Ru, Os, Re, Rh, Pt, Cu,[8-11] and acridine and flavin derivatives,[12-14] have been examined as the chormophores and sensitizers for light-driven processes.

$$2\ H_2O \rightarrow O_2 + 2\ H_2 \qquad \Delta G = +474\ kJ/mol = 4.92\ eV \qquad (1)$$

Of these sensitizers, Ru(II)-polypyridyl complexes are among the best explored due to their excellent chemical stability, broad absorption spectrum in the visible and favorable photophysical and redox properties. In this chapter, progress in the use of ruthenium polypyridyl-based molecular and supramolecular assemblies as sensitizers for 'artificial photosynthesis' is reviewed. Most of the comprehensive reviews on the application of such complexes towards artificial photosynthesis are over a decade old,[8,15,16] excepting an recent review by T. J. Meyers and co-workers[17] which was part of a 'Forum on Solar and Renewable Energy' thematic issue of Inorganic Chemistry (Oct 3, 2005).[18] Included in this thematic issue were additional reviews by G. J. Meyer[19] and M. Gratzel[20] on the application of ruthenium polypyridyl complexes as sensitizers for photovoltaics based on semiconducting nanoparticles such as TiO_2 which is a topic covered in other chapters of this book. Included in this issue is also a review of ruthenium polypyridyl complexes sensitizers for O_2 evolution in heterogeneous systems.[21] The focus of this chapter is largely on advances in homogeneous molecular systems that can be applied to the water-splitting problem.

2 Fundamental Principles

In photosynthesis and also as elaborated further in Chapter 8, nature has developed a molecular-based system able to capture light and transiently store it as a reducing potential (Eq. 2) and as high energy molecules (Eq. 3).[26] This energy is either quickly used or stored in the form of reduced CO_2

$$NADP^+ + 2e\text{-} + H^+ \rightarrow NADPH \tag{2}$$

$$ADP + P_i \rightarrow ATP + H_2O \tag{3}$$

$$6\ CO_2 + 6\ H_2O \rightarrow C_6H_{12}O_6 + 6\ O_2 \tag{4}$$

products (e.g., glucose, see Eq. 4) which are more stable forms of *stored energy.* Water is the ultimate source of electrons for reactions 2 and 4 and dioxygen is the byproduct that is lost to the atmosphere. The development of artificial photosynthetic systems that would mimic the natural process, at least in basic function, is a challenging yet realistic chemical problem with obvious long-term benefits for mankind.

As seen in Eq. 1, the water-splitting reaction has an overall energy requirement of 4.92 eV per O_2 molecule formed (or +474.7 kJ/mol O_2 formed). The most abundant solar radiation to strike the earth's surface falls in the visible range (750–400 nm) and fortunately, these photons are energetic enough (1.65–3.1 eV)[27] so that as little as two photons are required to drive this process thermodynamically. When broken down into redox half-reactions (5 and 6), the multi-electron nature of reaction 1 is readily apparent.

$$2\ H^+ + 2\ e^- \rightarrow H_2 \qquad\qquad + 79.9\ kJ/\ mol\ H_2,\ pH\ 7 \tag{5}$$

$$2\ H_2O \rightarrow O_2 + 4\ H^+\ + 4\ e^- \qquad + 314.9\ kJ/\ mol\ O_2,\ pH\ 7 \tag{6}$$

These two reactions are often referred to as the hydrogen evolving reaction (HER) and oxygen evolving reaction (OER), respectively. The problem of driving these two half-reactions with light is two-fold. First, water does not absorb light in the visible and a chromophore is needed to capture and concentrate the solar energy. Second, the absorption of a photon by a chromophore is typically associated with the excitation of a single electron in the molecule. Even if this electron has sufficient reducing potential for reaction 5, the electron stoichiometry is not met. Similar arguments apply to the holes generated and their ability to drive reaction 6. Nature has addressed these problems by developing enzymes proficient at stepwise storage of multiple redox equivalents (electrons or holes) until the appropriate redox stoichiometry is met. For example, the oxygen-evolving center (OEC) in photosystem II is composed of a tetramanganese cofactor that does not evolve O_2 until 4 electrons have been removed.[28] In general, we observe that the transformation of many small molecule substrates into desirable products, such as H_2O to O_2 and H_2, CO_2 to $C_6H_{12}O_6$, N_2 to NH_3, etc. are multi-electron processes and require not only the appropriate driving force but cofactors that enable the appropriate redox stoichiometry to be met. Artificial photosynthetic systems will similarly require entities/catalysts capable of driving multi-electron transfer (MET) reactions and proton-coupled electron transfer (PCET) reactions.[29-31]

3 Nature's Photosynthetic Machinery

Natural photosystems have been extensively studied and the machinery of photosynthesis found to be highly modular, organized on multiple scale levels and compartmentalized.[32] Importantly, all natural photosystems are molecule-based and their function can be understood both at the schematic level and, in many cases, at the molecular-level. Highly specialized components interact in controlled manner to ultimately deliver a product (reduction equivalents) and to effectively deal with waste (oxidation equivalents).

The bacterial photosystem functions without dioxygen production which simplifies the task at hand. Namely, electrons are obtained from more easily oxidized terminal electron donors such as H_2S instead of water. Nonetheless, the basic design needed to transform solar energy into stored chemical energy is present. The protein subunits and cofactors that comprise the photosystem in purple bacteria, such as *Rhodobacter (Rb.) sphaeroides* and *Rhodopseudomonas (Rps.) viridis*,[33] are shown schematically in Fig. 1 which is based on a crystal structure of this assembly.[34]

The sensitizer in this photosystem, P_{865}, is a symmetric bacteriochlorophyll dimer (labeled D_M and D_L in Fig. 1) which has a strong absorption at 865 nm corresponding to 1.43 eV. In the trans-cellular membrane assembly, the photoexcited $P_{865}*$ initiates charge separation by electron transfer down just one side of the photosynthetic assembly (the L side) as indicated by the arrows in Fig. 1. The reason for this asymmetry in electron transfer is still unclear, however, it is clear that electron transfer through a series of acceptor units: B_L to ϕ_L to Q_A and finally Q_B, leads of a charge separated complex with very slow (~ 1 sec) back rates. The P_{865}^+ cation is a modest oxidant, (E_{red} = +0.45 V, and is reduced to the starting state by external reductants such as H_2S. The reducing potential stored in Q_B is utilized in the cell (localized on the cytoplasmic side of the photosynthetic complex) to convert $NADP^+$ to NADPH. Thus, the net chemical reaction is:

$$H_2S + NAD^+ \rightarrow S + NADH + H^+ \tag{7}$$

The structure and function of this bacterial photosystem reveals important principles for the design of artificial photosystems. First, the sensitizer needs to be positioned close to secondary acceptors and donors which themselves are spatially isolated from each other such that photoexcitation leads to rapid spatial separation of the electron-hole pair. Second, compartmentalization of the photosynthetic assembly is likely to be necessary so as to prevent wasteful back reactions. For water-splitting, a system in which H_2 and O_2 are generated in separate compartments would have both safety and efficiency advantages.

In green plants and certain algae, the photosynthetic machinery is elaborated over those found in purple bacteria and is now able to reach the high potentials needed to oxidize water to dioxygen. This oxygenic photosystem is comprised of two photosynthetic reaction centers (sensitizer assemblies), photosystem I (PS I) and

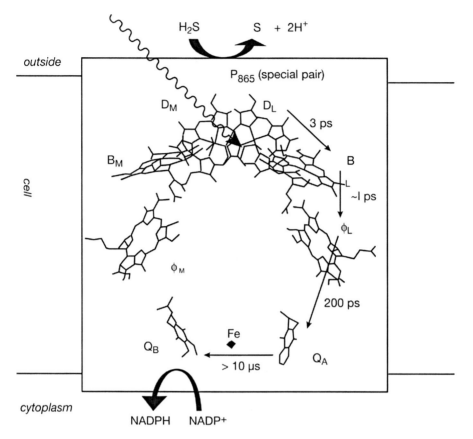

Fig. 1. Schematic of photosystem in purple bacteria

photosystem II (PS II) and thus requires more photons to drive the overall reaction.While the details have changed upon moving from the bacterial photosystem to PS I, the principle remains that PS I is unable to reach the potentials required to oxidize water. Figure 2 shows the classic Z scheme, first proposed in 1960 by Hill and Bendall[35] and subsequently elaborated many times,[36-38] which indicates how the two photosystems work together. The two sensitizers or reaction centers for PS I and PS II are again chlorophyll dimers such as found in bacteria, however these chlorophylls absorb at higher energies, 700 and 680 nm, and transiently store more energy. The special pairs for PSI and PSII are denoted P_{700} and P_{680}, respectively, and excitation of these reaction centers either by direct light absorption or by energy-transfer from an antenna assembly results in charge separation.

A schematic view of the cofactor arrangement in PSII is shown in Fig. 3.[39] The reducing potential of the P_{680}^* complex (~ –0.58 V) is used to shuttle an electron down a chain of redox acceptors in the PS II complex (pheotyphytin (phe/phe$^-$ at –0.42 V) to quinone A (Q_A/Q_A^- at –0.08 V) to quinone B (Q_B/Q_B^-))[40] and onto an

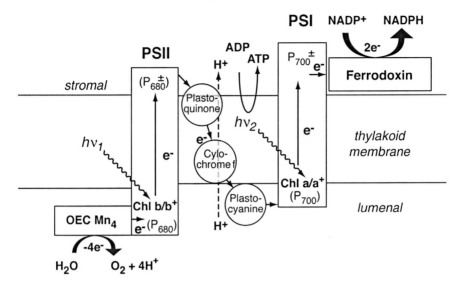

Fig. 2. The classic *Z scheme* for PSI and PSII.

external protein acceptor, plastoquinone. Reduced plastoquinone feeds another redox chain in which the reducing potential of these electrons are used to translocate protons across the membrane and thus generate a proton gradient to be used by the cell to generate ATP from ADP.[41] The oxidizing potential of the P_{680}^+ cation is estimated at +1.25 V[40,42] which is highly oxidizing and rapidly oxidizes the nearby tyrosine Z (Y_z/Y_z^+ at +1.21 V). Ten microseconds after excitation, the electron-hole pair is mostly localized as [Y_z^+/Q_A^-]. The tyrosine radical cation is positioned near the tetra-nuclear manganese oxygen evolving center (OEC) and possibly even coordinates to the Ca^{2+} site based on recent X-ray structural data.[43] Four sequential excitations (the 'S-state cycle') removes four electrons from the OEC with the last oxidation con-commitent with the evolution of O_2 and regeneration of the starting state of the OEC.[44] The structure of the OEC in PSII has recently been determined at resolutions as low as 3.2 A[43,45,46] which is where the structural 'cubane' model[43] shown in Figure 3 originates. The structure must be viewed with caution as there is some criticism of the co-factor structure due to the resolution of the data[47] and the possibility of X-ray damage to the OEC cofactor during data collection.[48] Nonetheless, this *cubane* structure incorporates most of the spectroscopic and compositional data known for PSII and serves as a useful model for the active site of oxygen evolution.

Photosystem I forms the second light-absorbing component in the Z scheme for green plants and algae and like PS II, the structure of the protein complexes has been determined by X-ray crystallography.[34,49-51] PS I is similar in function to that of the purple bacterial photosystem in that the oxidation potential generated is modest (P_{700}^+/P_{700} at ~ +0.5 V), however, the primary function of this photosystem is to

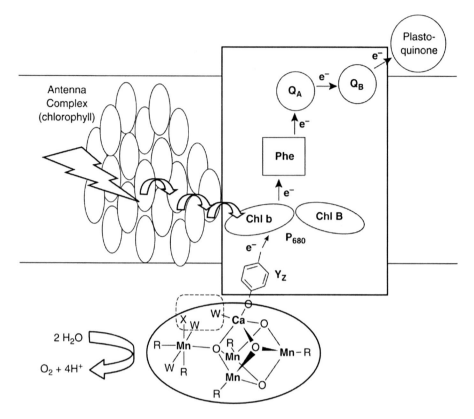

Fig. 3. Schematic view of co-factor arrangement in PS II.

generate a reducing potential for the generation of NADPH. The electrons for this process come from plastocyanin that is the terminal reduced product from PS II.

Both PSI and PSII are membrane proteins which span the thylakoid membranes in chloroplasts.[52] As indicated in Fig. 2, the luminal side of the membrane is where water is oxidized to O_2 and protons are generated. The energy dependent transloca-tion of these protons to the stromal side occurs as the electrons flow down the redox gradient indicated in the Z scheme. Ultimately, the protons are either consumed in the production of NADPH on the stromal side or their energy (in the form of a pro-ton gradient) is used to produce ATP.[38] The membrane is also the site in which the antenna complex, consisting of hundreds of chlorophylls are organized such that light excitation of any chlorophyll quickly results in energy transfer to the special pair in PS I or PS II for charge-separation.[32]

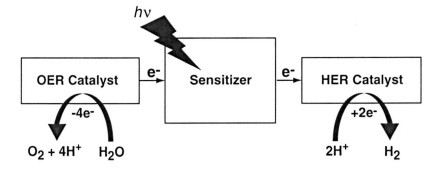

Fig. 4. Schematic of artificial photosynthetic system.

4 Design of Artificial Photosystems

From the natural photosystems, it is clear that a number of essential molecular components must be present and be organized in a supramolecular fashion such that the photogenerated electron-hole pair is quickly and efficiently separated and their potential energy delivered to functional co-catalysts. An artificial water-splitting photosystem will also require the controlled interaction of multiple subunits.[17] A schematic of such a system containing the minimum number of necessary components is given in Fig. 4. An efficient and robust sensitizer absorbs photons and generates long-lived electron-hole pairs that are coupled to appropriate multi-electron co-catalysts for oxygen and hydrogen evolution.

As mentioned previously, this chapter will focus largely on the use of Ru(II)-polypyridyl complexes as sensitizers to drive the water-splitting reaction. While we will examine the various properties of 'Rubpy' complexes in more depth, an initial analysis of the energetic of water splitting with $[Ru(bpy)_3]^{2+}$ will set the stage for the subsequent material. As seen in the modified Latimer diagram,[17,53] shown in Fig. 5, the photoexcited complex, denoted $[Ru(bpy)_3]^{2+*}$ or more simply Ru^{2+*}, is both a good oxidant and a good reductant. Electron transfer either to or from the excited state complex traps the ruthenium complex as $[Ru(bpy)_3]^+$ or $[Ru(bpy)_3]^{3+}$, respectively.[54] These photoproducts are potent reductants ($[Ru(bpy)_3]^+$) and oxidants ($[Ru(bpy)_3]^{3+}$) which can drive subsequent redox reactions.

These various species can be used to drive a catalytic reaction such as that shown in Eq. (8),

$$A + D \rightarrow A^- + D^+ \qquad (8)$$

by either an oxidative or reductive quenching pathway. These two catalytic manifolds are shown in Fig. 6, where A and D are generic acceptor and donor molecules.

While the two manifolds can store the same amount of energy, they encompass different potential ranges and therefore differ with respect to the types of acceptor and donor molecules that may be used or the driving force that may be applied to any

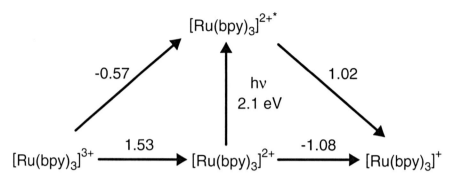

Fig. 5. Latimer diagram of the excited- and ground-state redox processes for [Ru(bpy)₃]²⁺.
Potentials given in V vs. NHE reference electrode.

given step in the cycle. If we analyze the energetics of the water-splitting reaction within these two manifolds, as shown in reactions 9 to 17, we can see that by designing a system properly we can deliver more driving force where it is most needed. All the reactions are normalized to a 4-photon, 4-electron process as presumably (though not necessarily) would be required and a pH of 7.0 is assumed. The first three photoreactions, 9–11, summarize the capture of light energy (at ~ 450 nm or 2.75 eV) to form the excited singlet molecule [Ru(bpy)₃]²⁺**. This molecule rapidly loses some energy upon intersystem-crossing to the triplet state and relaxation to vibronically cooled E_{00} triplet state, [Ru(bpy)₃]²⁺*.[17] The useable energy of this species is ~ 2.1 eV (~ 200 kJ/mol).

Light-capture:

$$4 Ru^{2+} (\lambda \sim 450 \text{ nm}) \rightarrow 4 Ru^{2+**} \qquad \Delta E = +1060 \text{ kJ/mol} \qquad (9)$$

$$\underline{4 Ru^{2+**} \rightarrow 4 Ru^{2+*}} \qquad \Delta E = -250 \text{ kJ/mol} \qquad (10)$$

$$4 Ru^{2+} + 4 h\nu \rightarrow 4 Ru^{2+*} \qquad \Delta E = +810 \text{ kJ/mol} \qquad (11)$$

Reductive quenching water-splitting path:

$$4 Ru^{2+*} + 2 H_2O \rightarrow 4 Ru^+ + O_2 + 4 H^+ \qquad \Delta G = -76 \text{ kJ/mol} \qquad (12)$$

$$\underline{4 Ru^+ + 4 H^+ \rightarrow 4 Ru^{2+} + 2 H_2} \qquad \Delta G = -260 \text{ kJ/mol} \qquad (13)$$

$$2 H_2O + 4 Ru^{2+*} \rightarrow O_2 + 2 H_2 + 4 Ru^{2+} \qquad \Delta G = -336 \text{ kJ/mol} \qquad (14)$$

Oxidative quenching water-splitting path:

$$4 Ru^{2+*} + 4 H^+ \rightarrow 4 Ru^{3+} + 2 H_2 \qquad \Delta G = -62 \text{ kJ/mol} \qquad (15)$$

$$\underline{4 Ru^{3+} + 2 H_2O \rightarrow 4 Ru^{2+} + O_2 + 4 H^+} \qquad \Delta G = -274 \text{ kJ/mol} \qquad (16)$$

$$2 H_2O + 4 Ru^{2+*} \rightarrow O_2 + 2 H_2 + 4 Ru^{2+} \qquad \Delta G = -336 \text{ kJ/mol} \qquad (17)$$

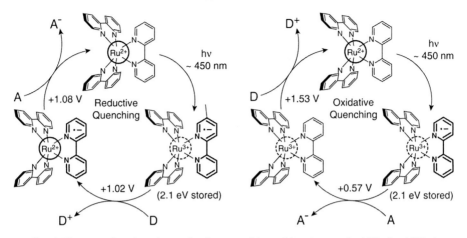

Fig. 6. Photoreactions based on reductive quenching of the photoexcited [Ru(bpy)3]2+*.

As can be seen both pathways are exothermic by −336 kJ/mol, however the exo-thermicity of the HER, Eq. (13), can be favored in the reductive quenching path at the expense of the OER, Eq. (12). The reverse is true for the oxidative quenching pathway as may be expected.

Of course, the solution pH is a major factor that can be tuned to favor hydrogen reduction or water oxidation. Application of the Nernst equation for the two mani-folds yields expressions 18–21:

reductive quenching pathway:

$$\Delta G_{OER} = \quad 84 - 22.85(pH) \qquad (kJ/mol) \qquad (18)$$

$$\Delta G_{HER} = -420 + 22.85(pH) \qquad (kJ/mol) \qquad (19)$$

oxidative quenching pathway:

$$\Delta G_{OER} = -116 - 22.85(pH) \qquad (kJ/mol) \qquad (20)$$

$$\Delta G_{HER} = -220 + 22.85(pH) \qquad (kJ/mol) \qquad (21)$$

From these expressions, we find the OER and HER are isoenergetic at pH 11 and 2.3 for the reductive and oxidative quenching pathways, respectively. More importantly, the pH range in which water splitting can occur can be determined, assuming the two half-reactions are not energetically coupled, i.e., the free energy of HER is not used to help drive the OER (or vice versa). Following the reductive quenching pathway, both the OER and HER are only spontaneous above pH 3.7. Conversely, the oxida-tive quenching pathway requires a pH below 9.6 for both to be spontaneous. Thus the quenching pathway plays an important role in determining optimum pH condi-tions. These energies and pH ranges are specific for [Ru(bpy)3]2+ assuming a 4 pho-

Table 1. Room temperature electrochemical and photophysical data for homoleptic ru-polypyridyl complexes.

Complex	$E(Ru^{3+/2+})^a$	$E(Ru^{2+/+})$	$E(Ru^{3+/2+*})$	$E(Ru^{2+*/+})$	λ_{em}, nm (τ, ns)^b	ϕ_{em}
$[Ru(bpz)_3]^{2+}$	2.10	−0.56	−0.06	1.60	627 (720)	0.024
$[Ru(4,4'-Cl_2bpy)]^{2+}$	1.64	−0.83	−0.42	1.23	632 (480)	−
$[Ru(bpy)_3]^{2+}$	1.53	−1.09	−0.57	1.02	610 (890)	0.073
$[Ru(phen)_3]^{2+}$	1.51	−1.11	−0.68	1.08	604 (460)	0.028
$[Ru(terpy)_2]^{2+}$	1.50	−1.05	−0.58	1.03	(<250 ps)	
$[Ru(4,4'- Me_2bpy)_3]^{2+ \, d}$	1.10	−1.37	−0.94	0.69	631 (335)	0.014
$[Ru(4,4'-(Et_2N)_2bpy)_3]^{2+}$	0.84		−1.07		700 (130)^c	0.010^c

^{a}As measured in acetonitrile unless otherwise noted. Electrochemical measurements were done in the presence of 0.1-M $Bu^n_4NPF_6$ or 0.1-M $Bu^n_4NClO_4$. All potentials are quoted relative to NHE using the correction factor of +0.247 V for SCE and +0.225 for Ag/Ag^+ where required.
^{b}degassed.
^{c}in MeOH/EtOH.
^{d}in H_2O.
bpz = 2,2'-bipyrazine. Data obtained from Ref. 54.

ton process but are tunable by modifying the complex structure. This is most commonly achieved by modification of the diimine ligand with electron withdrawing or donating groups. As seen in Table 1, ligand substitution can shift the relevant redox potentials by nearly a volt in either direction, however any gain in the oxidizing potential is offset by a loss in reducing potential or vice-versa. An extensive listing of the ground and excited state redox data for Ru(II) and Os(II) diimine complexes can be found in a review by Balzani and coworkers.[54] Also, Vlcek et. al. have reported that in many cases the excited state energies can be predicted simply from ligand redox parameters.[55]

This above analysis assumes that the free-energies of the HER and OER cannot be coupled which need not be true. As we observe in the natural photosystems, the free energy of one reaction may be used to change local pH by the translocation of protons across a membrane. The resulting pH change could easily be applied to increase the exothermicity of both O_2 and H_2 evolution in a straightforward manner and would be highly desirable. However, such a system requires an additional layer of sophistication in the design of the photosynthetic machinery. That said, artificial photodriven proton pumps are a reality as Gust, Moore and coworkers have shown that a membrane-bound carotenoid-porphyrin-quinone triad can actively transport protons against a gradient upon visible irradiation.[56]

There are additional aspects that must be considered for artificial or man-made photosystems. Nature's photosynthetic machinery is easily repaired or regenerated by the living organism. Artificial systems are likely to be harder to repair and therefore should be constructed of the most robust components available. The various components are likely to be subject to relatively harsh environmental conditions including high photon flux, thermal stress, aqueous solutions with variable pH and ionic strengths and oxygen. Only a few molecular species can meet all, or even most, of these conditions and while it is feasible to encapsulate or otherwise protect less stable molecular components, it is preferable that they be as robust as possible. No doubt the popularity of metalloporhyrins and ruthenium polypyridyl complexes as sensitizers is due, in large part, to their exceptional stability and favorable photophysical properties. Below we review the properties of ruthenium polypyridyl complexes that warrant all this attention, the current state of the art in 'wiring-up' these sensitizer to efficiently undergo energy or electron-transfer over large distances, the current ability to store multiple redox equivalents, and our ability to generate long-lived charge-separated states. Finally the development of efficient co-catalysts for water oxidation (OER) and hydrogen evolution (HER) are reviewed as are aspects of linking these catalysts with the sensitizer as indicated in Fig. 4.

5 The Ideal Sensitizer: Does *Rubpy* Come Close?

Rubpy is an often used shorthand for the well-studied $[Ru(bpy)_3]^{2+}$ complex cation and is also used more loosely to refer to any Ru(II) polypyridyl complex with similar properties. As shown in Fig. 7, the three best studied members of the Rubpy family are the parent complex, $[Ru(bpy)_3]^{2+}$, the phenanthroline analogue, $[Ru(phen)_3]^{2+}$, and the bis terpy complex, $[Ru(terpy)_3]^{2+}$, which has less optimal photophysical properties but has a number of synthetic and stereochemical advantages. The synthesis, stability and photophysics of Rubpy complexes has been reviewed many times[17,54,57] and the highlights of these properties as related to artificial photosynthetic assemblies is summarized here.

5.1 Stability

The stability of Rubpy complexes is a result of a number of factors including:

1. the large electronic stabilization found in these *octahedral* low spin d^6 second row transition metal complexes (while the symmetry of the complexes in Fig. 7 is not O_h, to a first approximation an octahedral coordination environment adequately describes many of the electronic properties);
2. the considerable covalent character of the bond as reflected in the similar electronegativities for Ru (χ 2.2) and N (χ 3.0), and
3. the multiple-bond character associated with π bond formation between the full Ru t_{2g} orbitals and the π-accepting orbitals on the polypyridyl ligands.

These factors and the associated kinetic inertness of the low spin d^6 electronic configuration make these complexes exceptionally stable with respect to ligand loss or decomposition.

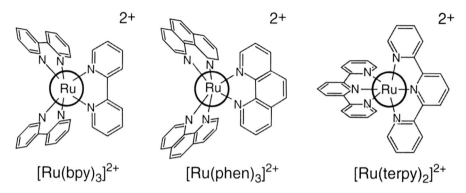

Fig. 7. Chemical structures of three commonly used Ru(II) polypyridyl complexes.

[Ru(bpy)₃]²⁺ and [Ru(phen)₃]²⁺ are modestly sensitive to photochemical degradation via photolabilization of the ligands upon excitation of the MLCT band.[58,59] The ³MLCT state that is utilized in most light-to-energy conversion schemes, is populated with essentially a 100% quantum yield.[60] At room temperature, however, thermal population of ruthenium-based d-d states, only ~ 43 kJ/mol higher in energy than the ³MLCT state, can occur resulting in decomposition via ligand dissociation.[58,59] Quantum yields for photochemical decomposition are generally small (ϕ ~ 0.04) and highly dependent on the solvent conditions and temperature.[58] Aqueous solutions of varying ionic strength and acidity show quantum decomposition yields between 0.3% to 0.001% at 343 K.[59] Dichloro methane solutions are considerably more reactive with decomposition quantum yields between 6 and 1 % at 298 K with the nature of the counterion playing an important role as small coordinating anions, such as halides, increasing the decomposition yield.[58] In dichloromethane, the increased decomposition is attributed to more significant ion pairing in this non-polar solvent, which leads to more effective *trapping* of the 5-coordinate intermediate, and stabilization of the product by anation. Photoracemization of optically pure Δ-[Ru(bpy)₃]²⁺ in aqueous solutions was also observed with a similar activation energy to the photodecomposition reaction and is presumed to occur via the same 5-coordinate intermediate.[61]

Photochemical decomposition of the ruthenium polypyridyl chromophore has not been significant obstacle to their use in energy conversion schemes because in most cases the ³MLCT state is rapidly quenched and the resulting photoproduct is stable with respect to ligand dissociation. For example, addition of the oxidative quencher, methylviologen (MV²⁺), greatly reduces the photodecomposition of [Ru(bpy)₃]²⁺* as the MV²⁺ rapidly reacts with the ³MLCT state by electron transfer yielding [Ru(bpy)₃]³⁺ and MV⁺.[62]

[Ru(bpy)₃]³⁺ and [Ru(bpy)₃]⁺ are commonly produced as photoproducts in the photochemistry of [Ru(bpy)₃]²⁺ and, as noted, are potent oxidants and reductants, respectively. Barring a redox reaction, these complexes show good stability towards both substitution and racemization,[63] however Ru(III) complex is not indefinitely stable in aqueous solutions. Over time spontaneous reduction of [Ru(bpy)₃]³⁺ to [Ru(bpy)₃]²⁺ is observed with some sacrificial degradation of the bpy ligands

(~ 10%).[64] Reductions are generally considered to be ligand-based and up to three reductions are often possible. Both the monocationic [Ru(bpy)₃]⁺ [65] and the neutral Ru(bpy)₃ [66] have been isolated and structurally characterized by X-ray diffraction.

The presence O₂ either from the atmosphere or as a product in water oxidation suggests its reactions with [Ru(bpy)₃]²⁺* bear close attention. Dioxygen is a well-known quencher of [Ru(bpy)₃]²⁺* and acts either by energy transfer or oxidative electron transfer quenching to yield singlet O₂ or superoxide radical O₂⁻, respectively.[67-69] As the ³MLCT state is quickly quenched, dioxygen typically protects the complex from photodecomposition via ligand loss.[58] However, when the O₂ and [Ru(bpy)₃]²⁺ are not free to diffuse apart as in viscous solution or zeolite matrices, the activated singlet O₂ or other reactive oxygen species can oxidize the diimine ligands irreversibly and thus damage the complex.[62,70]

Elaborate structures including those with up to three different bidentate ligands[71] are readily prepared due to the substitutional inertness of the Ru(II) ion. A number of synthetic procedures exist to build such complexes with most dealing with different strategies to coordinate select diimine or triimine ligands in a specific order. Some useful and often used starting complexes include Ru(bpy)₂Cl₂,[72] {Ru(bpy)Cl₃}ₓ,[73] Ru(bpy)(CO)₂Cl₂,[71,74] and Ru(terpy)Cl₃.[75] For most of these syntheses, it is trivial to substitute another diimine for the bpy ligand or triimine for the terpy ligand.

The tris-diimine and bis-triimine complexes, once formed, are tolerant of any number of ligand-based functional groups and the transformations typical of these functional groups can be performed on the metal complex with virtually no effect on the structural core integrity and stereochemistry. For example, the base catalyzed cross-coupling reactions of arylhalides with Ru(II) complexes, bearing di- or triimine ligands with peripheral phenols or hydroxymethyl (benzylic alcohol) substituents, gives the corresponding ethers in high yield and no reported degradation of the Ru complex.[76-79]

Palladium-catalyzed cross-coupling of bromoaryl-containing complexes with ethynylaryl-containing complexes is readily accomplished without degradation or racemization.[80-82] Barton and coworkers,[83,84] MacDonnell and coworkers[85-88] and others[89,90] have shown that the condensation of Ru(II) coordinated 1,10-phenanthroline-5,6-diones with ortho-phenylenediamines occurs without degradation of the complex or affecting the ruthenium stereochemistry. Electroploymerization of Ru(II) complexes bearing vinylbpy ligands has been demonstrated by Meyers and coworkers.[91] One illustrative example of the robustness of such complexes is shown in Figure 8. The coordinatively saturated Ru(II)trisphenanthroline complex can be oxidized to the tris-phenanthroline-5,6-dione complex in high yield (ca. 80 %) under extremely harsh chemical conditions.[88,92] The oxidation occurs in a mixture of concentrated sulfuric acid, nitric acid and NaBr at reflux, yet not only is the basic structural core retained, but if enantiomerically pure starting material is used, e.g., Λ-[Ru(phen)₃]²⁺, the quinone product is also optically pure. This reaction has been carried out on larger dendritic assemblies of [Ru(phen)₃]²⁺ units with similar success indicating the stability of this structural core to acidic and oxidizing conditions.[85]

The shapes of the three complexes in Figure 7 have important consequences in their use as sensitizers in multi-component assemblies. The tris-bpy and tris-phen complexes have three-fold symmetry (D₃ point group) while the bis-terpy complex

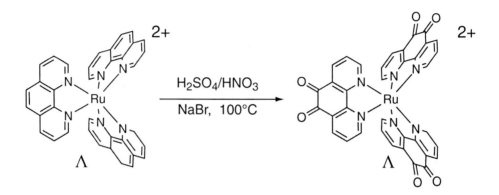

Fig. 8. Oxidation of [Ru(phen)₃]²⁺ with retention of stereochemistry.

has two-fold symmetry (D_{2h} point group). The stereochemical issues related to the tris complexes complicates their use in multi-component assemblies. As can be seen, the two tris complexes lack a convenient attachment site to build highly symmetric multi-component assemblies. Furthermore, these complexes are chiral (existing as Δ and Λ enantiomers) and the incorporation of multiple chiral centers in any supramolecular or covalent assembly leads to diastereomers which are very difficult to isolate pure. A number of research groups have addressed the stereochemical issues that arise when multiple ruthenium tris-diimine centers are covalently joined.[81,86,93-97] The stereochemistry of the terpy complexes conversely is well-suited to appending other components along the major two-fold axis such that stereochemically pure, linear arrays are prepared in a relatively straightforward manner.[98]

5.2 Photophysics and Photochemistry

All three Ru(II) complexes in Fig. 7 have strong absorption bands ($\varepsilon \sim 15{,}000 - 20{,}000$ M⁻¹cm⁻¹) in the visible region from 450 to 480 nm which are assigned to Ru dπ → bpy (π*) MLCT transitions. As described previously, [Ru(bpy)₃]²⁺* refers to the thermally equilibrated ³MLCT excited-state which is the state responsible for luminescence. This ³MLCT state is converted from the initial ¹MLCT state (often denoted [Ru(bpy)₃]²⁺**) with near unit efficiency in less than 300 fs.[99,100] The ³MLCT state decays by both luminescent and non-radiative decay pathways. As shown in Table 1, emission quantum yields (ϕ_{em}) in MeCN at room temperature are on the order of 0.03 to and 0.07 for [Ru(phen)₃]²⁺ and [Ru(bpy)₃]²⁺, respectively,[54] showing that non-radiative decay processes dominate this temperature. A survey of tris-diimine Ru(II) compounds reveals that such yields for ϕ_{em} are commonly between 0.10 and 0.001 however exceptions as high as 0.40 and as low as 5x10⁻⁶ are known.[54] One major non-radiative decay pathway for [Ru(bpy)₃]²⁺ is via thermal population of ligand field (d-d) states which are approximately 3000 cm⁻¹ (0.37 eV) higher in energy that the ³MLCT state.[101] At low temperature, this pathway is attenuated and the complexes often show dramatically longer luminescent lifetimes. Nonetheless, these d-d states are sufficiently high in energy that at room temperature that the lifetimes (τ) of the ^{3}MLCT state range from 400 to 1200 ns for many simple Rubpy analogues. Furthermore, because the decay pathways are well understood, several strategies can be employed to considerably lengthen the excited-state lifetimes.

For example, the excited-state lifetime of [Ru(terpy)₂]²⁺ is less than 250 ps at room temperature[102] however, synthetic modifications of the terpy ligand have lead to bis-terpy complexes with lifetimes as long as 200 ns.[103,104] As with the trisdiimine complexes, the non-radiative decay for [Ru(terpy)₂]²⁺ is attributed thermal population of low lying d-d states. However, in this case the tridentate terpy ligand causes a larger distortion of the O$_h$ ligand field which further lowers the energy of some d-d states to energies close to or below those of the ^{3}MLCT state. At room temperature, thermal population of these d-d states provides a rapid, non-radiative decay pathway to the ground state. At low temperature (77 K), thermal population of the d-d states is lessened and vibronic coupling between the ^{3}MLCT and d-d states can be frozen out to an appreciable degree, leading to excited state lifetimes on the order of 10 μs.[105]

As might be expected, simply lowering the energy of the ^{3}MLCT state relative to the d-d states results in longer excited state lifetimes. Balzani and coworkers found that modifying the terpy ligands with peripheral electron withdrawing methylsulfonate groups, lengthened the room temperature lifetime of the complex to 25 ns.[105] Similarly, the cyano-derivatized terpy complex **1** has a τ = 75 ns at 298 K, however, there is a limit to this approach.[106] The lifetime for the homoleptic complex **2** is only 50 ns indicating that if the ^{3}MLCT is lowered too much, direct non-radiative decay to the GS from the ^{3}MLCT state can become an important factor, as expected from the energy gap law.[107] Hanan and coworkers showed that stabilizing the LUMO of the terpy ligand by extending the planar aromatic system also extends the luminescent lifetime.[103,104] For the heteroleptic [Ru(terpy)(terpy-cyano pyrimidine)]²⁺ complex **3**, shown in Fig. 7, τ is 200 ns at 298 K which is attributed to a combination of the extended planar aromatic system and the EWG cyano substituent.

In general, when the excited-state is sufficiently long-lived (typically > 1 ns), bimolecular reactions in which the excited state acts as a reactant become possible. Typical reactions for quenching the excited state of Rubpy occur via one of the three following mechanisms. Schemes showing reductive and oxidative electron-transfer were shown in Fig. 6. Reaction 18 shows a third quenching mechanism known as energy transfer whereby relaxation of the ^{3}MLCT to the ground state is coupled to

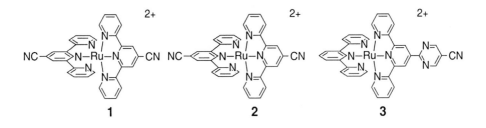

formation of an excited state in the quencher. While other quenching mechanisms are known these three are the dominant ones observed with Rubpy complexes.

$$[Ru(bpy)_3]^{2+*} + Q \rightarrow [Ru(bpy)_3]^{2+} + Q* \tag{22}$$

Literally, hundreds of inorganic and organic compounds have been examined as quenchers for Ru^{2+*} and its analogues. An extensive list of quenchers for excited state metal complexes (including Rubpy complexes), their rate constants (k_q) and mechanism of quenching (reductive, oxidative or energy transfer) was compiled by Hoffman and coworkers.[108] Quenching rate constants (k_q) are typically obtained via a Stern-Volmer analysis.[109] Typical rate constants for reductive and oxidative quenching of photoexcited Rubpy complexes usually range from 10^5 L·mol^{-1}s^{-1} to the diffusion limit of $\sim 10^{10}$ L·mol^{-1}s^{-1}. Energy transfer rate constants typically fall between 10^7 L·mol^{-1}s^{-1} and the diffusion limit. Some of the more commonly used reductive quenchers are organic amines (triethylamine (TEA), triethanolamine (TEOA), dimethylaniline (DMA)), organics (phenathiazine, phenols, hydro-quinones), coordination complexes ($[Fe(CN)_6]^{4-}$, $[Co(bpy)_3]^{2+}$), and simple inorganic ions (I^-, Eu^{2+}, $S_2O_6^{2-}$). Commonly oxidative quenchers include organics, such as methylviologen (MV^{2+}), quinones, and nitrobenzene, coordination complexes ($[Co(NH_3)_5Cl]^{2+}$, $Cr(acac)_3$) and simple inorganic ions ($S_2O_8^{2-}$, Cu^{2+}) and molecules (O_2). A number of donors (e.g., trialkylamines) and acceptors (e.g., $[Co(NH_3)_5Cl]^{2+}$ and $S_2O_8^{2-}$) are known to rapidly decompose upon electron-transfer which thereby prevents any back-reaction between quencher and Rubpy photoproduct.[110-112] These donors and acceptors are considered sacrificial and are useful when one desires to trap the reduced $[Ru(bpy)_3]^{1+}$ or oxidized $[Ru(bpy)_3]^{3+}$ photoproduct.

6 Supramolecular Assemblies: Dyads, Triads and Beyond

As illustrated by nature and shown schematically in Fig. 4, an artificial photosynthetic system will require several molecular components that are spatially juxtaposed to favor the various tasks at hand, i.e., light-harvesting, charge separation, catalysis for HER and OER. This next section will cover some of the molecular dyads, triads and larger assemblies prepared to address some of these challenges.

6.1 Energy Transfer Quenching: Antenna Complexes

Energy transfer quenching is the mechanism that natural photosystems use to harvest light over large areas using an 'antenna system' and subsequently funnel this energy to a specific site. In green plants, the antenna is composed of thousands of chlorophyll molecules which, when photoexcited to the triplet state (^{3}Chl), rapidly pass their energy to neighboring molecules eventually resulting in the electronic excitation of the special pair of chlorophyll molecules in the photochemical reaction center of PSI or PSII.[113] The whole process occurs on the order of a few tens of picoseconds, thus by using an antenna system, a relatively dilute photon flux that may typically excite a single reaction center chromophore on the order of a few excitations per second is dramatically enhanced.[113]

Similarly, efficient energy transfer between ruthenium(II) and osmium(II) poly-pyridyl complexes has been the basis for numerous artificial antenna complexes which act to harvest visible light energy and funnel it to of specific metal site.[114,115] Energy-transfer from $[Ru(bpy)_3]^{2+*}$ to $[Os(bpy)_3]^{2+}$ is found to occur at a rate close to the diffusion limit ($k_q = 1 \times 10^9$ M^{-1}s^{-1}) for the bimolecular reaction[116] and covalently linking these complexes is a common strategy for the construction of antenna complexes. Energy transfer rates in these multi-metallic assemblies are reasonably fast even over large internuclear distances provided that the bridging ligand can provide some electronic coupling. The role of the bridging ligand and the distance dependence of energy transfer in donor-acceptor dyads has been extensively investigated with bridges constructed of non-conjugated hydrocarbons,[117-120] polyalkyny-lenes,[121] polyenenes,[122] oligothienylenes,[123] and oligophenylenes,[124-129] and others.[8,130-132] For example, energy transfer rate constants were measured in acetonitrile for a series dimers in which a $Ru(bpy)_2^{2+}$ unit is bridged to a $Os(bpy)_2^{2+}$ unit by rigid conjugated ligands of the type bpy-(ph)$_n$-bpy (where n = 3, 5 and 7).[133] In this case, the rate constants for energy transfer from the Ru ^{3}MLCT to Os ^{3}MLCT were practically temperature independent and decreased with increasing length of the oligophenylene spacer on the order of $k_{en} = 6.7 \times 10^8$ s^{-1} for $n = 3$; $k_{en} = 1.0 \times 10^7$ s^{-1} for $n = 5$; $k_{en} = 1.3 \times 10^6$ s^{-1} for $n = 7$ at 293 K. Typically, the energy-transfer process in these conjugated systems occurs via a superexchange (Dexter) mechanism where the structure of the bridge plays an integral role in the ET process. Importantly, such systems show that vectoral energy transfer is possible at reasonable rates over large distances or as more commonly referred to 'molecular wires' are possible. From the above example, the Ru-Os distance is 4.2 nm in the $[(bpy)_2Ru(bpy-(ph)_7-bpy)Os(bpy)_2]^{4+}$ dimer and the rate constant is still 1.3×10^6 s^{-1}.

In addition to simple dyads, dendritic assemblies of Ru(II) and Os(II) polypyridyl complexes have been extensively investigated as the branching pattern of a dendrimer is ideally suited for orienting numerous chromophores to funnel energy to a single site.[114,134-137] For example, the Ru$_3$Os tetramer 4 shows quantitative energy transfer from the three peripheral Ru sites to the central Os(II) center as was expected as the Os(II) site has the lowest lying excited state.[138] When more complex dendrimers such as the decametallic 5 (shown schematically having the same bridging motif as 4) are examined, the energy migration pathways were less straightforward with energy transfer from the intermediate Ru sites (Ru$_i$) going to both the central Os(II) site and the peripheral Ru sites (Ru$_p$).[139] The bidirectionality is due to the relative energy ordering of the local ^{3}MLCT states which is Ru$_i$>Ru$_p$>Os, thus energy transfer from Ru$_i$ is downhill whether it goes towards the Os or the Ru$_p$. While this has limited the utility of these complexes as antenna which funnel the energy to a single site, recent developments suggest this can be overcome. Campagna, Scandola and co-workers showed that complex 6 which contains seven Ru chromophores (4 Ru$_p$, 2 Ru$_i$, 1 Ru(dpp)$_2$Cl$_2$ apical) undergoes energy transfer exclusively to the apical site even though the Ru$_i$ is again the highest energy ^{3}MLCT state.[140] The result is understood to involve an unusual two-step electron transfer process, importantly suggesting that indeed such complexes can funnel captured solar energy to a single site!

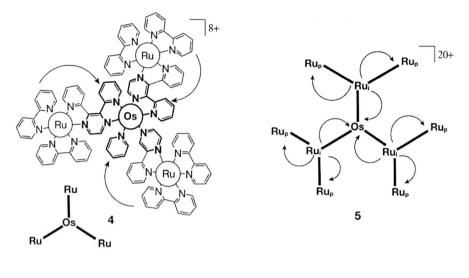

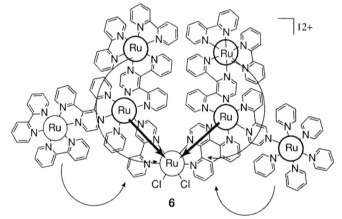

6.2 Bichromophores: Increasing Excited-State Lifetimes

Reversible or *equilibrated* energy-transfer between two chromophores with similar excited-state energies has also been used as a strategy for extending excited-state lifetimes of both Rubpy and Ruterpy analogues and has been the topic of several excellent recent reviews.[141-143] Conjugates of a ruthenium-polypyridyl chromophore with an organic chromophore that has a triplet excited state similar in energy to the ^{3}MLCT state can show reversible triplet-triplet energy transfer between the two chromophores, in essence leading to a ligand-centered triplet state (^{3}LC) that is thermally equilibrated with the ^{3}MLCT state.[144] A particularly striking example of this effect is seen by comparing the luminescent lifetime of $[Ru(bpy)_3]^{2+}$ and compounds **7** and **8**. Luminescent lifetimes at 298 K in deaerated MeCN increase from 0.920 μs

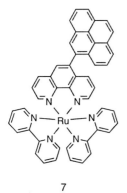

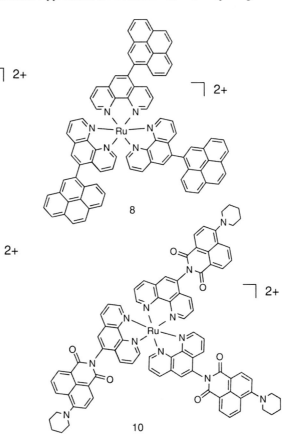

7

8

9

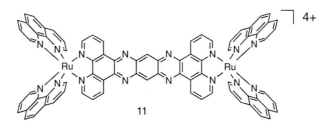

10

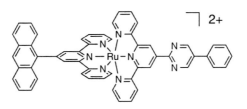

11

12

for [Ru(bpy)$_3$]$^{2+}$ to 23.7 μs and 148 μs for **7** and **8**, respectively.[145,146] Pyrene is by far the most organic chromophore used for such applications as its lowest triplet state is nearly isoenergetic with Rubpy ^{3}MLCT state. Recently, Castellano and co-workers reported that the piperindinyl-naphthalimide (PNI) chormophore is also functional in this capacity with the complexes **9** and **10** exhibiting lifetimes on the order of 47 and 115 μs, respectively.[147] This mechanism of transiently 'storing' the excited state as an organic triplet can greatly enhance the lifetime and therefore the utility of the inorganic chromophore, however, it should be noted that the equilibrium between the ^{3}MLCT state and the ^{3}LC state is easily perturbed by the solvent polarity and therefore the lifetimes are usually very sensitive to the solvent.

Compound **11** presents an interesting case in which the ^{3}LC chromophore in equilibrium with the ^{3}MLCT state is not so easily discerned. Chiorboli et. al. report a increase in excited state lifetime for compound **11** from 5 ns in MeCN to 1.3 μs in CH$_2$Cl$_2$.[148,149] In this case the organic chromophore is the tetraazapentacene backbone which is fused to two bipyridines to form the bridging tatpp ligand. Such intimate linking between the two chromophores typically leads to strong electronic coupling,[150] however in this case the electronic coupling between the tetra-azapentacene-like component and the bpy-like components is surprisingly weak. While unusual for most fused aromatic systems, this weak coupling between the bpy-like and phenazine-like components is commonly observed in the complexes of closely related and well-studied dipyrido-[3,2-a:2',3'-c]-phenazine (dppz)[151-156] and tetrapyrido-[3,2-a:2',3'-c:3'',2''-h:2''',3'''-j]-phenazine (tpphz) ligands.[157-159]

Hanan and coworkers have applied the bichromophoric approach with considerable success to the Ruterpy complexes.[103,160] Compound **12** shows excited state equilibrium between an anthracene ^{3}LC state and a 'Ruterpy' ^{3}MLCT state leading to room temperature luminescent lifetime of 1052 ns which is compared to 250 ps for [Ru(terpy)$_2$]$^{2+}$.[102]

6.3 Reductive and Oxidative Quenching: Dyads and Triads with Donors and Acceptors

Once light energy has been harvested in the form of a localized excited state, spatial separation of the electron-hole pair becomes important to prevent wasteful recombination reactions and to spatially direct the oxidizing and reducing equivalents to the appropriate co-catalysts. Dyads are usually composed of a sensitizer (S) and a donor (D) or acceptor (A) moiety whereas and triads are typically composed of S with both a D and A covalently attached. As opposed to antenna assemblies, the mode of action here is light-driven endogonic electron transfer from the donor to the acceptor as mediated by the Ru sensitizer. The linked assembly favors fast intramolecular reductive or oxidative quenching of the excited state and in the case of the D-S-A triads, subsequent ET to regenerate the ground state sensitizer (e.g., [Ru(bpy)$_3$]$^{2+}$). This area has been the subject of several excellent reviews over the past 10 years.[98,161-164]

Some of the more donors that have incorporated into dyads and triads are ferrocene[165], phenathiazine (PTZ),[166-169] phenols[170] and, more recently some tethered Mn

Table 2. Commonly used donor and acceptor moieties in dyads and triads.

Compound	Couple	E_{red} (vs NHE)	Ref.
	Donors		
Phenothiazine (PTZ)	0/+	1.08	294
Ferrocene (Fc)	0/+	0.55	
Phenol	0/1–	0.65	
	Acceptors		
Methylviologen (MV^{2+})	2+/1+	–0.11	166
Naphthalene diimide (NDI)		–0.42	170
C_{60}	0/1–	–0.16	175,295
		$–0.22^a$	
		$–0.38^b$	
p-benzoquinone (BQ)	0/1–	–0.36	173
Anthroquinone (AQ)	0/1–	–0.52	294
Nitrobenzene	0/1–		53

As measured in acetonitrile unless otherwise noted. Electrochemical measurements were done in the presence of 0.1 M $Bu^n_4NPF_6$ or 0.1 M $Bu^n_4NClO_4$. All potentials are quoted relative to NHE using the correction factor of +0.548 Fc, +0.236 SSCE, +0.247 V for SCE and +0.225 for Ag/Ag^+ where required.[296]
^{a}in THF.
^{b}in CH_2Cl_2.

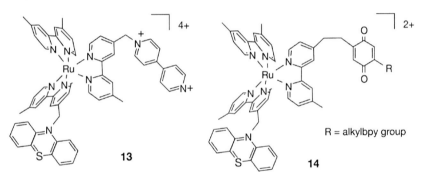

R = alkylbpy group

13   **14**

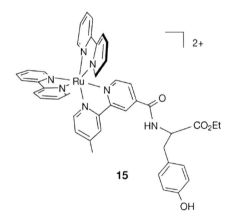

15

complexes.[171-174] Common acceptors include C_{60},[175-177] methylviologen (MV^{2+}),[166-169] naphthalenediimide (NDI),[170,178], quinones[179-181], nitroaromatics[53] and other transition metals complexes.[126] Although most dyads and triads are covalently linked an increasing number of assemblies based on supramolecular interactions (e.g., H-bonding, host-guest, salt-bridge) or are appearing.[182-186] Many of these donors and acceptors are listed in Table 2 along with the relevant redox couple and potential.

The function of a typical triad is seen in complex **13**. This D-S-A triad yields charge separated (CS) state characterized as [Ru^{II}(dmb)(bpyCH$_2$PTZ$^{\cdot+}$)(bpyCH$_2$MV$^{\cdot+}$)]$^{4+}$ upon photoexcitation of the Ru chromophore.[166] The half-life of the CS state is only ~160 ns, however it transiently stores 1.1 eV of energy. Complex **14** is similar in function however a benzoquinone function, is used in place of the MV^{2+}.[173] Like complex **13**, a CS state ($\tau \sim 90$ ns) is formed in **14** upon photoirradiation. In this complex both the quantum yield (90% vs. 22%) and amount of energy stored (1.32 eV) in the CS state are greater. In both **13** and **14**, it appears that oxidative quenching is favored and subsequent electron transfer from the PTZ group reduces the Ru^{III} site.

In an important step to mimic the natural photosystem, tyrosine residues tethered to a Rubpy sensitizer as in **15** have been shown to reduce the Ru^{III} center obtained after oxidative quenching with methylviologen or [Co(NH$_3$)$_5$Cl]$^{2+}$.[187] Formation of the resulting tyrosyl radical is a proton-coupled process and it has been shown to be a concerted process in which the reorganization energy associated with deprotonation can be tuned by H-bonding and pH.[188-191] Similar results are observed for tyrosyl residues tethered to Re(I)diimine based chromophores.[192]

The Uppsala group has shown that a photogenerated tyrosyl radical can oxidize a Mn^{III}Mn^{III} dimer to the Mn^{III}Mn^{IV} state in an intermolecular reaction.[193] This same group has gone on to incorporated the tyrosyl group into a manganese chelating ligand as in **16** (see Fig. 9).[194] These results are particularly significant given that in PSII, tyrosine Z plays a pivotal role in reducing the oxidized P$_{680}$ cofactor and then oxidizing the nearby tetranuclear Mn-cofactor.[195] Importantly, complex **16** has been shown to undergo multiple light-induced oxidations as indicated in Fig. 9 to form the Mn^{III}Mn^{IV} complex **17** [196,197] which is just one 'hole' short of the 4 electron stoichiometry needed for water oxidation. Moreover, this manganese centered oxidation requires water as the acetate groups are replaced by bridging oxo groups suggesting that the development of dyads which are functional in light driven oxidation may not be long in coming. In general, electron transfer rate constants for these Ru/Mn dyads vary between 1×10^5 to 2×10^7 s^{-1} and are related to the internuclear distance and believed to be limited primarily by the large inner reorganization energy of the Mn complex.[198,199] Wiegahardt and coworkers have also explored dyads which couple Rubpy sensitizers to a mononuclear Mn^{IV} complex containing phenolate ligands and with a Mn^{II} trimer assembled within the same ligand system.[171,172] Tethered [Ru(bpy)$_3$]$^{3+}$ centers are generated by photoexcitation and oxidative quenching with Co^{III}. Subsequently, electron transfer from the phenolate group in the Mn^{IV} complex or from the Mn^{II} ions in the trimer is observed to reduce the Ru. Like the Uppsala groups results, ET rates were on the order of 5×10^7 s^{-1}. This result is particularly notable in that the phenolate radical is directly coordinated to the Mn^{IV} ion and this unusual complex survives up to 1 ms.

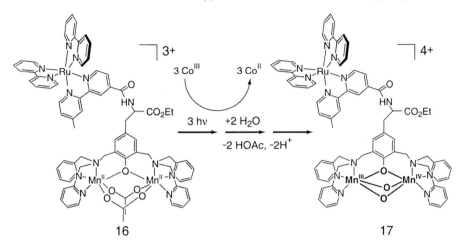

Fig. 9. Photodriven multi-electron oxidation of a manganese dimer using a covalently attached Rubpy sensitizer (see Ref. 193).

While numerous donor-Rubpy-acceptor triads have been studied,[98,161,200] triad **18** is notable for several reasons.[201] First, the complex shows a remarkably long-lived charge separated state of ~ 600 μS in solution at room temperature. This is at least two orders of magnitude longer than previous triads based on Rubpy sensitizers. Second, the charge separated state localizes the electrons on the NDI acceptors (as NDI⁻) and the holes on the Mn dimer as a $Mn^{II}Mn^{III}$ complex. The authors believe the unusually long lifetime is due, in part, to the large inner sphere reorganization that occurs in the Mn dimer, which makes the back reaction strongly activated. Third, the donor 'cofactor' is has the potential to directly act as both a multi-electron donor and ultimately as a water oxidation catalyst. It is also interesting to note that photoexcitation in this triad is first followed by oxidative quenching to give [$Mn^{II}Mn^{II}$-Ru^{III}-NDI^-]³⁺ which then undergoes intramolecular ET to yield the long-lived charge separated product [$Mn^{II}Mn^{III}$-Ru^{II}-NDI^-]³⁺.

6.4 Single versus Multi-Electron Processes

Photoexcitation is almost always a one-photon, one-electron process.[202] As seen in reactions 5 and 6, water splitting is a multi-electron process. One significant challenge in the development of artificial photosystems has been addressing this mismatch. As seen in reactions 19 and 20, one-electron routes towards water splitting involve the formation of highly reactive intermediates and require considerably higher potentials[202,203] than needed for the corresponding multi-electron reactions 5 at –0.414 V and –0.816 V, respectively. The redox potentials required to drive reactions 18 and 19 are not attainable by Ru-polypyridyl systems and are similarly out of the reach for many organic sensitizers. Furthermore, controlling and directing these highly reactive intermediates (e.g., ·OH, ·H, ·OOH) towards the desired products (H₂ and O₂) is an extremely challenging task which is best avoided if possible:

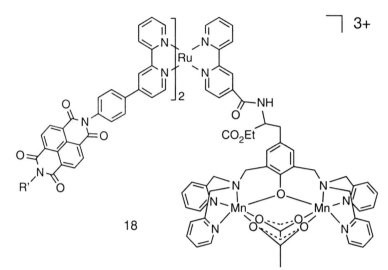

18

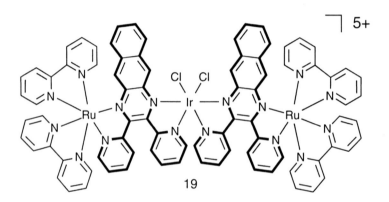

19

$$H^+ + e^- \rightarrow H^. \qquad -2.69\ V \qquad (+243\ kJ/\ mol,\ pH\ 7) \qquad (23)$$

$$OH^- \rightarrow OH^. + e^- \qquad -2.33\ V \qquad (+224.8\ kJ/\ mol,\ pH\ 7) \qquad (24)$$

It is clear that the electron stoichiometry strongly affects the reaction mechanism and therefore the potentials required. As demonstrated by the natural photosystems, properly designed co-catalysts can circumvent these high-energy one-electron path-

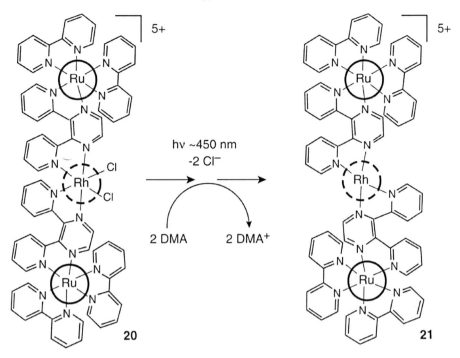

Fig. 10. The photoreduction of the trimeric Ru^{II}-Rh^{III}-Ru^{II} complex (**20**) at the Rh(III) site yields the Ru^{II}-Rh^{I}-Ru^{II} complex (**21**).

ways, however this is still a tremendous challenge for man-made systems. The first photoinduced charge separation event typically changes the physical properties of the acceptor and donor such that successive electron transfer events are not favorable.[204,205]

One of the first complexes capable of storing more than a single electron was the Ru^{II}-Ir^{III}-Ru^{II} trimer **19**. In this case, reductive quenching of the photoexcited complex with dimethylaniline led to the storage of two electrons, one in each of the π^* orbitals of the two bridging polypyridyl ligands (highlighted in bold).[206] More recently, the same group reported a trimeric Ru^{II}-Rh^{III}-Ru^{II} complex **20** (see Fig. 10), structurally similar to **19**, that undergoes photoreduction at the Rh(III) site to yield the Ru^{II}-Rh^{I}-Ru^{II} complex **21**. In this case, the photoproduct **21** has undergone a considerable structural rearrangement at the Rh site with loss of two chloride ligands. This doubly-reduced Rh^{I} center is an attractive catalytic site for heterolytic reactions such as oxidative addition or the formation of hydrides. In fact, the complex is reported to catalytically evolve hydrogen under photochemical conditions.[207]

MacDonnell and coworkers have shown that compounds **11** and **23** undergo two and four electron reductions to yield **22** and **24**, respectively, under photochemical conditions in both MeCN and water (see Fig. 11).[208-210] In this system, the photoreductions are ligands based and are seen to occur as stepwise one-electron processes under basic conditions. At lower pH's (~6-8), protonation of the reduced cen-

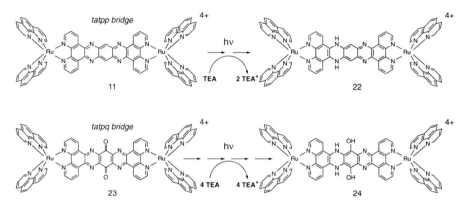

Fig. 11. Under photochemical conditions compounds **11** and **23** undergo two- and four-electron reductions to yield compunds **22** and **24**, respectively, in both, MeCN and water.

tral bridging ligands becomes evident and the individual one-electron reductions merge such that only two electron reductions are observed at the relatively slow timescale of the photochemical reaction (minutes). The process is reversible and air oxidation reoxidizes the reduced complexes back to **11** and **23**. Visible light irradiation of a acetonitrile solution of **23**, Pd(bpy)Cl$_2$, NH$_4$PF$_6$ and triethanolamine slowly produces H$_2$ at a rate of ~ 3 turnovers (per **23**) a day.[211] As before when discussing the unusual bichromophore behavior of **11**, the unusual acceptor capabilities of tatpp and tatpq bridging ligands seems to arise from the weak electronic coupling of the tetraazapentacene-like orbitals for tatpp (or tetraazapentacene-like and quinone-like orbitals for tatpq) and the bipyridine-like orbitals. As the bpy-like orbitals are the ones initially populated upon excitation, subsequent 'intramolecular electron-transfer' reduces the central portion of the ligand leaving the bpy-like portion open for another reductive cycle (after reductive quenching of the Ru(III) center with a sacrificial reductant).

As shown previously in Fig. 9, compound **16** is a promising addition to this family of complexes capable of photodriven multi-electron processes. Flash photolysis reveals a stepwise three electron oxidation at the Mn^{II}Mn^{II} center to yield the Mn^{III}Mn^{IV} complex **17**.[196,197] Unlike the preceding examples, multiple oxidizing equivalents or 'holes' are stored during the photochemical reaction making this system complimentary to those that collect multiple electrons.

Bocarsly, Pfennig and co-workers reported interesting multi-electron photoreactions for the trimeric M^{II}-Pt^{IV}-M^{II} complexes **25a-c**.[212-215] In this system, a single photon excitation into the intervalence charge transfer band results dissociation of the trimer into [Pt(NH$_3$)$_4$]$^{2+}$ and two equivalents of a M^{III}. The initial photoexcited complex is though to dissociate first to a M^{III} complex and Pt^{III}-M^{II} intermediate. The latter dimer subsequently undergoes a thermal electron transfer reaction to yield the final products.

In a related fashion, Haga and coworkers showed that the tetranuclear [Ru$_4$]$^{8+}$ complex **26** is doubly-reduced by a combination of photo- and thermal-induced re-

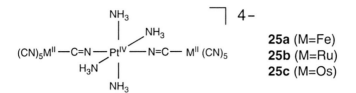

25a (M=Fe)
25b (M=Ru)
25c (M=Os)

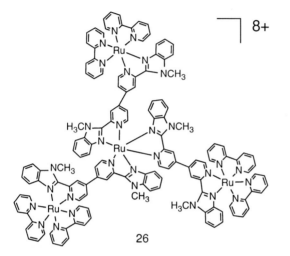

26

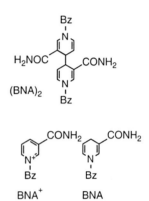

BNA⁺   BNA

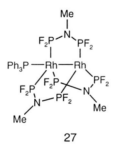

27

duction.[216] Initial photoexcitation of the Ru-diimine chromophores leads to reductive quenching of the photoexcited complex by N-benzyldihydronicotinomide, (BNA)₂. The oxidized (BNA)₂⁺ dimer then breaks into BNA⁺ and BNA·; the latter radical species subsequently reduces the ruthenium tetramer to form a [Ru₄]⁶⁺ complex in which the electrons are stored in the π* orbitals of the bridging ligands. In this case, the sacrificial reductant (BNA)₂ is 'non-innocent' in that the formation of a stable neutral radical leads to the second thermal reduction.

It is worth noting that similar processes could be in occurring with compounds **11**, **19**, **20** and **23** in which the oxidized sacrificial donor (e.g., TEA, DMA) deprotonate to form a neutral radical species with good reducing power. While it is difficult to rule such a possibility out, MacDonnell and co-workers have shown that the singly-reduced version of compound **11**, $[(phen)_2Ru^{II}(tatpp^{\cdot-})Ru^{II}(phen)_2]^{3+}$ can be isolated and, when subject to photochemical reduction, cleanly undergoes the second reduction. Thus, while the overall reduction of **11** to **22** could include a thermal redox reaction, it does not require one.

Nocera and coworkers have shown that a two-electron catalytic manifold is accessible in the Rh^0Rh^0 dimer **27** which is formed in solution by photolabilization of the CO adduct.[31,217,218] Earlier Gray and coworkers showed that certain Rh^IRh^I dimers stoichiometrically form X-Rh^{II}-Rh^{II}-X and H_2 when irradiated in the presence HX.[219,220] Under photolytic conditions, the dimer **27** is thought to oxidatively add hydrohalides (HX) to form the mixed-valent dimer $Rh^0Rh^{II}HX$ which then aggregates to form higher nuclearity species and H_2. These aggregates absorb at visible wavelengths (~580 nm) and undergo a photochemical reaction to form $Rh^0Rh^{II}X_2$ with the evolution of additional H_2. In the presence of halogen atom traps, UV photolysis regenerates **27** and thus it is possibly to photocatalytically evolve H_2.

7. OER and HER Co-Catalysts

7.1 Mimicking the Oxygen Evolving Center: Water Oxidation Catalysts

Developing good molecular catalysts for water oxidation remains one of the most challenging aspects of artificial photosynthesis. The oxidation process requires a number of difficult steps including the binding of at least two waters in close proximity, the removal of 4 electrons from these two waters, the formation of an O-O bond, the release of the product O_2, and the regeneration of the catalyst starting state. Several comprehensive reviews on molecular water oxidation catalysts have appeared[221-223] including one that focuses on the bioinorganic chemistry of the OEC in PSII.[224] Heterogeneous, polyoxometallate and colloidal catalysts for oxygen evolution continue to be explored by a number of groups[21,225-227] and are not covered here. Complexes which mimic the structure and spectroscopy of the OEC in PSII have been recently reviewed[228,229] and this section will focus on some of the newer or most notable complexes that show functional similarities.

A collection of functional molecular oxygen-evolving catalysts is shown in Fig. 12. Of these, the oldest and best studied is the ruthenium oxo-bridged dimer, **31**,[230-233] which was first reported to oxidize water in the presence of strong oxidants by Meyers and co-workers in 1982.[234] It is known to proceed via a $[(bpy)_2Ru^V(O)(\mu\text{-}O)Ru^V(O)(bpy)_2]^{4+}$ intermediate[235] and although the formation of a peroxo bridge between adjacent oxo groups is appealing, recent isotopic labeling studies suggest the O_2 is at least partially derived from solvent H_2O.[236] Yamada et. al propose oxo attack on water H-bonded to the catalyst to give a hydroperoxo intermediate prior to O_2 evolution.[236] The catalytic activity is attenuated by an anation side reaction which nonetheless is reversible by substitution of the anion with water.[235] Overall the num-

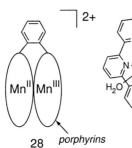

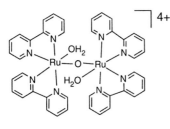

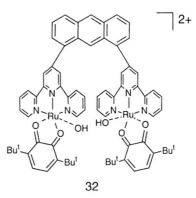

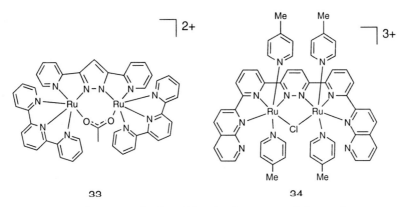

Fig. 12. A collection of functional oxygen-evolving catalysts.

ber of catalytic turnovers is limited (< 25) due to eventual decomposition of the complex. Catalyst **28** consists of two cofacial Mn(III) tetraarylporphyrins and recently it has been shown that a $Mn^V(O)$-$Mn^V(O)$ intermediate is formed prior to oxygen evolution.[237,238] This result is particularly interesting in that a terminal $Mn^V(O)$ group is implicated in many O_2 evolving mechanisms for PS II. [239-243] The $Mn^{III}(\mu$-$O)_2Mn^{IV}$ dimer **29** was the first bis μ-oxo Mn dimer shown to evolve O_2 catalytically under homogeneous conditions.[244,245] The disposition of the two open coordination sites makes it difficult to see how the two putative $Mn^V(O)$ groups could form an intramolecular peroxo intermediate and thus suggests that water plays an active role in the O_2 forming step. This is supported by isotopic labeling studies, however these results are not conclusive in that water may exchange with the oxo species bound to the Mn ions. Turnover numbers for this catalysts are modest (~30) with Oxone ($KHSO_5$) as the terminal oxidant presumably as the complex is seen to decompose with the formation of permanganate.[245]

The $Mn^{II}Mn^{II}$ dimer **30** has an unusual structure with two 7 coordinate Mn ions forming a bis μ-oxygen dimer via an η^1carboxylato oxygen ligand.[246] Upon mixing **30** with Bu^tOOH or Ce(IV), O_2 is evolved with a catalytic mechanism invoking dissociation of the dimer to form monomeric $Mn^{III}(OH)$ complexes, reassociation to a $Mn^{III}(μ\text{-}O)Mn^{III}$ dimer, further oxidation to the $Mn^{IV}(μ\text{-}O)_2Mn^{IV}$ dimer and formation of $Mn^{III}(μ\text{-}OO)Mn^{III}$ dimer which then loses O_2 to reform **30**.[247] Unfortunately, the data is not reported in a way that turnover numbers are easily compared with other catalysts but it appears that at least 10 turnovers are possible. Interestingly, isotopic labeling studies show the O_2 evolved consistently incorporates one oxygen atom from the solvent water and one from the Bu^tOOH. Curiously, the same labeling studies show that when $(NH_4)_2[Ce(NO_3)_6]$ is used as the oxidant, some of the nitrate oxygen atoms are incorporated into the dioxygen product. [247] The mechanism by which this occurs is unclear but it is notable in that many groups use this oxidant because it is 'oxygen-atom' free. It is worth mentioning that O_2 evolution has been observed from a model complex containing a $Mn_4O_4^{6+}$ cubane core, however this result is only seen in the gas phase and the relevancy to condensed phase systems is unclear.[248]

The remaining water oxidation catalysts **31-34** are $Ru^{II}Ru^{II}$ dimers with complexes **33** and **34** sharing similar bridging ligands. While these catalysts are less relevant as models of the natural photosystem, they are the most active and robust molecular water oxidation catalysts. Dimer **32** was reported by Tanaka and coworkers in 2000 and utilizes an unusual bridging ligand based on two terpyridine ligands which are held in a cofacial manner by an anthracene spacer.[249,250] Coordination of an ortho-quinone to each Ru(II) center leaves one accessible substrate site per metal ion. The ortho-quinone ligand appears to have an integral role in the catalytic cycle as replacement of this ligand with bpy results in markedly diminished O_2 evolving activity. In fact, one postulated reaction mechanism involves a base-catalyzed internal disproportionation of the quinone and hydroxo ligands in **32** resulting in the formation of a peroxo bridged dimer as indicated in reaction 21:

$$Ru^{II}Q(OH)\text{-}Ru^{II}Q(OH) \rightarrow Ru^{II}SQ\text{-}(OO)\text{-}Ru^{II}SQ + 2\,H^+ \qquad (21)$$

Controlled potential electrolysis of **32** in10% water in CF_3CH_2OH gave 21 turnovers (O_2 per **32**) whereas the activity jumps to 33,500 turnovers for the catalyst immobilized on an ITO electrode in aqueous solution. Unfortunately, no attempts to use chemical oxidants in a completely homogeneous system for **32** are reported.

Complex **33** consists of two Ru^{II} ions are bridged by a pyrazol-based chelating ligand and an acetate group.[251] Terpy ligands are use to tie-up three coordination sites on each Ru(II) ion leaving the complex coordinatively saturated but with an acid labile acetate group. Displacement of the acetate group with water gives the active catalyst which is though to proceed through a $Ru^{IV}(O)\text{-}Ru^{IV}(O)$ species as the highest oxidation state intermediate. At pH 1, the highest oxidation potential revealed by CV was at 1.05 V($E_{p,a}$ vs. SSCE). If this irreversible process is due to formation of the $Ru^{IV}(O)$-$Ru^{IV}(O)$ intermediate then the overpotential for O_2 evolution is nearly nil. Using Ce^{IV} as the oxidant, catalytic turnover numbers on the order of 19 after 48 h are observed.

Zong and Thummel recently reported on the water oxidation ability of complex **34** and some closely related derivatives.[252] Complex **34** shows turnover numbers as high as 3200 in aqueous solution (pH 1, CF_3SO_3H) using Ce^{IV} as the oxidant. Interestingly, they also show that a closely related Ru^{II} monomer can also catalyze O_2 evolution albeit with only 580 turnovers compared to the dimer. The CV of **34** shows an oxidation at 1.66 V vs SCE in acetonitrile which is more typical of the uppermost redox process in these water oxidation catalysts. The authors do not speculate on the mechanism at this time. Not included in Fig. 12 is the dinuclear complex $[(NH_3)_3Ru(\mu\text{-}Cl)_3Ru(NH_3)_3]^{2+}$ which has also been reported to catalyze water oxidation using Ce^{IV} as the oxidant.[253] Immobilizing the catalysts in a Nafion membrane was shown to significantly enhance the catalyst lifetime.

Of the water oxidation catalysts mentioned, all use either a powerful chemical oxidant (e.g., Ce^{IV}, OCl^-, Oxone, Bu^tOOH) or an electrode to drive the reaction. Efforts to couple the oxidation to a photoprocess have not yielded an active photocatalyst but nonetheless are beginning to yield some promising results as shown previously in Fig. 9 for the Ru-Mn_2 dyad.[254-256]

7.2 The Hydrogen Evolving Reaction (HER): Hydrogen Evolution Catalysts

The vast majority of hydrogen evolving catalysts are still mostly limited to noble-metal colloids or solids.[257-260] Early studies on solar or photo-driven hydrogen production used these colloids along with a sensitizer such as Rubpy and electron relays (e.g., MV^{2+}) with modest success.[15,16,261-268] The development of good molecular catalysts for HER remains a surprisingly elusive goal excepting the biologically produced iron-only and iron-nickel hydrogenases.[269] These enzymes are able to catalyze proton-reduction or hydrogen oxidation essentially at the thermodynamic potential and do so with rapid turnover.[270] With the recent X-ray crystallographic structural elucidation of the active sites of these enzymes,[271-273] shown schematically in Fig. 13, considerable progress has been made in understanding their mechanism of action. Numerous groups have models the active site of the $FeNi$[274,275] and Fe-only[276-279] hydrogenases in small molecules, however, the functional activity of these model complexes is limited by the large overpotentials required for function.[280,281] Similarly, dyads in which a Rubpy sensitizer is covalently linked to a biomimetic diiron complex have been prepared but photodriven H_2 is still elusive.[282,283]

An early review by Koelle on transition metal catalyzed proton reduction nicely developed the various chemical steps involved in hydrogen evolution including metal hydride formation, hydride acidity (basicity) and protonation and requisite redox potentials.[284] The complexes review here have little structural relevance to the hydrogenase active sites but many show   promising catalytic activity.   More recently

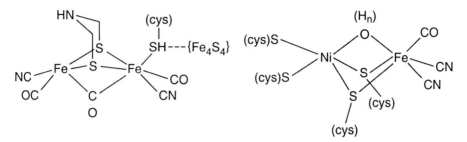

Fig. 13. Biologically produced iron-only and iron-nickel hydrogenases used in the catalysis for HER.

(2005), Artero and Fontecave reviewed the most efficient catalysts for hydrogen evolution from an applications perspective and nicely defined the general principles for the design of robust and economically viable catalysts for the HER.[285] A number of the catalysts highlighted by Artero and Fontecave are shown in Fig. 14 and while a detailed description of their action would be repetitive, the general properties desired are good basicity of the metal center (either as is or after reduction), open sites for hydride coordination, and finally an accessible redox n+2 state for heterolytic cleavage upon protonation. Homolytic pathways are also possible but require close juxtaposition of two metal centers or unfavorable bimolecular reactions between catalytic metal centers. Collman and coworkers prepared co-facial Ru porphyrins to address this possibility but saw little improvement over monomeric porphyrin complexes.[286-289]

The cobalt diglyoximate complexes like **35a-c** were first investigated as HER catalysts in the early 1980's by Espenson and coworkers[290-292] and have recently been revisited by Peters, Lewis and coworkers.[293] In particular, the difluoroboryl-dioxime complexes **36a** and **36b** seem promising as the macrocylic structure imparts good stability in acidic solution and the potential at which H_2 is evolved electrochemically is surprisingly high. Complex **36b** evolves H_2 at -0.28 V vs. SCE in acetonitrile solutions with HCl·Et$_2$O as the proton source.

8. Future Outlook and Concluding Remarks

Over the past decade a number of significant advances have been made in the development of viable artificial photosystems for water splitting. Antenna complexes offer systems that can collect solar energy over large spatial regions and funnel it to specific sites for charge-separation. New methods exist for extending the excited state lifetimes for Rubpy based sensitizers and numerous studies have shown that the charge separation event can be directed towards specific donor and acceptor 'cofactors'. Advances in our ability to collect and store multiple electrons or holes, suggests that the catalytically desirable multi-electron reactions for HER and OER can be accessed. Molecular dyads and triads containing potentially catalytic active reaction centers have been constructed and shown to be active towards photochemi-

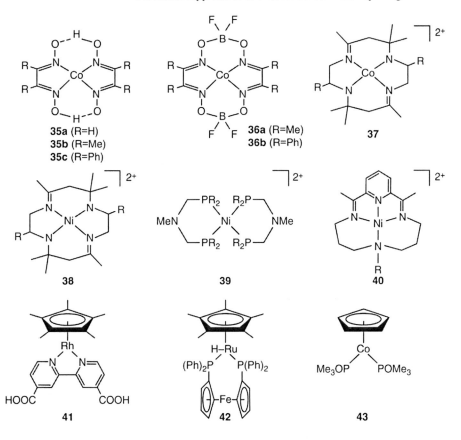

Fig. 14. Chemical structure of some of the more efficient hydrogen evolving catalysts.

cally induced electron transfer reactions. This and the development of new and better catalysts for both OER and HER suggests that photocatalytic assemblies are not long in coming.

The current state of the art is evolving however some notable problems remain. Current catalysts are slow and inefficient compared to their biological counterparts. Challenges remain in coupling the multi-electron equivalency in the OER and HER with the Rubpy sensitizers. Back electron transfer or related wasteful back reactions still account for the serious losses in efficiency. These issues will be difficult if not impossible to address in a single compartment molecular system. Compartmentalization coupled to vectoral electron transfer remains a largely unexplored aspect of this work (if heterogeneous systems are excluded) yet is likely to be critical for success for a purely molecular system.

Acknowledgements

The author thanks the US National Science Foundation (CHE-0518649) and the Robert A. Welch Foundation (Y-1301) for continued financial support.

References

1. Wasielewski, M. R. *Chem. Rev.* **1992**, *92*, 435-461.
2. Cole-Hamilton, D. J.; Bruce, D. W. In *Comp. Coord. Chem. V. 6*; Wilkinson, G., Gillard, R. D., McCleverty, J. A., Eds.; Pergamon Books Ltd: Maxwell House, NY, 1987; Vol. 6.
3. Holten, D.; Bocian, D. F.; Lindsey, J. S. *Acc. Chem. Res.* **2002**, *35*, 57-69.
4. Kodis, G.; Liddell, P. A.; de la Garza, L.; Clausen, P. C.; Lindsey, J. S.; Moore, A. L.; Moore, T. A.; Gust, D. *J. Phys. Chem. A.* **2002**, *106*, 2036 - 2048.
5. Fukuzumi, S.; Imahori, H. In *Electron Transfer in Chemistry* Balzani, V. Ed.,   Wiley-VCH Verlag GmbH, Weinheim, Germany, 2001, 2 927-975.
6. You, C.-C.; Dobrawa, R.; Saha-Moeller, C. R.; Wuerthner, F. *Top. Curr.   Chem.* **2005**, *258*, 39-82
7. Lin, V. S.; DiMagno, S. G.; Therien, M. J. *Science* **1994**, *264*, 1105-11.
8. Balzani, V.; Juris, A.; Venturi, M.; Campagna, S.; Serroni, S. *Chem. Rev.* **1996**, *96*, 759-833.
9. Willner, I.; Kaganer, E.; Joselevich, E.; Dorr, H.; David, E.; Gunter, M. J.; Johnston, M. R. *Coord. Chem. Rev.* **1998**, *171*, 261-285.
10. Hissler, M.; McGarrah, J. E.; Connick, W. B.; Geiger, D. K.; Cummings, S. D.; Eisenberg, R. *Coor. Chem. Rev.* **2000**, *208*, 115-137.
11. Dempsey, J. L.; Esswein, A. J.; Manke, D. R.; Rosenthal, J.; Soper, J. D.; Nocera, D. G. *Inorg. Chem., 44 (20)*, , **2005**, *44*, 6879 -6892.
12. Krasna, A. I. *Photochem. Photobiol.* **1980**, *31*, 75-82.
13. Nenadovic, M. T.; Micic, O. I.; Kosanic, M. M. *Rad. Phys. Chem.* **1981**, *17*, 159-161.
14. Eng, L. H.; Lewin, M. B. M.; Neujahr, H. Y. *Photochem. Photobiol.* **1993**, *58*, 594-599.
15. Amouyal, E. *Solar Energy Materials and Solar Cells* **1995**, *38*, 249-276.
16. Bard, A. J.; Fox, M. A. *Acc. Chem. Res.* **1995**, *28*, 141-145.
17. Alstrum-Acevedo, J. H.; Brennaman, M. K.; Meyer, T. J. *Inorg. Chem.* **2005**, *44*, 6802-6827.
18. Eisenberg, R.; Nocera, D. G. *Inorg. Chem.* **2005**, *44*, 6779-6781.
19. Meyer, G. J. *Inorg. Chem.* **2005**, *44*, 6852 - 6864.
20. Grätzel, M. *Inorg. Chem.* **2005**, *44*, 6841 - 6851.
21. Hoertz, P. G.; Mallouk, T. E. *Inorg. Chem.* **2005**, *44*, 6828 -6840.
22. Turner, J. A. *Science* **1999**, *285*, 687.
23. Jacoby, M. In *Chemical and Engineering News*, May 30, 2003; Vol. 33, pp 35-37.
24. Lewis, N. S. *Nature* **2001**, *414*, 589-590.
25. Dresselhaus, M. S.; Thomas, I. L. *Nature* **2001**, *414*, 332-337.
26. Danks, S. M. *Photosynthetic Systems : Structure, Function, and Assembly*; Wiley: New York, 1983.
27. Serway, R.; Beichner, R. *Physics for Scientists and Engineers with Modern Physics*; 5th ed.; Saunders College Publishing: Orlando, 2000.
28. Manchanda, R.; Brudvig, G. W.; Crabtree, R. H. *Coord. Chem. Rev.* **1995**, *144*, 1-38.
29. Cukier, R. I.; Nocera, D. G. *Ann. Rev. Phys. Chem.* **1998**, *49*, 337-369.
30. Heyduk, A. F.; Macintosh, A. M.; Nocera, D. G. *J. Am. Chem. Soc.* **1999**, *121*, 5023-5032.
31. Heyduk, A. F.; Nocera, D. G. *Science* **2001**, *293*, 1639-1641.

32. Ke, B. *Photosynthesis: Photobiochemistry and Photobiophysics*; Kluwer Academic Publishers: Dordrecht, The Netherlands, 2001; Vol. 10.
33. Lancaster , C. R. D.; Michel, H. In *Handbook of Metalloproteins*; Messerschmidt, A., Ed.; John Wiley & Sons Ltd.: Chichester, UK, 2001; Vol. 1, pp 119-135.
34. Deisenhofer, J.; Michel, H. In *Molecular Mechanisms in Bioenergetics*; Ernster, L., Ed.; Elsevier: Amsterdam, 1992, pp 103-120.
35. Hill, R.; Bendall, F. *Nature* **1960**, *186*, 136-137.
36. Katz, E.; Shipway, A. N.; Willner, I. In *Electron Transfer in Chemistry V. 4*; V. Balzani, P. P., M.A.J. Rodgers, Ed.; Wiley-VCH: Weinheim, Germany, 2001; Vol. 4, pp 127-201.
37. Arnon, D. I. *Photosynth. Res.* **1995**, *46*, 47-71.
38. Allen, J. F. *Cell* **2002**, *110*, 273-276.
39. Barber, J. *Curr. Opin. Struct. Biol.* **2002**, *12*, 523-530.
40. Grabolle, M.; Dau, H. *Biochim. Biophys. Acta* **2005**, *1708*, 209-218.
41. Moser, C. C.; Page, C. C.; Cogdell, R. J.; Barber, J.; Wraight, C. A.; Dutton, P. L. *Adv. Protein Chem.* **2003**, *63*, 71-109.
42. Rappaport, F.; Guergova-Kuras, M.; Nixon, P. J.; Diner, B. A.; Lavergne, J. *Biochemistry* **2002**, *41*, 8518-8527.
43. Ferreira, K. N.; Iverson, T. M.; Maghlaoui, K.; Barber, J.; Iwata, S. *Science* **2004**, *303*, 1831 - 1838.
44. Barber, J. *Q. Rev. Biophys* **2003**, *36*, 71-89.
45. A. Zouni, A.; Witt, H. T.; Kern, J.; Fromme, P.; Krauss, N.; W., S.; Orth, P. *Nature* **2001**, *409*, 739–743.
46. Kamiya, N.; Shen, J.-R. *Proc. Natl. Acad. Sci. USA* **2003**, *100*, 98-103.
47. Biesiadka, J.; Lolla, B.; Kernb, J.; K.-D., I.; Zouni, A. *Phys. Chem. Chem. Phys.* **2004**, *20*, 4733-4736.
48. Yano, J.; Kern, J.; Irrgang, K.-D.; Latimer, M. J.; Bergmann, U.; Glatzel, P.; Pushkar, Y.; Biesiadka, J.; Loll, B.; Sauer, K.; Messinger, J.; Zouni, A.; Yachandra, V. K. *Proc. Natl. Acad. Sci. USA* **2005**, *102*, 12047-12052.
49. Deisenhofer, J.; Epp, O.; Miki, K.; Huber, R.; Michel, H. *J. Mol. Bio.* **1984**, *180*, 385-398.
50. Byrdin, M.; Jordan, P.; Krauss, N.; Fromme, P.; Stehlik, D.; Schlodder, E. *Biophys. J.* **2002**, *83*, 433-457.
51. Jordan, P.; Fromme, P.; Witt, H. T.; Klukas, O.; Saenger, W.; Krauss, N. *Nature* **2001**, *411*, 909–917.
52. Armond, P. A.; Staehelin, L. A.; Arntzen, C. J. *J. Cell Biol.* **1977**, *73*, 400-418.
53. Bock, C. R.; Connor, J. A.; Gutierrez, A. R.; Meyer, T. J.; Whitten, D. G.; Sullivan, B. P.; Nagle, J. K. *J. Am. Chem. Soc.* **1979**, *101*, 4815-4824.
54. Juris, A.; Balzani, V.; Barigelletti, F.; Campagna, S.; Belser, P.; Von Zelewsky, A. *Coord. Chem. Rev.* **1988**, 85-277.
55. Vlcek, A. A.; Dodsworth, E. S.; Pietro, W. J.; Lever, A. B. P. *Inorg. Chem.* **1995**, *34*, 1906-1913.
56. Gust, D.; Moore, T. A.; Moore, A. L. *Acc. Chem. Res.* **2001**, *34*, 40-48.
57. Balzani, V.; Juris, A. *Coordination Chemistry Reviews* **2001**, *211*, 97-115.
58. Durham, B.; Caspar, J. V.; Nagle, J. K.; Meyer, T. J. *J. Am. Chem. Soc.* **1982**, *104*, 4803-4810.
59. Van Houten, J.; Watts, R. J. *Inorg. Chem.* **1978**, *17*, 3381 - 3385.
60. Demas, J. N.; Taylor, D. G. *Inorg. Chem.* **1979**, *18*, 3177 - 3179.
61. Porter, G. B.; Sparks, R. H. *J. Photochem.* **1980**, *13*, 123-131.
62. Vaidyalingam, A.; Dutta, P. K. *Anal. Chem.* **2000**, *72*, 5219-5224.
63. Dwyer, F. P.; Gyarfas, E. C. *J. Proc. Roy. Soc. New South Wales* **1949**, *83*, 174-176.
64. Ghosh, P. K.; Brunschwig, B. S.; Chou, M.; Creutz, C.; Sutin, N. *J. Am. Chem. Soc.* **1984**, *106*, 4772-4783.

65. Pérez-Cordero, E.; Brady, N.; Echegoyen, L.; Thummel, R.; Hung, C.-Y.; Bott, S. G. *Chem. Eur. J.* **1996**, *2*, 781-788.
66. Pérez-Cordero, E.; Campagna, C.; Echegoyen, L. *Angew. Chem. Intl. Ed. Engl.* **1997**, *36*, 137-140.
67. Demas, J. N.; Harris, E. W.; McBride, R. P. *J. Am. Chem. Soc.* **1977**, *99*, 3547-3551.
68. Buell, S. L.; Demas, J. N. *J. Phys. Chem.* **1983**, *87*, 4675-4681.
69. Hartmann, P.; Leiner, M. J. P.; Kohlbacher, P. *Sensors and Actuators, B: Chemical* **1998**, *B51*, 196-202.
70. Fuller, Z. J.; Bare, W. D.; Kneas, K. A.; Xu, W.-Y.; Demas, J. N.; DeGraff, B. A. *Anal. Chem.* **2003**, *75*, 2670-2677.
71. Anderson, P. A.; Deacon, G. B.; Haarmann, K. H.; Keene, F. R.; Meyer, T. J.; Reitsma, D. A.; Skelton, B. W.; Strouse, G. F.; Thomas, N. C. *Inorg. Chem.* **1995**, *34*, 6145-6157.
72. Sullivan, B. P.; Salmon, D. J.; Meyer, T. J. *Inorg. Chem.* **1978**, *17*, 3334 - 3341.
73. Kelch, S.; Rehahn, M. *Macromolecules* **1997**, *30*, 6185-6193.
74. Strouse, G. F.; Anderson, P. A.; Schoonover, J. R.; Meyer, T. J.; Keene, F. R. *Inorg. Chem.* **1992**, *31*, 3004-3006.
75. Sullivan, B. P.; Calvert, J. M.; Meyer, T. J. *Inorg. Chem.* **1980**, *19*, 1404-1407.
76. Constable, E. C.; Harverson, P.; Oberholzer, M. *Chem. Commun.* **1996**, 1821-1822.
77. Constable, E. C.; Harverson, P. *Chem. Commun.* **1996**, 33-34.
78. Constable, E. C.; Harverson, P. *Inorg. Chim. Acta.* **1996**, *252*, 9-11.
79. Worl, L. A.; Strouse, G. F.; Younathan, J. N.; Baxter, S. M.; Meyer, T. J. *J. Am. Chem. Soc.* **1990**, *112*, 7571-7578.
80. Tzalis, D.; Tor, Y. *Angew. Chem. Int. Ed. Engl.* **1997**, *36*, 2666-2668.
81. Tzalis, D.; Tor, Y. *J. Am. Chem. Soc.* **1997**, *119*, 852-853.
82. Tzalis, D.; Tor, Y. *Chem. Commun.* **1996**, 1043-1044.
83. Friedman, A. E.; Chambron, J.-C.; Sauvage, J.-P.; Turro, N. J.; Barton, J. K. *J. Am. Chem. Soc.* **1990**, *112*, 4960-4961.
84. Hartshorn, R. M.; Barton, J. K. *J. Am. Chem. Soc.* **1992**, *114*, 5919-5925.
85. Kim, M.-J.; MacDonnell, F. M.; Gimon-Kinsel, M. E.; DuBois, T.; Asgharian, N.; Griener, J. C. *Angew. Chem., Intl. Ed.* **2000**, *39*, 615-619.
86. MacDonnell, F. M.; Kim, M.-J.; Wouters, K. L.; Konduri, R. *Coord. Chem. Rev.* **2003**, *242*, 47-58.
87. MacDonnell, F. M.; Wouters, K. L.; Zeng, H. *Polymer Preprints* **2003**.
88. Torres, A. S.; Maloney, D. J.; Tate, D.; MacDonnell , F. M. *Inorg. Chim. Acta.* **1999**, *293*, 37-43.
89. Hiort, C.; Lincoln, P.; Nordén, B. *J. Am. Chem. Soc.* **1993**, *115*, 3448-3454.
90. Lincoln, P.; Norden, B. *Chem. Comm.* **1996**, *18*, 2145-2146.
91. Denisevich, P.; Abruna, H. D.; Leidner, C. R.; Meyer, T. J.; Murray, R. W. *Inorg. Chem.* **1982**, *21*, 2153-2161.
92. Gillard, R. D.; Hill, R. E. E. *J. Chem. Soc., Dalton. Trans.* **1974**, 1217.
93. MacDonnell, F. M.; Ali, M. M.; Kim, M.-J. *Comm. Inorg. Chem.* **2000**, *22*, 203-225.
94. MacDonnell, F. M.; Kim, M.-J.; Bodige, S. *Coord. Chem. Rev.* **1999**, *185-186*, 535-549.
95. Fletcher, N. C.; Keene, F. R.; Viebrock, H.; von Zelewsky, A. *Inorg. Chem.* **1997**, *36*, 1113-1121.
96. Keene, F. R. *Chem. Soc. Rev.* **1998**, *27*, 185.
97. Morgan, O.; Wang, S.; Bae, S.-A.; Morgan, R. J.; Baker, A. D.; Strekas, T. C.; Engel, R. *J. Chem. Soc., Dalton Trans.* **1997**, 3773-3776.
98. Baranoff, E.; Collin, J.-P.; Flamigni, L.; Sauvage, J.-P. *Chem. Soc. Rev.* **2004**, *33*, 147-155.
99. McCusker, J. K. *Acc. Chem. Res.* **2003**, *36*, 876-887.

100. Damrauer, N. H.; Cerullo, G.; Yeh, A.; Boussie, T. R.; Shank, C. V.; McCusker, J. K. *Science* **1997**, *275*, 54-57.

101. Van Houten, J.; Watts, R. J. *J. Am. Chem. Soc.* **1976**, *98*, 4853-4858.

102. Winkler, J. R.; Netzel, T. L.; Creutz, C.; Sutin, N. *J. Am. Chem. Soc.* **1987**, *109*, 2381-2392.

103. Medlycott, E. A.; Hanan, G. S. *Chem. Soc. Rev.* **2005**, *34*, 133-142.

104. Fang, Y.-Q.; Taylor, N. J.; Hanan, G. S.; Loiseau, F.; Passalacqua, R.; Campagna, S.; Nierengarten, H.; Van Dorsselaer, A. *J. Am. Chem. Soc.* **2002**, *124*, 7912-7913.

105. Maestri, M.; Armaroli, N.; Balzani, V.; Constable, E. C.; Thompson, A. M. W. C. *Inorg. Chem.* **1995**, *34*, 2759-2767.

106. Wang, J.; Fang, Y.-Q.; Hanan, G. S.; Loiseau, F.; Campagna, S. *Inorg. Chem.* **2005**, *44*, 5-7.

107. Englman, R.; Jortner, J. *Mol. Phys.* **1970**, *18*, 145-164.

108. Hoffman, M. Z.; Bolletta, F.; Moggi, L.; Hug, G. L. *J. Phys. Chem. Ref. Data* **1989**, *18*, 219-543.

109. Turro, N. J. *Modern Molecular Photochemistry*; University Science Books: Sausalito, CA, 1991.

110. DeLaive, P. J.; Foreman, T. K.; Giannotti, C.; Whitten, D. G. *J. Am. Chem. Soc.* **1980**, *102*, 5627-5631.

111. Demas, J. N.; Addington, J. W. *J. Am. Chem. Soc.* **1976**, *98*, 5800-5806.

112. Bolletta, F.; Juris, A.; Maestri, M.; Sandrini, D. *Inorg. Chim. Acta.* **1980**, *44*, L175-L176.

113. Blankenship, R. E.; Olson, J. M.; Miller, M. A. *Advances in Photosynthesis* **1995**, *2*, 399-435.

114. Balzani, V.; Campagna, S.; Denti, G.; Juris, A.; Serroni, S.; Venturi, M. *Acc. Chem. Res.* **1998**, *31*, 26-34.

115. Balzani, V.; Credi, A.; Scandola, F. In *Transition Metals in Supramolecular Chemistry*; Fabbrizzi, L., Poggi, A., Eds.; Kluwer Academic: Amsterdam, 1994, pp 1-32.

116. Creutz, C.; Chou, M.; Netzel, T. L.; Okumura, M.; Sutin, N. *J. Am. Chem. Soc.* **1980**, *102*, 1309-1319.

117. Closs, G. L.; Miller, J. R. *Science* **1988**, *240*, 440-442.

118. Klan, P.; Wagner, P. J. *J. Am. Chem. Soc.* **1998**, *120*, 2198-2199.

119. Closs, G. L.; Johnson, M. D.; Miller, J. R.; Piotrowiak, P. J. *J. Am. Chem. Soc.* **1989**, *111*, 3751-3753.

120. Paddon-Row, M. N. *Acc. Chem. Res.* **1994**, *27*, 18-25.

121. Harriman, A.; Ziessel, R. *Coord. Chem. Rev.* **1998**, *171*, 331-339.

122. Effenberger, F.; Wolf, H. C. *New J. Chem.* **1991**, *15*, 117-123.

123. Vollmer, M. S.; Würthner, F.; Effenberger, F.; Emele, P.; Meyer, D. U.; Stümpfig, T.; Port, H.; Wolf, H. C. *Chem. Eur. J.* **1998**, *4*, 260-269.

124. Barigelletti, F.; Flamigni, L.; Balzani, V.; Collin, J.-P.; Sauvage, J.-P.; Sour, A.; Constable, E. C.; Cargill-Thompson, A. M. W. *J. Am. Chem. Soc.* **1994**, *116*, 7692-7699.

125. Barigelletti, F.; Flamigni, L.; Guardigli, M.; Juris, A.; Beley, M.; Chodorowsky-Kimmes, S.; Collin, J.-P.; Sauvage, J.-P. *Inorg. Chem.* **1996**, *35*, 136-142.

126. Indelli, M. T.; Scandola, F.; Collin, J.-P.; Sauvage, J.-P.; Sour, A. *Inorg. Chem.* **1996**, *35*, 303-312.

127. Osuka, A.; Satoshi, N.; Maruyama, K.; Mataga, N.; Asahi, T.; Yamazaki, I.; Nishimura, Y.; Onho, T.; Nozaki, K. *J. Am. Chem. Soc.* **1993**, *115*, 4577-4589.

128. Osuka, A.; Maruyama, K.; Mataga, N.; Asahi, T.; Yamazaki, I.; Tamai, N. *Inorg. Chem.* **1990**, *112*, 4958-4959.

129. Helms, A.; Heiler, D.; McLendon, G. *J. Am. Chem. Soc.* **1991**, *113*, 4325-4327.

130. De Cola, L.; Belser, P. *Coord. Chem. Rev.* **1998**, *177*, 301-346.

131. Ziessel, R.; Hissler, M.; El-ghayoury, A.; Harriman, A. *Coord. Chem. Rev.* **1998**, *178-180*, 1251-1298.

132. Tour, J. M. *Chem. Rev.* **1996**, 537-554.

133. Schlicke, B.; Belser, P.; De Cola, L.; Sabbioni, E.; Balzani, V. *J. Am. Chem. Soc.* **1999**, *121*, 4207-4214.

134. Balzani, V.; Ceroni, P.; Maestri, M.; Vicinelli, V. *Curr. Opin. Chem. Biol.* **2003**, *7*, 657-665.

135. Campagna, S.; Denti, G.; Serroni, S.; Ciano, M.; Balzani, V. *Inorg. Chem.* **1991**, *30*, 3728-3732.

136. Campagna, S.; Denti, G.; Serroni, S.; Juris, A.; Venturi, M.; Ricevuto, V.; Balzani, V. *Chem. Eur. J.* **1995**, *1*, 211.

137. Denti, G.; Campagna, S.; Sabatino, L.; Serroni, S.; Ciano, M.; Balzani, V. *Inorg. Chem.* **1994**, *29*, 4750-4758.

138. Campagna, S.; Denti, G.; Sebatino, L.; Serroni, S.; Ciano, M.; Balzani, V. *J. Chem. Soc. Chem. Commun.* **1989**, 1500-1501.

139. Denti, G.; Campagna, S.; Serroni, S.; Ciano, M.; Balzani, V. *J. Am. Chem. Soc.* **1992**, *114*, 2944-2950.

140. Puntoriero, F.; Serroni, S.; Galletta, M.; Juris, A.; Licciardello, A.; Chiorboli, C.; Campagna, S.; Scandola, F. *ChemPhysChem* **2005**, *6*, 129-138.

141. Wang, X.-y.; Del Guerzo, A.; Schmehl, R. H. *J. Photochem Photobio. C.* **2004**, *5*, 55-77.

142. McClenaghan, N. D.; Leydet, Y.; Maubert, B.; Indelli, M. T.; Campagna, S. *Coord. Chem. Rev.* **2006**, *249*, 1336-1350.

143. Baba, A. I.; Shaw, J. R.; Simon, J. A.; Thummel, R. P.; Schmehl, R. H. *Coord. Chem. Rev.* **1998**, *171*, 43-59.

144. Ford, W. E.; Rodgers, M. A. J. *J. Phys. Chem.* **1992**, *96*, 2917-2920.

145. Tyson, D. S.; Henbest, K. B.; Bialecki, J.; Castellano, F. N. *J. Phys. Chem. A.* **2001**, *105*, 8154-8161.

146. Tyson, D. S.; Bialecki, J.; Castellano, F. N. *Chem. Comm.* **2000**, *23*, 2355-2356.

147. Tyson, D. S.; Luman, C. R.; Zhou, X.; Castellano, F. N. *Inorg. Chem.* **2001**, *40*, 4063-4071.

148. Chiorboli, C.; Fracasso, S.; Scandola, F.; Campagna, S.; Serroni, S.; Konduri, R.; MacDonnell, F. M. *Chem. Comm.* **2003**, 1658-1659.

149. Chiorboli, C.; Sandro, F.; Ravaglia, M.; Scandola, F.; Campagna, S.; Wouters, K. L.; Konduri, R.; MacDonnell, F. M. *Inorg. Chem.* **2005**, *44*, 8368-8378.

150. Treadway, J. A.; Loeb, B.; Lopez, R.; Anderson, P. A.; Keene, F. R.; Meyer, T. J. *Inorg. Chem.* **1996**, *35*, 2242-2246.

151. Brennaman, M. K.; Alstrum-Acevedo, J. H.; Fleming, C. N.; Jang, P.; Meyer, T. J.; Papanikolas, J. M. *J. Am. Chem. Soc.* **2002**, *124*, 15094-15098.

152. Coates, C. G.; Olofsson, J.; Coletti, M.; McGarvey, J. J.; Oenfelt, B.; Lincoln, P.; Norden, B.; Tuite, E.; Matousek, P.; Parker, A. W. *J. Phys. Chem. B.* **2001**, *105*, 12653-12664.

153. Coates, C. G.; Callaghan, P.; McGarvey, J. J.; Kelly, J. M.; Jacquet, L.; Kirsch-De Mesmaeker, A. *J. Mol. Struc.* **2001**, *598*, 15-25.

154. Onfelt, B.; Lincoln, P.; Norden, B.; Baskin, J. S.; Zewail, A. H. *Proc. Natl. Acad. Sci. USA* **2000**, *97*, 5708-5713.

155. Fees, J.; Kaim, W.; Moscherosch, M.; Matheis, W.; Klima, J.; Krejcik, M.; Zalis, S. *Inorg. Chem.* **1993**, *32*, 166-174.

156. Amouyal, E.; Homsi, A.; Chambron, J.-C.; Sauvage, J.-P. *J. Chem. Soc., Dalt. Trans.* **1990**, 1841-1845.

157. Chiorboli, C.; Rodgers, M. A. J.; Scandola, F. *J. Am. Chem. Soc.* **2003**, *125*, 483-491.

158. Chiorboli, C.; Bignozzi, C.-A.; Scandola, F.; Ishow, E.; Gourdon, A.; Launay, J.-P. *Inorg. Chem.* **1999**, *38*, 2402-2410.

159. Campagna, S.; Serroni, S.; Bodige, S.; MacDonnell, F. M. *Inorg. Chem.* **1999**, *38*, 692-701.

160. Passalacqua, R.; Loiseau, F.; Campagna, S.; Fang, Y.-Q.; Hanan, G. S. *Angew. Chem. Int. Ed.* **2003**, *42*, 1608-1611.

161. Sauvage, J.-P.; Collin, J.-P.; Chambron, J.-C.; Guillerez, S.; Coudret, C.; Balzani, V.; Barigelletti, F.; De Cola, L.; Flamigni, L. *Chem. Rev.* **1994**, *94*, 993.

162. Scandola, F.; Chiorboli, C.; ndelli, M. T.; Rampi, M. A. In *Electron Transfer in Chemistry*; Balzani, V., Ed.; Wiley-VCH: Weinheim, Germany, 2001; Vol. 3, pp 337-408.

163. Huynh, M. H. V.; Dattelbaum, D. M.; Meyer, T. J. *Coord. Chem. Rev.* **2005**, *249*, 457-483.

164. Durr, H.; Bossmann, S. *Acc. Chem. Res.* **2001**, *34*, 905-917.

165. Siemeling, U.; Vor der Brueggen, J.; Vorfeld, U.; Neumann, B.; Stammler, A.; Stammler, H.-G.; Brockhinke, A.; Plessow, R.; Zanello, P.; Laschi, F.; Fabrizi de Biani, F.; Fontani, M.; Steenken, S.; Stapper, M.; Gurzadyan, G. *Chem. Eur. J.* **2003**, *9*, 2819-2833.

166. Treadway, J. A.; Chen, P.; Rutherford, T. J.; Keene, F. R.; Meyer, T. J. *J. Phys. Chem. A.* **1997**, *101*, 6824-6826.

167. Danielson, E.; Elliott, C. M.; Merkert, J. W.; Meyer, T. J. *J. Am. Chem. Soc.* **1987**, *109*, 2519-2520.

168. Cooley, L. F.; Larson, S. L.; Elliott, C. M.; Kelley, D. F. *J. Phys. Chem.* **1991**, *95*, 10694-10700.

169. Larson, S. L.; Elliott, C. M.; Kelley, D. F. *J. Phys. Chem.* **1995**, *99*, 6530-6539.

170. Johansson, O.; Wolpher, H.; Borgstroem, M.; Hammarstroem, L.; Bergquist, J.; Sun, L.; Kermark, B. *Chem. Comm.* **2004**, 194-195.

171. Burdinski, D.; Bothe, E.; Wieghardt, K. *Inorg. Chem.* **2000**, *39*, 105-116.

172. Burdinski, D.; Wieghardt, K.; Steenken, S. *J. Am. Chem. Soc.* **1999**, *121*, 10781-10787.

173. Borgstroem, M.; Johansson, O.; Lomoth, R.; Baudin, H. B.; Wallin, S.; Sun, L.; Aakermark, B.; Hammarstroem, L. *Inorg. Chem.* **2003**, *42*, 5173-5184.

174. Hammarström, L. *Curr. Opin. Chem. Biol.* **2003**, *7*, 666-673.

175. Maggini, M.; Guldi, D. M.; Mondini, S.; Scorrano, G.; Paolucci, F.; Ceroni, P.; Roffia, S. *Chem. Eur. J.* **1998**, *4*, 1992-2000.

176. Guldi, D. M.; Luo, C.; Da Ros, T.; Prato, M.; Dietel, E.; Hirsch, A. *Chem. Comm.* **2000**.

177. Guldi, D. M.; Maggini, M.; Menna, E.; Scorrano, G.; Ceroni, P.; Marcaccio, M.; Paolucci, F.; Roffia, S. *Chem. Eur. J.* **2001**, *7*, 1597-1605.

178. Johansson, O.; Borgstroem, M.; Lomoth, R.; Palmblad, M.; Bergquist, J.; Hammarstroem, L.; Sun, L.; Kermark, B. **2003**, *42*, 2908-2918.

179. Opperman, K. A.; Mecklenburg, S. L.; Meyer, T. J. *Inorg. Chem.* **1994**, *33*, 5295-5301.

180. Benniston, A. C.; Chapman, G. M.; Harriman, A.; Rostron, S. A. *Inorg. Chem.* **2005**, *44*, 4029-4036.

181. Mecklenburg, S. L.; McCafferty, D. G.; Schoonover, J. R.; Peek, B. M.; Erickson, B. W.; Meyer, T. J. *Inorg. Chem.* **1994**, *33*, 2974-2983.

182. Ballardini, R.; Balzani, V.; Clemente-Leon, M.; Credi, A.; Gandolfi, M. T.; Ishow, E.; Perkins, J.; Stoddart, J. F.; Tseng, H.-R.; Wenger, S. *J. Am. Chem. Soc.* **2002**, *124*, 12786-12795.

183. Schild, V.; Van Loyen, D.; Duerr, H.; Bouas-Laurent, H.; Turro, C.; Woerner, M.; Pokhrel, M. R.; Bossmann, S. H. *J. Phys. Chem. A* **2002**, *160*, 9149-9158.

184. Nelissen, H. F. M.; Kercher, M.; De Cola, L.; Feiters, M. C.; Nolte, R. J. M. *Chem. Eur. J.* **2002**, *8*, 5407-5414.

185. Roberts, J. A.; Kirby, J. P.; Wall, S. T.; Nocera, D. G. *Inorg. Chim. Acta.* **1997**, *263*, 395-405.

186. Roberts, J. A.; Kirby, J. P.; Nocera, D. G. *J. Am. Chem. Soc.* **1995**, *117*, 8051-8052.

187. Magnuson, A.; Berglund, H.; Korall, P.; Hammarstrom, L.; Aakermark, B.; Styring, S.; Sun, L. *J. Am. Chem. Soc.* **1997**, *119*, 10720-10725.
188. Sjoedin, M.; Styring, S.; Kermark, B.; Sun, L.; Hammarstrom, L. *J. Am. Chem. Soc.* **2000**, *122*, 3932-3936.
189. Sjodin, M.; Ghanem, R.; Polivka, T.; Pan, J.; Styring, S.; Sun, L.; Sundstroem, V.; Hammarstrom, L. *Phys. Chem. Chem. Phys.* **2004**, *6*, 4851-4858.
190. Sjodin, M.; Styring, S.; Akermark, B.; Sun, L.; Hammarstrom, L. *Phil. Trans. R. Soc. Lond. B.* **2002**, *357*, 1471-1479.
191. Sjodin, M.; Styring, S.; Wolpher, H.; Xu, Y.; Sun, L.; Hammarstrom, L. *J. Am. Chem. Soc.* **2005**, *127*, 3855-3863.
192. Reece, S. Y.; Nocera, D. G. *J. Am. Chem. Soc.* **2005**, *127*, 9448 - 9458.
193. Magnuson, A.; Frapart, Y.; Abrahamsson, M.; Horner, O.; Kermark, B.; Sun, L.; Girerd, J.-J.; Hammarstrom, L.; Styring, S. *J. Am. Chem. Soc.* **1999**, *121*, 89-96.
194. Sun, L.; Raymond, M. K.; Magnuson, A.; LeGourrierec, D.; Tamm, M.; Abrahamsson, M.; Kenez, P. H.; Martensson, J.; Stenhagen, G.; Hammarstrom, L.; Styring, S.; Akermark, B. *J. Inorg. Biochem.* **2000**, *78*, 15-22.
195. Vermaas, W.; Stryring, S.; Schroder, W.; Andersson, B. *Photosynth. Res.* **1993**, *38*, 249-260.
196. Huang, P.; Hogblom, J.; Anderlund, M. F.; Sun, L.; Magnuson, A.; Styring, S. *J. Inorg. Biochem.* **2004**, *98*, 733-745.
197. Huang, P.; Magnuson, A.; Lomoth, R.; Abrahamsson, M.; Tamm, M.; Sun, L.; van Rotterdam, B.; Park, J.; Hammarstrom, L.; Akermark, B.; Styring, S. *J. Inorg. Biochem.* **2002**, *9*, 159-172.
198. Berg, K. E. T., Anh; Raymond, Mary Katherine; Abrahamsson, Malin; Wolny, ; Berg, K. E.; Tran, A.; Raymond, M. K.; Abrahamsson, M.; Wolney, J.; Rendon, S.; Andersson, M.; Sun, L.; Styring, S.; Hammarstrom, L.; Toftlund, H.; Akermark, B. *Eur. J. Inorg. Chem.* **2001**, *4*, 1019-1029.
199. Abrahamsson, M. L. A.; Baudin, H. B.; Tran, A.; Philouze, C.; Berg, K. E.; Raymond-Johansson, M. K.; Licheng Sun, L.; Åkermark, B.; Styring, S.; Hammarström, L. *Inorg. Chem.* **2002**, *41*, 1534 -1544.
200. McCafferty, D. G.; Friesen, D. A. D., E.; Wall, C. G.; Saderholm, M. J.; Erickson, B. W.; Meyer, T. J. *Proc. Natl. Acad. Sci. USA* **1996**, *93*, 8200-8204.
201. Borgstrom, M.; Shaikh, N.; Johansson, O.; Anderlund, M. F.; Styring, S.; Akermark, B.; Magnuson, A.; L., H. *J. Am. Chem. Soc.* **2005**, *127*, 17504-17515.
202. Watts, R. J. *Comm. Inorg. Chem.* **1991**, *11*, 303-337.
203. Meyer, T. J. *J. Electrochem. Soc.* **1984**, 221C-228C.
204. Heyduk, A. F.; Nocera, D. G. *Science* **2001**, *293*, 1639-1641.
205. Sutin, N.; Creutz, C.; Fujita, E. *Comm. Inorg. Chem.* **1997**, *19*, 67-92.
206. Molnar, S. M.; Nallas, G.; Bridgewater, J. S.; Brewer, K. J. *J. Am. Chem. Soc.* **1994**, *116*, 5206-5210.
207. Brewer, K. J.; Elvington, M. "Supramolecular complexes as photocatalysts for the production of hydrogen from water", 2006
208. de Tacconi, N. R.; Lezna, R. O.; Konduri, R.; Ongeri, F.; Rajeshwar, K.; MacDonnell, F. M. *Chem. Eur. J.* **2005**, *11*, 4327-4339.
209. Konduri, R.; de Tacconi, N. R.; Rajeshwar, K.; MacDonnell, F. M. *J. Am. Chem. Soc.* **2004**, *126*, 11621-11629.
210. Konduri, R.; Ye, H.; MacDonnell, F. M.; Serroni, S.; Campagna, S.; Rajeshwar, K. *Angew. Chem. Int. Ed.* **2002**, *41*, 3185-3187.
211. MacDonnell, F. M.; McAllister, C.; Lane, T. J. *unpublished results.*
212. Chang, C. C.; Pfennig, B.; Bocarsly, A. B. *Coord. Chem. Rev.* **2000**, *2008*, 33-45.

213. Pfennig, B. W.; Lockard, J. V.; Cohen, J. L.; Watson, D. F.; Ho, D. M.; Bocarsly, A. B. *Inorg. Chem.* **1999**, *38*, 2941-2946.
214. Pfennig, B. W.; Bocarsly, A. B. *J. Phys. Chem.* **1992**, *96*, 226-233.
215. Zhou, M.; Pfennig, B. W.; Steiger, J.; Van Engen, D.; Bocarsly, A. B. *Inorg. Chem.* **1990**, *29*, 2456-2460.
216. Ali, M.; Sato, H.; Haga, M.; Koji, T.; Yoshimura, A.; Ohno, T. *Inorg, Chem.* **1998**, *37*, 6176-6180.
217. Esswein, A. J.; Veige, A. S.; Nocera, D. G. *J. Am. Chem. Soc.* **2005**, *127*, 16641-16651.
218. Rosenthal, J.; Bachman, J.; Dempsey, J. L.; Esswein, A. J.; Gray, T. G.; Hodgkiss, J. M.; Manke, D. R.; Luckett, T. D.; Pistorio, B. J.; Veige, A. S.; Nocera, D. G. *Coord. Chem. Rev.* **2005**, *249*, 1316-1326.
219. Mann, K. R.; Lewis, N. S.; Miskowski, V. M.; Erwin, D. K.; Hammond, G. S.; Gray, H. B. *J. Am. Chem. Soc.* **1977**, *99*, 5525-5526.
220. Mann, K. R.; Bell, R. A.; Gray, H. B. *Inorg. Chem.* **1979**, *18*, 2671-2673.
221. Rüettinger, W.; Dismukes, G. C. *Chem. Rev.* **1997**, *97*, 1.
222. Rüettinger, W.; Dismukes, G. C. *Concepts Photobiol.* **1999**, 330-363.
223. Yagi, M.; Kaneko, M. *Chem. Rev.* **2001**, *101*, 21.
224. Vrettos, J. S.; Brudvig, G. W. In *Comprehensive Coordination Chemistry II*; McCleverty, J. A., Meyer, T. J., Eds., 2004; Vol. 8, pp 507-547.
225. Kaschak, D. M.; Johnson, S. A.; Waraksa, C. C.; Pogue, J.; Mallouk, T. E. *Coor. Chem. Rev.* **1999**, *185-186*, 403-416.
226. Liu, H.; Nakamura, R.; Nakato, Y. *ChemPhysChem* **2005**, *6*, 2499-2502.
227. Howells, A. R.; Sankarraj, A.; Shannon, C. *J. Am. Chem. Soc.* **2004**, *126*, 12258-12259.
228. Mukhopadhyay, S. M., S. K.; Bhaduri, S.; Armstrong, W. H. *Chem. Rev.* **2004**, *104*, 3981-4026.
229. Yocum, C. F.; Pecoraro, V. L. *Curr. Opin. Chem. Biol.* **1999**, *3*, 182-187.
230. Gilbert, J. A.; Eggleston, D. S.; Murphy, W. R., Jr.; Geselowitz, D. A.; Gersten, S. W.; Hodgson, D. J.; Meyer, T. J. *J. Am. Chem. Soc.* **1985**, *107*, 3855-3864.
231. Schoonover, J. R.; Ni, J.; Roecker, L.; White, P. S.; Meyer, T. J. *Inorg. Chem.* **1996**, *35*, 5885-5892.
232. Morris, N. D.; Mallouk, T. E. *J. Am. Chem. Soc.* **2002**, *124*, 11114-11121.
233. Hurst, J. K. *Coord. Chem. Rev.* **2005**, *249*, 313-328.
234. Gersten, S. W.; Samuels, G. J.; Meyer, T. J. *J. Am. Chem. Soc.* **1982**, *104*, 4029-4030.
235. Binstead, R. A.; Chronister, C. W.; Ni, J.; Hartshorn, C. M.; Meyer, T. J. *J. Am. Chem. Soc.* **2000**, *122*, 8464-8473.
236. Yamada, H.; Siems, W. F.; Koike, T.; Hurst, J. K. *J. Am. Chem. Soc.* **2004**, *126*, 9786-9795.
237. Naruta, Y.; Sasayama, M.; Sadaki, T. *Angew. Chem. Int. Ed.* **1994**, *33*, 1839-1841.
238. Shimazaki, Y.; Nagano, T.; Takesue, H.; Ye, B.-H.; Tani, F.; Naruta, Y. *Angew. Chem. Int. Ed.* **2004**, *43*, 98-100.
239. Hoganson, C. W.; Babcock, G. T. *Science* **1997**, *277*, 1953-1956.
240. Messinger, J.; Badger, M.; Wydrzynski, T. *Proc. Natl. Acad. Sci. USA* **1995**, *92*, 3209-3213.
241. Pecoraro, V. L.; Baldwin, M. J.; Caudle, M. T.; Hsieh, W.-Y.; Law, N. A. *Pure Appl. Chem.* **1998**, *70*, 925-929.
242. Limburg, J. B., Gary W.; Crabtree, Robert H.. (2000), In *Biomimetic Oxidations Catalyzed by Transition Metal Complexes*; Meunier, B., Ed.; Imperial College Press: London, UK, 2000, pp 509-541.
243. McEvoy, J. P.; Brudvig, G. W. *Phys. Chem. Chem. Phys.* **2004**, *6*, 4754-4763.
244. Limburg, J.; Vrettos, J.; Liable-Sands, L.; Rheingold, A.; Crabtree, R. H.; Brudvig, G. W. *Science* **1999**, *283*, 1524-1527.

245. Limburg, J.; Vrettos, J. S.; Chen, H.; de Paula, J. C.; Crabtree, R. H.; Brudvig, G. W. *J. Am. Chem. Soc.* **2001**, *123*, 423-430.
246. Baffert, C.; Collomb, M.-N.; Deronzier, A.; Kjaergaard-Knudsen, S.; Latour, J.-M.; Lund, K. H.; McKenzie, C. J.; Mortensen, M.; Nielsen, L. P.; Thorup, N. *Dalton Trans.* **2003**, *9*, 1765-1772.
247. Poulsen, A. K.; Rompel, A.; McKenzie, C. J. *Angew. Chem. Int. Ed.* **2005**, *44*, 6916-6920.
248. Rüettinger, W.; Yagi, M.; Wolf, K.; Bernasek, S.; Dismukes, G. C. *J. Am. Chem. Soc.* **2000**, *122*, 10353-10357.
249. Wada, T.; Tsuge, K.; Tanaka, K. *Inorg. Chem.* **2001**, *40*, 329-337.
250. Wada, T.; Tsuge, K.; Tanaka, K. *Angew. Chem. Int. Ed.* **2000**, *39*, 1479-1482.
251. Sens, C.; Romero, I.; Rodríguez, M.; Llobet, A.; Parella, T.; Benet-Buchholz, J. *J. Am. Chem. Soc.* **2004**, *126*, 7798 - 7799.
252. Zong, R.; Thummel, R. P. *J. Am. Chem. Soc.* **2005**, *127*, 12802 - 12803.
253. Yagi, M.; Osawa, Y.; Sukegawa, N.; Kaneko, M. *Langmuir* **1999**, *15*, 7406-7408.
254. Lomoth, R.; Magnuson, A.; Sjodin, M.; Huang, P.; Styring, S.; Hammarstrom, L. *Photosynth. Res.* **2006**, *87*, 25-40.
255. Xu, Y.; Eilers, G.; Borgstrom, M.; Pan, J.; Abrahamsson, M.; Magnuson, A.; Lomoth, R.; Bergquist, J.; Polivka, T.; Sun, L.; Sundstrom, V.; Styring, S.; Hammarstrom, L.; Akermark, B. *Chem. Eur. J.* **2005**, *11*, 7305-7314.
256. Sun, L.; Hammarstrom, L.; Akermark, B.; Styring, S. *Chem. Soc. Rev.* **2001**, *30*, 36-49.
257. Grzeszczuk, M.; Poks, P. *Electrochimica Acta* **2000**, *45*, 4171-4177.
258. Bose, C. S. C.; Rajeshwar, K. *J. Electroanal. Chem.* **1992**, *333*, 235.
259. Brugger, P.-A.; Cuendet, P.; Gratzel, M. *J. Am. Chem. Soc.* **1981**, *103*, 2423.
260. Dhanalakshmi, K. B.; Latha, S.; Anandan, S.; Maruthamuthu, P. *Int. J. Hydrog. Energ.* **2001**, *26*, 669-674.
261. Kiwi, J.; Grätzel, M. *Angew. Chemie* **1979**, *91*, 659-660.
262. Lehn, J.-M.; Sauvage, J.-P.; Ziessel, R. *Nov. J. Chim.* **1981**, *5*, 291.
263. Grätzel, M. *Acc. Chem. Res.* **1981**, *14*, 376.
264. Grätzel, M., Ed. *Energy Resources Through Photochemistry and Catalysis*; Academic Press: New York, 1983.
265. Loy, L.; Wolf, E. E. *Solar Energy* **1985**, *34*, 455-461.
266. Harriman, A.; Mills, A. *J. Chem. Soc., Faraday Trans.* **1981**, 2111-2124.
267. Amouyal, E.; Keller, P.; Moradpour, A. *J. Chem. Soc., Chem. Commun.* **1980**, 1019-1020.
268. Amouyal, E.; Grand, D.; Moradpour, A.; Keller, P. *New. J. Chem.* **1982**, *6*, 241-244.
269. Jones, A. K.; Sillery, E.; Albracht, S. P. J.; Armstrong, F. A. *Chem. Comm.* **2002**, *8*, 866-867.
270. Cammack, R.; Frey, M.; Robson, R., Eds. *Hydrogen as a Fuel: Learning from Nature*; Taylor and Francis: New York, 2001.
271. Peters, J. W.; Lanzilotta, W. N.; Lemon, B. J.; Seefeldt, L. C. *Science* **1998**, *282*, 1853-1858.
272. Volbeda, A.; Charon, M.-H.; Piras, C.; Hatchikian, E. C.; Frey, M.; Fontecilla-Camps, J. C. *Nature* **1995**, *373*, 580-587.
273. Fontecilla-Camps, J.-C.; Frey, M.; Garcin, E.; Higuchi, Y.; Montet, Y.; Nicolet, Y.; Volbeda, A. *Molecular architectures* **2001**, *93-109*, 238-261.
274. Marr, A. C.; Spencer, D. J. E.; Schroder, M. *Coord. Chem. Rev.* **2001**, *219-221*, 1055-1074.
275. Bouwman, E.; Reedijk, J. *Coord. Chem. Rev.* **2005**, *249*, 1555-1581.
276. Liu, X.; Ibrahim, S. K.; Tard, C.; Pickett, C. J. *Coord. Chem. Rev.* **2005**, *249*, 1641-1652.

277. Capon, J.-F.; Gloaguen, F.; Schollhammer, P.; Talarmin, J. *Coord. Chem. Rev.* **2005**, *249*, 1664-1676.

278. Darensbourg, M. Y.; Lyon, E. J.; Zhao, X.; Georgakaki, I. P. *Proc. Natl. Acad. Sci. USA* **2003**, *100*, 3683-3688.

279. Gloaguen, F.; Lawrence, J. D.; Rauchfuss, T. B. *J. Am. Chem. Soc.* **2001**, *123*, 9476-9477.

280. Mejia-Rodriguez, R.; Chong, D.; Reibenspies, J. H.; Soriaga, M. P.; Darensbourg, M. Y. *J. Am. Chem. Soc.* **2004**, *126*, 12004-12014.

281. Borg, S. J.; Behrsing, T.; Best, S. P.; Razavet, M.; Liu, X.; Pickett, C. J. *J. Am. Chem. Soc.* **2004**, *126*, 16988-16999.

282. Sun, L.; Aakermark, B.; Ott, S. *Coord. Chem. Rev.* **2005**, *249*, 1653-1663.

283. Ott, S.; Borgstrom, M.; Kritikos, M.; Lomoth, R.; Bergquist, J.; Akermark, B.; Hammarstrom, L.; Sun, L. *Inorg. Chem.* **2004**, *43*, 4683-4692.

284. Koelle, U. *New J. Chem.* **1992**, *16*, 157.

285. Artero, V.; Fontecave, M. *Coord. Chem. Rev.* **2005**, *249*, 1518-1535.

286. Collman, J. P.; Wagenknecht, P. S.; Hutchison, J. E. *Angew. Chem. Int. Ed.* **1994**, *33*, 1537-1556.

287. Collman, J. P.; Ha, Y.; Wagenknecht, P. S.; Lopez, M. A.; Guilard, R. *J. Am. Chem. Soc.* **1993**, *115*, 9080-9088.

288. Collman, J. P.; Wagenknecht, P. S.; Hutchison, J. E.; Lewis, N. S.; Lopez, M. A.; Guilard, R.; L'Her, M.; Bothner-By, A. A.; Mishra, P. K. *J. Am. Chem. Soc.* **1992**, *114*, 5654-5664.

289. Collman, J. P.; Wagenknecht, P. S.; Lewis, N. S. *J. Am. Chem. Soc.* **1992**, *114*, 5665-5673.

290. Chau, T.-H.; Espenson, J. H. *J. Am. Chem. Soc.* **1978**, *100*, 129-133.

291. Connolly, P.; Espenson, J. H. *Inorg. Chem.* **1986**, *25*, 2684-2688.

292. Bakac, A.; Espenson, J. H. *J. Am. Chem. Soc.* **1984**, *106*, 5197-5202.

293. Hu, X.; Cossairt, B. M.; Brunschwig, B. S.; Lewis, N. S.; Peters, J. C. *Chem. Comm.* **2005**, *37*, 4723-4725.

294. Coe, B. J.; Friesen, D. A.; Thompson, D. W.; Meyer, T. J. *Inorg. Chem.* **1996**, *35*, 4575-4584.

295. Chaignon, F.; Torroba, J.; Blart, E.; Borgstroem, M.; Hammarstroem, L. *New J. Chem.* **2005**, *29*, 1272-1284.

296. Bard, A. J.; Faulkner, L. R. *Electrochemical Methods*; Wiley: New York, 1980.

7

Hydrogen Generation from Irradiated Semiconductor-Liquid Interfaces

Krishnan Rajeshwar

University of Texas at Arlington, Arlington, TX

1 Introduction and Scope

This Chapter explores the possibility of using sunlight in conjunction with semiconductor/electrolyte interfaces for the production of hydrogen from water and other suitable solvents. The underlying principles of solar energy conversion using semiconductor/electrolyte interfaces have been discussed in several review articles, book chapters and books,[1-46] and will not be repeated here. This field of *photoelectrochemistry* had its early origins in attempts to use inorganic semiconductor/electrolyte interfaces in electronic devices.[47-52] Subsequently, it was found that an electrochemical cell made from a n-TiO$_2$ photoanode and a Pt counterelectrode evolved H$_2$ and O$_2$ from water under UV irradiation or even sunlight.[53-57] In a historical sense, it is worth noting that H$_2$ evolution had been observed at semiconductor (e.g., Ge)/electrolyte interfaces several years prior to the Japanese discovery, although the hydrogen evolution reaction (HER) occurred as a result of semiconductor photocorrosion.[58] A flurry of activity ensued in the 1970s and 1980s on the photoelectrolysis of water; indeed, attempts to split water using sunlight and inorganic semiconductor(s) have continued in an unabated manner to the present time.

Table 1 contains a chronological listing of review articles summarizing the developments on this topic. Several books contain discussions on the photoelectrolysis of water[4-8] and the reader also is referred to entire journal issues or book chapters (e.g., Refs. 57, 81–96) devoted to this subject. The present discussion builds on this body of literature by summarizing the state-of-the-art and future challenges. Every attempt is made here to cite the majority of articles that have appeared in the literature on this topic, although, inevitably, space constraints preclude an all-inclusive compilation. The interested reader can gain entry into the specialized literature using either the books, the reviews, or the selected articles cited below to highlight a par-

Table 1. Chronological listing of review articles on the photoelectrolysis of water.

Entry number	Title of review article	Comments	Reference
	Period 1975-1985		
1	Photocatalytic Hydrogen Production: A Solar Energy Conversion Alternative?	This early review classifies solar energy conversion methods according to photosensitizer type. Review concludes that photoelectrolysis of water is the most promising scheme meriting further consideration.	59
2	p-n Photoelectrolysis Cells	The concept of combining both n- and p-type semiconductor electrodes for water splitting first discussed.	60
3	Solar Energy-Assisted Electrochemical Splitting of Water	Kinetic, energetic, and solid-state considerations in the search for suitable electrodes for water splitting elaborated.	61
4	Semiconducting Oxide Anodes in Photoassisted Electrolysis of Water	A variety of binary and ternary oxide anodes for the photoassisted OER discussed.	62
5	Photoelectrolysis of Water with Semiconductors	Appears to be the first comprehensive review article on the topic.	63
6	Design and Evaluation of New Oxide Photoanodes for the Photoelectrolysis of Water with Solar Energy	Review in a similar vein as Ref. 62 (Entry 4 above) but with new data from the authors' laboratory.	64
7	Oxide Semiconductors in Photoelectrochemical Conversion of Solar Energy	Once again, oxide electrodes are examined but with applicability directed toward the conversion of solar energy into either electrical power or H_2.	65
8	Conversion of Sunlight into Electrical Power and Photoassisted Electrolysis of Water in Photoelectrochemical Cells	Advances in the development of efficient regenerative photoelectrochemical cells reviewed with a brief discussion of p-type semiconductor photocathodes for the HER.	29
9	Artificial Photosynthesis: Water Cleavage into Hydrogen and Oxygen by Visible Light	Review deals mostly with micro-heterogeneous (particulate) systems for water splitting.	30
10	Hydrogen Evolving Solar Cells	Principles in the design of semiconductor electrodes, surface modification strategies, p-n junction cells, and photoelectrolysis by suspended semiconductor particles, discussed.	66
11	The Energetics of p/n Photoelectrolysis Cells	The interfacial aspects of combining both photoanodes and photocathodes discussed using both theory and experiment.	67

Table 1. Continuation.

Entry number	Title of review article	Comments	Reference
	Period 1985-2005		
12	Solar Hydrogen Production through Photobiological, Photochemical and Photoelectrochemical Assemblies	A general review on the subject with a section on semiconductor- and dye-based systems.	68
13	Overall Photodecomposition of Water on a Layered Niobate Catalyst	Review mostly dealing with developments in the authors' laboratory on particulate systems.	69
14	Artificial Photosynthesis: Solar Splitting of Water to Hydrogen and Oxygen	Various approaches based on the use of semiconductors discussed with a look at future prospects.	40
15	Solar Photoproduction of Hydrogen	Review mainly addresses potential and experimental efficiencies for four types of systems of which one comprises photoelectrolysis cells with one or more semiconductor electrodes.	70
16	Recent Progress of Photocatalysts for Overall Water Splitting	See Entry 13 above.	71
17	Photocatalytic Decomposition of Water	Principles of water-splitting reviewed including a section on semiconductor-based approaches.	72
18	Photo- and Mechano- Catalytic Overall Water Splitting Reactions to Form Hydrogen and Oxygen on Heterogeneous Catalysts	See Entry 13 above.	73
19	Multiple Band Gap Semiconductor/Electrolyte Solar Energy Conversion	The strategy of stacking semiconductors with variant E_g's discussed with a goal to enhance the overall process efficiency.	44
20	Development of Photocatalyst Materials for Water Splitting with the Aim at Photon Energy Conversion	See Entry 13 above.	74
21	New Aspects of Heterogeneous Photocatalysts for Water Decomposition	See Entry 13 above.	75
22	Photoelectrochemical Hydrogen Generation from Water Using Solar Energy. Materials-Related Aspects	As in Entry 5, a comprehensive review which outlines the principles, R&D progress, impact on hydrogen economy, and cost issues.	76
23	Photocatalytic Materials for Water Splitting	See Entry 13 above.	77
24	Photocatalytic Water Splitting into H_2 and O_2 over Various Tantalate Photocatalysts	See Entry 13 above.	78
25	Strategies for the Development of Visible-Light-Driven Photocatalysts for Water Splitting	See Entry 13 above.	79
26	Metal Oxide Photoelectrodes for Hydrogen Generation Using Solar Radiation Driven Water Splitting	Topics reviewed include preparation of oxide electrodes, sensitization of wide band gap oxides, tandem cells, solid solutions of oxides and porous/nano-crystalline materials.	80

ticular development. An annotated bibliography does exist for the period, 1975–1983, for photoelectrochemical cell studies with semiconductor electrodes.[97]

2 Types of Approaches

Figure 1 illustrates the interfacial energetics involved in the photoelectrochemical evolution of H_2. Thus, the electronic energy levels in the semiconductor and in the contacting solution are shown on a common diagram. In a semiconductor, the filled electronic levels (valence band or VB) and the empty levels (conduction band or CB) are separated by a *forbidden* gap, namely, the band gap energy, E_g.[98-100] Photoexcitation of the semiconductor with light of energy equal to or exceeding E_g (i.e., with wavelengths corresponding to or shorter than that corresponding to the energy gap) elicits electron-hole pairs, a fraction of which (as defined by the quantum yield) escape recombination and find their way to the semiconductor/ solution interface. For the photosplitting of water (Figure 1a), the CB and VB edges at the semiconductor surface (E_{CB} and E_{VB} respectively) must bracket the two redox levels corresponding to the HER and the oxygen evolution reaction (OER) respectively. This is tantamount to stating that the photogenerated electrons have sufficient energy to reduce protons and the photogenerated holes have sufficient energy to oxidize water (Figure 1a).

This is a stringent requirement indeed as further elaborated in the next Section. Instead of actually photosplitting water, sacrificial agents may be added to the solution such that the HER and OER steps may be separately optimized and studied (Figures 1b and 1c). It must be borne in mind that now the overall photoreaction becomes thermodynamically *down hill* and is more appropriately termed: photocatalytic (see below). Examples of sacrificial agents include sulfite for the photo-driven HER case (Figure 1b) or Ag^+ as the electron acceptor for the photocatalytic oxidation of water (Figure 1c).

Instead of using the semiconductor in the form of electrodes in an *electrochemical* cell, a *wireless* water splitting or HER system could be envisioned where particle suspensions are used (instead of electrodes) in a *photochemical* reactor. Two points regarding such an approach must be noted. First, unlike in the case of a semiconductor electrode, a bias potential cannot be applied in the suspension case. Second, the sites for the HER and OER are not physically separated as in the electrochemical case. Thus, the potential exists in a photochemical system for a highly explosive stoichiometric (2:1) mixture of H_2 and O_2 to be evolved. Nonetheless, strategies have been devised for immobilizing the semiconductor particles in a membrane so that the HER and OER sites are properly separated (see for example, Refs. 101–108). These include the so-called semiconductor septum photoelectrochemical cells where the n- and p- type semiconductor particles are embedded, for example, in a bilayer lipid membrane.[105] The OER and HER sites are thus compartmentalized in this approach. However, claims of enhanced solar conversion efficiency in such devices have been questioned[109] on the basis that in many of the cases reported, a galvanic cell (i.e., a sacrificial battery system) was used to drive the photoproduction of H_2.

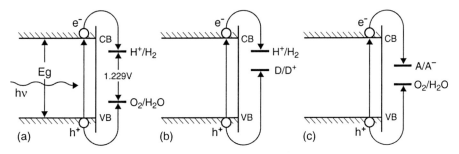

Fig. 1. Interfacial energetics at semiconductor-liquid junctions. D is an electron donor and A is an electron acceptor.

Bifunctional redox catalysts have been investigated in terms of their applicability for the solar-assisted splitting of water.[30,110-118] In this approach, Pt (an excellent catalyst for the HER) and RuO_2 (an excellent catalyst for the OER) are loaded onto colloidal TiO_2 particles. But unlike in the approaches discussed earlier, the oxide semiconductor is not used as a light absorber; instead, an inorganic complex [e.g., amphiphilic $Ru(bpy)_3^{2+}$ derivative, bpy = 2,2'-bipyridyl ligand] is used as the sensitizer.[30,110] Claims of cyclic and sustained water cleavage by visible light in this system, however, have not been independently verified. Other colloidal systems have also been reported for the OER.[119-123] Since these *microheterogeneous* assemblies do not involve photoexcitation of a semiconductor, they are not further discussed in this Chapter.

A photoelectrochemical (photoelectrolysis) system can be constructed using a n-type semiconductor electrode, a p-type semiconductor, or even mating n- and p-type semiconductor photoelectrodes as illustrated in Figs. 2a-c respectively. In the device in Fig. 2a, OER occurs on the semiconductor *photoanode* while the HER proceeds at a catalytic counterelectrode (e.g., Pt black). Indeed, the classical n-TiO_2 photocell alluded to earlier,[53-57] belongs to this category. Alternately, the HER can be photodriven on a p-type semiconductor while the OER occurs on a "dark" *anode*.

Unlike the single *photosystem* cases in Figs. 2a and 2b, the approach in Fig. 2c combines two photosystems. Both heterotype (different semiconductors) or homotype (same semiconductor) approaches can be envisioned, and it has been shown[60] that the efficiency of photoelectrolysis with solar radiation can be enhanced by using simultaneously illuminated n- and p- type semiconductor electrodes (Fig. 2c). It is interesting to note that this twin-photosystems approach mimics the plant photosynthesis system, intricately constructed by nature, albeit operating at rather low efficiency. The approach in Fig. 2c has at least two built-in advantages. First, the sum of two photopotentials can be secured in an additive manner such that the required threshold for the water splitting reaction (Chapter 2) can be more easily attained than in the single photoelectrode cases in Figs. 2a and 2b. Second, different segments of the solar spectrum can be utilized in the heterotype approach, and indeed, many semiconductors (with different E_g's) can be stacked to enhance the overall solar conversion efficiency of the device.[44] However, the attendant price to be paid is the concomitant increase in the device complexity. Further, the photocurrents through

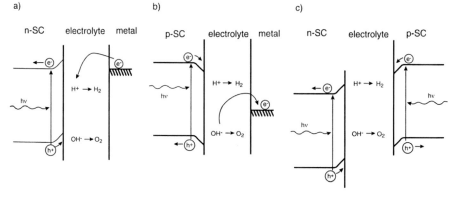

Fig. 2. Photoelectrolysis cell configurations (refer to text).

the two interfaces will have to be carefully matched since the overall current flowing in the cell must obviously be the same.

Finally, hybrid approaches for water photosplitting can be envisioned. As illustrated in Fig. 3a, a water electrolyzer can be simply hooked up to a solar panel that delivers the needed photovoltage.[40,70,124,125] A conceptually more appealing scenario deploys a p-n junction directly in ohmic (electronic) contact with the electroactive surface where the HER (or less commonly, the OER) occurs (Fig. 3b). A variety of such *monolithic* configurations have been discussed.[126–129] A p/n photochemical diode consisting of p-GaP and n-Fe$_2$O$_3$ has been assembled in a monolithic unit and studied for its capability to evolve H$_2$ and O$_2$ from seawater.[130] Silicon spheres comprising of p- and n- regions in electronic contact (forming p/n diodes) embedded in glass with a conductive backing have been used for photosplitting HBr into H$_2$ and Br$_2$.[131] These and other hybrid approaches are further elaborated in a subsequent Section.

Next, we define an ideal semiconductor photoanode and photocathode for the solar electrolysis of water. We also briefly examine *real world* issues related to charge-transfer kinetics at semiconductor/electrolyte interfaces and the need for an external bias to drive the photolysis of water.

3   More on Nomenclature and the Water Splitting Reaction Requirements

A bewildering array of terms have been deployed in this field; thus, a few clarifying remarks appear to be in order. The term *photoelectrochemical* refers to any scenario wherein light is used to augment an electrochemical process. This process could be either *uphill* (Gibbs free energy charge being positive) or *downhill* (negative ΔG) in a thermodynamic sense. In the former case, the process is called *photosynthetic* (the reaction H$_2$O $\rightarrow$ H$_2$ + ½ O$_2$ being an example) while the latter would be a photocatalytic process (e.g., the oxidation of hydrocarbons at an illuminated n-TiO$_2$/solution interface in an oxygenated medium). The term *photoelectrolysis* is correctly applied

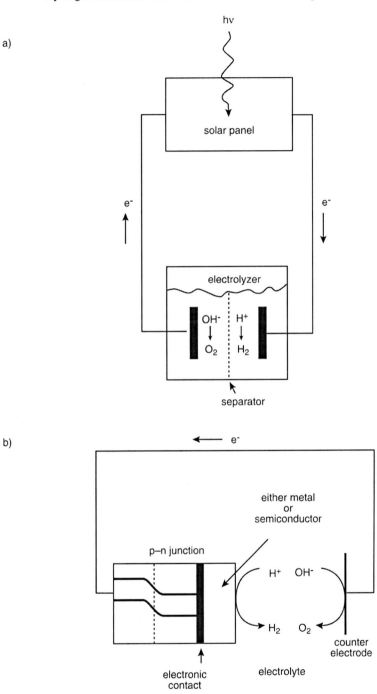

Fig. 3. Two hybrid photoelectrolysis cell configurations.

to a case involving semiconductor photoelectrode(s) in an electrochemical cell. On the other hand, the term *photolysis* is more general and also includes the case of semiconductor suspensions. The term *photoassisted splitting* should be reserved for the cases wherein the excitation light energy only *partially* furnishes the voltage needed for the electrolysis process, the rest being accommodated by an applied external bias (see below). Finally, the term *solar* should be reserved for cases where sunlight (or at least simulated sunlight) was used for the semiconductor excitation. In all the cases, the more general term (or prefix) *photo* should be used. For example, if water is split (into H_2 or O_2) in a photochemical reactor containing a UV light source and semiconductor particulate suspensions, the process descriptor that is appropriate here would be: *UV photolysis of water.*

What photovoltage and semiconductor bandgap energy (E_g) would be minimally needed to split water in a single photosystem case (c.f., Figa. 2a or 2b)? We have seen (Chapter 2) that, to split water into H_2 and O_2 with both products at 1 atm, a thermodynamic potential of 1.23 V would be needed. To this value would have to be added all the losses within an operating cell mainly related to resistive (Ohmic components) and the overpotentials (kinetic losses) required to drive the HER and OER at the two electrode/electrolyte interfaces. This would translate to a semiconductor E_g value of ~ 2 eV if the splitting of water to H_2 and O_2 is the process objective. On the other hand, photovoltaic theory[3] tells us that the photovoltage developed is nominally only ~ 60% of E_g. Taking all this into account, an E_g value around 2.5 eV would appear to be optimal.

What about a twin-photosystem configuration as in Fig. 2c? Optimal efficiency (we will define efficiency soon) is reached in such a configuration when one semiconductor has an E_g value of ~ 1.0 eV and the second ~ 1.8 eV.[66] On the other hand, it has been pointed out[64] that an optimal combination would be two matched electrodes of equal 0.9-eV band gap, since, in the absence of other limitations, the photocurrent would have been dictated by the higher E_g electrode of a pair.

An irradiated semiconductor particle in a microheterogeneous system can be regarded as a short-circuited electrochemical cell where that particle is poised at a potential (ΔV) such that the anodic and cathodic current components are precisely balanced (i.e., no *net* current obviously is flowing through that particle.[132] This photovoltage obviously has to attain a value around ~ 2 V for the water splitting reaction to be sustained. Given the need to reduce the kinetic losses and move the photovoltage value down to one around the thermodynamic (ideal) limit of 1.23 V, it is therefore not surprising that many of the studies on semiconductor particle suspensions have utilized (partially) metallized surfaces – the metals being selected to be catalytic toward the HER. The prototype here is the platinized semiconductor particle (e.g., Pt/TiO_2) and the platinum islands are deposited on the oxide surface using photolysis in a medium containing the Pt precursor (e.g., $PtCl_6^{2-}$) and a sacrificial electron donor (e.g., acetate).[133,134] Obviously, the bifunctional catalyst assemblies discussed in the preceding Section, are motivated by considerations to make the HER and OER processes more facile.

Very detailed studies also have appeared on catalytic modification of semiconductor *electrode* surfaces to improve the HER performance; the reader is referred to the many review articles and book chapters on this topic.[22,29,88,135,136,136a]

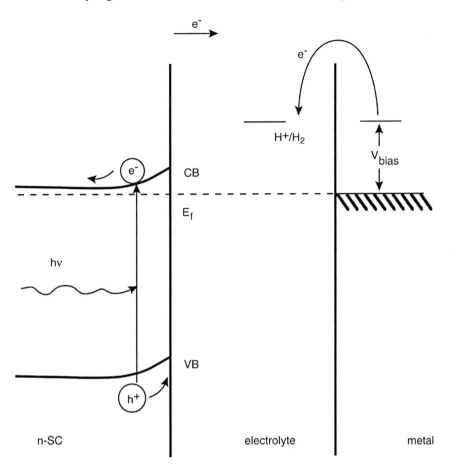

Fig. 4. An interfacial energetic situation in a photoelectrolysis cell where the flat-band potential of the n-type semiconductor photoanode lies positive of the HER potential. V_{bias} is the external bias potential needed in this case to drive the photoelectrolysis process.

The earlier discussion related to Fig. 1a should have indicated that it is simply not the magnitude of E_g (and the ΔV generated) alone that is the sole criterion for sustaining the water photosplitting process. Where the CB and VB levels lie on the energy diagram for the semiconductor at the interface is crucial. Assuming that we are dealing with thermalized electrons here (i.e., no *hot carrier* processes), the CB edge for the n-type semiconductor has to be higher (i.e., be located at a more negative potential) relative to the H_2/H^+ redox level in the solution (c.f., Fig. 1a). In the event that this is not true (see Fig. 4), an external bias potential would be needed to offset the deficit energy content of the photogenerated electrons. Other equivalent statements can be made for the requirements for an n-type semiconductor, namely that the semiconductor has to have low electron affinity or that the flat-band potential for that particular semiconductor/electrolyte interface has to be more negative than the H_2/H^+ redox level.

Interestingly, rutile TiO_2 electrodes have an interfacial situation similar to that schematized in Fig. 4. Thus, the authors in the classical n-TiO_2 water splitting study[53-57] circumvented this problem via a *chemical* bias in their electrochemical cell by imposing a pH gradient between the photoanode and cathode chambers. On the other hand, photogenerated holes in TiO_2 are generated at a very positive potential (because of its low-lying VB edge at the interface) so that they have more than enough energy to oxidize water to O_2. Not too many semiconductor surfaces are stable against photocorrosion under these conditions; i.e., the photogenerated holes attack the semiconductor itself rather than a solution species such as OH^- ions. Cadmium sulfide (E_g = 2.4 eV) is a particularly good example of a semiconductor that undergoes photocorrosion instead of evolving O_2 from H_2O. Thus, the requirements for a single photosystem for splitting water should have semiconductor energy levels that straddle the two redox levels (H_2/H^+ and OH^-/O_2), have an E_g value of ~ 2.5 eV for the semiconductor, and with a semiconductor surface that is completely immune to photocorrosion under OER (or HER) conditions. Additionally, the semiconductor surface has to be made catalytically active toward OER or HER.

Interfacial energetics in two-photosystem cells combining n- and p-type semiconductor electrodes respectively (Fig. 2c) have been discussed.[67] Stability issues in photoelectrochemical energy conversion systems have been reviewed.[31a]

Given the above, it is hardly surprising that the search for satisfactory semiconductor candidates has continued at an unabated rate to the time of writing of this Chapter. In a historical sense, it is interesting that the shift of the research objective from initially photoelectrolysis toward regenerative photoelectrochemical cells (which generate electricity rather than a fuel such as H_2) in the early years (1980s) is undoubtedly a consequence of the many challenges involved in the discovery (and optimization) of a semiconductor for the solar water splitting application. The field is now coming full circle with realization of the importance of a renewable H_2 economy,[138-141] and researchers are once again turning their attention to the use of semiconductor/electrolyte interfaces for solar H_2 generation.

One approach to circumvent the semiconductor stability problem is to simply remove the photoactive junction from physical contact with the liquid. An alternate approach is to reduce the activity of water at the interface (and thus the proclivity to corrosion) by using a hydrophobic environment. In the first category, a variety of coatings have been deployed to protect the semiconductor surface (see Table 2). In the latter category, ionic liquids (such as concentrated lithium halide electrolytes) have been used[173,190,191] so that instead of splitting water, compounds such as HCl or HI can be photodecomposed to H_2 and Cl_2 (a value-added chemical) or I_2 respectively. Note that many of the examples in Table 2 are really hybrid systems where the photovoltaic junction (consisting of the semiconductor and a metal, polymer, or a transparent conducting oxide) simply biases an electrochemical interface.

Table 2. Types of coatings for protecting semiconductor surfaces against photocorrosion.[a]

Entry number	Semiconductor(s)	Type of coating	Specific coating(s) employed	References
1	GaP	metal	Au	142
2	GaP, Si	metal	Pt, Pd, Ni or Cu	143
3	Si	metal	Pt	144
4	GaAs	metal	Au, Pt, Rh	145
5	Si	e.c.p.[b]	polypyrrole	146–148
6	Si	e.c.p.	polypyrrole	149
7	GaAs	e.c.p.	polypyrrole	150
8	CdS	e.c.p.	polypyrrole	89, 151–153
9	CdTe	e.c.p.	polypyrrole	154
10	Si	e.c.p.	polypyrrole	154a
11	Si	e.c.p.	polyacetylene	154b
12	CdS, CdSe	e.c.p./catalyst hybrid	polypyrrole/ polybithiophene/RuO_2	155
13	Si	e.c.p./catalyst hybrid	Au/polypyrrole	156
14	Si	redox layer/e.c.p hybrid	ferrocene/polypyrrole	157
15	Si, InP	redox layer	ferrocene	158
16	Si, GaAs, GaP, InP and CdS	wide band gap sc[c]	TiO_2	159
17	GaAs, GaAlAs	wide band gap sc or insulator	TiO_2, SnO_2, Nb_2O_3, Al_2O_3 or Si_3N_4	160
18	CdSe	wide band gap sc	ZnSe	161
19	CdSe	sc (photoconductor)	Se	162, 163
20	Si	insulator/catalyst hybrid	silicide/Pt	164, 165
21	Si	macrocycle film	Phthalocyanine (Cu- and metal-free)	166
22	Si	self-assembled monolayer	alkane-thiol	167
23	GaAs	self-assembled monolayer		168
24	InP	self-assembled monolayer	"	169–171
25[a]	Si	transparent conducting oxide	Sn-doped indium oxide (ITO)	172–174
26	Si	transparent conducting oxide	Sb-doped SnO_2	175, 176
27	Si	transparent conducting oxide	SnO_2	177
28	Si	conducting oxide	Tl_2O_3	178
29	CdS	conducting oxide	RuO_2	179, 180
30	Si	redox polymer/catalyst	N,N'dialkyl- 4,4'bipyridinium polymer/ Pt or Pd	181–184
31	Si	redox polymer/catalyst	poly(benzyl viologen)/Pt	185
32	GaS	redox polymer	as in Entry 30 without the metal catalyst	186
33	Si	redox polymer	[4,4'-bipyridinium]- 1,1'-diylmethylene- 1,2-phenylenemethylene dibromide polymer (poly-oXV^{2+}) and other viologen-based polymers	187
34	GaAs	redox polymer	polystyrene with pendant Ru(bpy)$_3^{2+}$ complex (bpy=2,2'-bipyridyl ligand)	188
35	GaAs	redox polymer/catalyst	polymer as in Entry 34/ RuO_2	189

[a]Not all these cited studies have focused on photodriven HER and OER applications for the coated semi-conductor/electrolyte interfaces. [b]e.c.p.= electronically-conducting polymer. [c]sc = semiconductor

4 Efficiency of Photoelectrolysis

In a regenerative solar energy conversion system, the device efficiency (η) is simply given by the ratio of the power delivered by the photovoltaic converter and the incident solar power (P_s in W/m^2 or mW/cm^2). However, we are concerned here with devices producing a fuel (H$_2$) and several expressions exist for the device efficiency. Thus, this efficiency can be expressed in kinetic terms:[70,192]

$$\eta_1 = \frac{\Delta G^0_{H_2} R_{H_2}}{P_s A} \times 100 \tag{1}$$

In Eq. 1, $\Delta G^0_{H_2}$ is the standard Gibbs energy for the water splitting reaction generating H$_2$, R_{H_2} is the rate (mol s^{-1}) of generation of H$_2$, and A is the irradiated area (m^2 or cm^2). In the above (as well as in the expressions below), it is assumed that the H$_2$ gas is evolved at 1 atm in its standard state. (Corrections have been discussed for cases where the gas is not evolved at 1 atm, see Ref. 70.) Another equation[31a] for the efficiency refers to the standard (Nernstian) voltage for the water splitting reaction, 1.23 V (see Chapter 2):

$$\eta_2 = \frac{(1.23 - V_{bias}) i_t}{P_s A} \times 100 \tag{2}$$

The bias voltage, that is needed in some cases (c.f., Fig. 3), is V_{bias} and i_t is the current corresponding to the maximum power point[29,66] of the cell.

In some cases, ΔH values are used in place of the free energy, and then the term, 1.23 in Eq. 2, must be replaced with 1.47. This assumes that the products will be burned (i.e., in a thermal combustion process) to recover the stored energy as heat rather than as electrical energy in a fuel cell. Other efficiency expressions have been proposed that take into account the energy throughput or the polarization losses at the photoelectrode(s) and the *dark* counterelectrode where relevant (see for example, Ref. 193). The shortcomings of these alternate expressions have been pointed out.[31a] In cases where the energy storage system generates a multitude of products rather than just H$_2$, the free energy term in the numerator in Eq. 1 becomes a summation of all the free energies stored in the various products.[194]

Let us now examine the ideal limit for the process efficiency as derived by various authors. Table 3 contains a compilation of these estimates dating back to 1980. These ideal limits range from ~ 3% all the way to ~ 72%! Admittedly, the scenarios pertaining to these estimates are variant and both single- and multi-photosystem configurations have been considered. Taken as a whole, a 10–12% process efficiency (under, say, AM 1.0-solar irradiation) for a solar photoelectrochemical water splitting system based on a *single* photoconverter, appears to be a reasonable target. Higher efficiencies can be realized in a multi-photosystem or even a tandem (i.e., hybrid, see above) configuration although attendant increase in costs associated with increased system complexity may have to be taken into account here. The sensitivity

Table 3. Ideal limits for water photoelectrolysis efficiencies as estimated by various authors.

Entry number	Efficiency (%)	Comments	Reference(s)
1	28	Assumptions behind estimate not known.	59
2	3.4–6.3	For AM 1.0 insolation; estimate depends on semiconductor E_g.	65
3	7	AM 1.2-for one-photon system.	195
	~ 10	For a two-photon system.	195
4	41	Two-photosystem configuration with two different (optimized) E_g values.	192
5	30.7–42.4	For AM 1.5-insolation. Estimated efficiency depends on type of solar conversion system; i.e., whether single- or twin-photosystem and whether one- or two-photon driven.	70, 195
6	38–72	Twin-photosystem at AM 1.0 for the lower limit and a 36 band gap solar cell for the upper limit.	44, 196
7	9–12	For AM 1.0-insolation, the estimate depends on whether single- or twin-photosystem configuration. Values cited based on original work in Ref. 198.	76, 197

of η to parameters such as the semiconductor band gap (E_g) has been analyzed by several authors.[76,197,198] The efficiency peaks at ~ 1.5 eV and ~ 2.2 eV for a twin- and a one-photosystem respectively[197,198] and at ~ 1.8 eV for a tandem cell combining a solar photovoltaic cell with a single photoanode-based electrochemical cell.[70]

Experimental data on photo- or solar-electrolysis efficiencies are contained in Table 4. Other compilations are available.[31a,76] In most of the instances, the attained laboratory efficiencies are well below the ideal limit (c.f., Table 3) although some of the numbers claimed (see entries 1, 9, 14, 15, for example, in Table 4) are indeed impressive. However, some of the values claimed have been questioned by others.[214] Further independent verification is undoubtedly warranted, and the losses associated with process scale-up are, as yet, unknown.

In closing this Section, two other types of efficiencies must be mentioned. The first is the quantum efficiency, Φ—a parameter well known to the photochemistry community. This pertains to the ratio of either a number (cumulative) or equivalently, a rate of useful events and the number (or rate) of photons incident on the cell. Thus in the case of the HER, the numerator could be either R_{H_2} or the number of moles of H₂ or even the number of electrons contributing to the HER—the denominator in each case being expressed either cumulatively or in terms of the rate of incidence of the photons. If deleterious carrier-recombination processes are carefully minimized, Φ could approach 100% in an optimized system. Note that this parameter automatically enters into *measured* estimates of η (e.g., because of the R_{H_2} and i_t terms in Eqs. 1 and 2, respectively.)

Finally, researchers also quote another efficiency parameter closely related to Φ, namely, the IPCE or incident photon-to-electron conversion efficiency.

Table 4. Examples of experimentally obtained efficiencies for the photoelectrolysis of water and other solvents.

Entry number	Efficiency (%)	Comments	Reference
1	13.3	p-InP photocathodes with catalytic modification were used in 1-M HClO$_4$ under sunlight at 81.7 mW/cm^2.	199
2	~ 1.8	The quoted value only a projection based on available data on a LaCrO$_3$-TiO$_2$ photoanode.	200
3	8.6	Efficiency for a p-InP modified with Ru electrocatalyst in 1-M HCl.	201
4	0.05	Measured for a polycrystalline p/n diode assembly based on Fe$_2$O$_3$. Poor efficiency attributed to the non-optimal charge transfer properties of the oxide.	202
5	8.2	Measured for a p-n junction Si electrode coated with noble metal and for the photoelectrolysis of HI under simulated AM 1solar radiation.	203
6	7.8	Measured for a twin-photosystem cell configuration based on p-InP photocathode and n-MoSe$_2$ (or n-WSe$_2$) photoanode under monochromatic radiation (632.8 nm) and for the photoelectrolysis of HBr.	204
7	~ 1–2.8	Bipolar CdSe/CdS panels used under 52 mW/cm^2 effective solar flux. Upper limit after correction for light absorption by the electrolyte.	205–207
8	1.84	Measured for a cell with n-Fe$_2$O$_3$ photoanode under 50.0 mW/cm^2 Xe arc lamp irradiation and at a bias potential of 0.2 V/SCE at pH 14.	208
9	18.3	Bipolar AlGaAs/Si/RuO$_2$/Pt configuration under illumination with a 50-W tungsten-halogen lamp.	209
10	4.8	Titania nanotubes were used under UV irradiation at 365 nm and at an intensity of 146 mW/cm^2.	210
11	0.6–2.2	A tandem monolithic configuration used with WO$_3$ films being biased by a PV junction.	211
12	2.5	An amorphous Si based triple-junction cell coated with HER and OER catalysts.	212
13	4.5	A WO$_3$ photoanode is coupled with a dye-sensitized TiO$_2$ solar cell under *standard* solar light.	46
14	12.4	A monolithic configuration with a p-GaInP$_2$ photocathode biased by a GaAs p-n junction is used.	126
15	7.8–16.5	Lower value measured for a triple junction p-i-n a-Si(Pt)/KOH hybrid PV/electrolysis cell. The higher value for a cell similar to that in Entry 14.	125
16	8.35	Measured for carbon-doped TiO$_2$ films with Xe-lamp irradiation at 40 mW/cm^2. However, see text.	213

5 Theoretical Aspects

With reference to Fig. 5, the so-called quasi-Fermi level formalism[3,7,11,12] is useful for understanding the interfacial energetics at illuminated semiconductor-electrolyte interfaces. Thus, in the *dark*, at equilibrium:

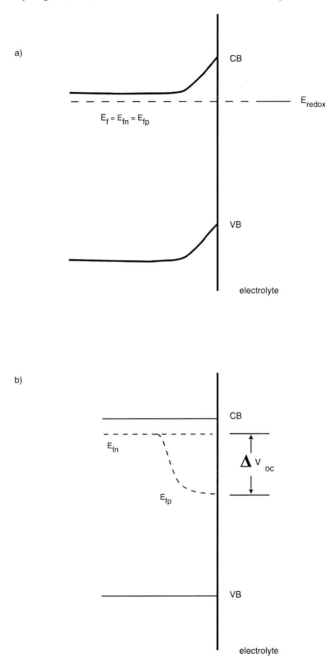

Fig. 5. Diagrams for a semiconductor-electrolyte interface (a) at equilibrium and (b) under irradiation showing the quasi-Fermi levels for electrons and holes.

$$E_f = E_{fn} = E_{fp} \tag{3}$$

where E_f stands for the Fermi level energy and the subscripts n and p denote the two types of carriers, electrons and holes in the semiconductor respectively. This situation is schematized in Fig. 5a. Under illumination, at open-circuit, a non-equilibrium concentration of electronic carriers is created, and separate quasi-Fermi levels (E_{fn} and E_{fp}) are required to describe the electron and hole concentrations (Fig. 5b):

$$n(x) = N_c \exp[(E_{CB} - E_{fn(x)})/kT] \tag{4a}$$

$$p(x) = N_v \exp[(E_{fp(x)} - E_{VB})/kT] \tag{4b}$$

In Eqs. 4a and 4b, x is a position variable since the values of n, p, E_{fn} and E_{fp} are position dependent to varying degrees, N_c and N_v are the densities of states in the conduction and valence bands respectively, k is the Boltzmann constant and T is the absolute temperature. The splitting of the electron and hole quasi-Fermi levels under illumination (Fig. 5b) defines the magnitude of the photovoltage developed, ΔV (ΔV_{oc} in the specific open-circuit case in Fig. 5b).

On the basis that the position of E_{fn} at the semiconductor surface is dependent on the photon flux and that E_{fn} has to lie above the HER redox level, a threshold in light intensity has been proposed[217] for the sustained photoelectrolysis of water to occur. However, as discussed by other authors,[218] no such threshold has been reported in the literature. It has been pointed out[218] that the driving force for the photoinduced electron transfer process is related to the difference in *standard* potentials of the *donor* (say, an electron at the semiconductor CB edge) and the *acceptor* (say, protons in solution). This is independent of the carrier concentration and photon flux and thus a light intensity threshold for *incipient* product formation through photoelectrolysis should not occur.[218]

The experimental observations[217] of an *apparent* light intensity threshold for the photocurrent onset have been rationalized[218] on the basis that a critical photon flux must be exceeded to counteract the dark current of opposite polarity flowing through the cell. Thus, there appears to be confusion between alternate definitions of a light intensity threshold: a threshold for *incipient* product (say H_2) formation and a threshold for product formation in a *specific* (e.g., *standard*) *state*.[218]

Other fundamental considerations for a solar photoelectrolysis system have been discussed.[219–223] Theoretical formulations for photocurrents at p- and n-type semiconductor electrodes have been presented[221] on the assumption that the rate-determining step is charge transfer across the interface. A theory for the light-induced evolution of H_2 has been presented by the same group for semiconductor electrodes.[222] The effect of an oxide layer of tunneling dimensions has been considered for photoelectrochemical cells designed for fuel (e.g., H_2) production.[223] It is fair to say that these theoretical developments have occurred fairly early on in the evolution of this field (before 1985). The work in the subsequent two decades has largely focused on the discovery of new semiconductor materials for the photosplitting of water.

We consider these materials aspects in the next few Sections.

6 Oxide Semiconductors

General reviews of the use of oxide semiconductors for the photoelectrolysis of water are contained in Refs. 62 and 65 (see Table 1). Eleven binary and ternary oxides were examined in Ref. 62. Linear correlations were presented between the flat band potential, V_{fb} of these oxides and their band gap energy (E_g); and between V_{fb} and the heat of formation of the oxide per metal atom per metal-oxygen bond. Aligning all the oxide energy levels on a common scale, these authors noted[62] that the position of the conduction band varies much more than those of valence bands – a trend expected from the cationic (d-band) character of the conduction band in the oxide while the valence band is mainly of O(2p) character. The latter should be relatively independent of the oxide parentage in terms of the metal.

A similar correlation between V_{fb} and E_g was presented in Ref. 65 but for a much more extensive collection of oxides including oxides with or without partially filled d levels and oxides formed anodically on metals. Only oxides with partially filled d levels (Type "a" in the author's notation, Ref. 65) yielded a straight-line correlation between the two parameters. This plot was used by the authors of the two studies[62,65] for predictive purposes to assess the efficacy of a given oxide for the photoelectrolysis of water. As seen earlier, the V_{fb} of the material has to be at a negative enough potential to drive the HER, and E_g has to be ~ 2 eV and yet bracket the HER and OER redox levels (c.f., Figure 1a).

The possibility of introducing new d-bands for Type "b" oxides (with filled d-bands) by introducing dopants into the host lattice was also discussed in Ref. 65 with examples. Other authors have advocated this approach as well (Ref. 26, for example). The review in Ref. 32 contains further examples of this approach for effectively *shrinking* the original E_g and sensitizing the oxide to visible portions of the solar spectrum. We shall return to this aspect for the specific case of TiO_2 later in this Chapter.

We now turn to discussions of individual oxide semiconductor materials for the photoelectrolysis of water, starting with the *mother* of all oxides, namely, TiO_2.

6.1 Titanium Dioxide: Early Work

Historically, this is the material which really sparked interest in the solar photoelectrolysis of water. Early papers on TiO_2 mainly stemmed from the applicability of TiO_2 in the paint/pigment industry[255] although fundamental aspects such as current rectification in the dark (in the reverse bias regime) shown by anodically formed valve metal oxide film/ electrolyte interfaces was also of interest (e.g., Ref. 52). Another driver was possible applications of UV-irradiated semiconductor/electrolyte interfaces for environmental remediation (e.g., Refs. 256, 257).

Representative early work on this remarkable material is presented in chronological order in Table 5, with all these studies aimed toward the photoelectrolysis of water. Further summaries of this early body of work are contained in Refs. 5, 6, 17–20, 25 and 32.

Table 5. Representative examples of early work (i.e., prior to ~1985) on the use of TiO_2 for the photoelectrolysis of water.

Entry number	Title of article	Comments	Reference
1	Electrochemical Photolysis of Water at a Semiconductor Electrode	First demonstration of the feasibility of water splitting.	54
2	The Quantum Yield of Photolysis of Water on TiO_2 Electrodes	Very low quantum yields (~10^{-3}) were measured when no external bias was applied. The effect of photon flux also explored.	224
3	Photoelectrolysis of Water Using Semiconducting TiO_2 Crystals	Study shows the necessity of a bias potential for rutile photoanodes.	225
4	Photoelectrolysis of Water in Cells with TiO_2 Anodes	Both single crystal and polycrystalline TiO_2 used and external quantum efficiency measured.	226
5	A Photo-Electrochemical Cell with Production of Hydrogen and Oxygen by a Cell Reaction	Cell configuration also employs an illuminated p-GaP photocathode (c.f. Ref. 60).	227
6	Photoassisted Electrolysis of Water by Irradiation of a Titanium Dioxide Electrode	The initial claim in Ref. 54 supported along with data on the wavelength response and the correlation of product yield and current.	228
7	Semiconductor Electrodes 1. The Chemical Vapor Deposition and Application of Polycrystalline n-Type Titanium Dioxide Electrodes to the Photosensitized Electrolysis of Water	Comparison of the behavior of CVD and single crystal n-TiO_2 presented.	229
8	Formation of Hydrogen Gas with an Electrochemical Photo-cell	See text.	55
9	Hydrogen Production under Sunlight with an Electrochemical Photo-cell	See text.	56
10	Photoproduction of Hydrogen: Potential Dependence of the Quantum Efficiency as a Function of Wavelength	—	230
11	Photoelectrolysis of Water with TiO_2-Covered Solar-Cell Electrodes	A hybrid structure, involving a p-n junction Si cell coated with a TiO_2 film by CVD, is studied.	231
12	Electrochemical Investigation of an Illuminated TiO_2 Electrode	Two types of TiO_2 films studied, namely, anodically formed layers on Ti sheets and those prepared by plasma jet spraying of TiO_2 powder.	232
13	Intensity Effects in the Electrochemical Photolysis of Water at the TiO_2 Anode	Quantum efficiency observed to approach unity at low light intensities.	233
14	Improved Solar Energy Conversion Efficiencies for the Photocatalytic Production of Hydrogen via TiO_2 Semiconductor Electrodes	Heat treatment of Ti metal found to influence performance.	234

Table 5. Continuation.

Entry number	Title of article	Comments	Reference
15	Near-UV Photon Efficiency in a TiO₂ Electrode: Application to Hydrogen Production from Solar Energy	—	235
16	Novel Semiconducting Electrodes for the Photosensitized Electrolysis of Water	Appears to be the first study on doping TiO₂ to extend its light response into the visible range of the electromagnetic spectrum.	236
17	Photoelectrolysis of Water in Sunlight with Sensitized Semiconductor Electrodes	Similar observations as in Ref. 236 for Al³⁺-doped TiO₂.	237
18	Photoelectrolysis	The behavior of single crystals of two different orientations (⊥ and ∥ to the C axis) and polycrystalline TiO₂ reported.	238
19	The Quantum Yields of Photoelectric Decomposition of Water at TiO₂ Anodes and p-Type GaP Cathodes	A more detailed study as in Ref. 227 by the same research group.	239
20	Anomalous Photoresponse of n-TiO₂ Electrode in a Photoelectrochemical Cell	The behavior of surface states at the TiO₂-electrolyte interface is focus of this study.	240
21	An Effect of Heat Treatment on the Activity of Titanium Dioxide Film Electrodes for Photosensitized Oxidation of Water	Heat treatment in argon atmosphere found to improve performance of both anodic and pyrolytically prepared TiO₂ films.	241
22	Preparation of Titanium Dioxide Films as Solar Photocatalysts	Low-cost polyimide plastic used as film substrate.	242
23	Photoelectrochemical Behaviour of TiO₂ and Formation of Hydrogen Peroxide	Other than the OER, reduction of O₂ to H₂O₂ also observed.	243
24	Photodeposition of Water over Pt/TiO₂ Catalysts	Powdered photocatalyst is employed.	244
25	Photocatalytic Decomposition of Gaseous Water over TiO₂ and TiO₂-RuO₂ Surfaces	As above but gaseous water used at room temperature.	245
26	Photoelectrolysis of Water with Natural Mineral TiO₂ Rutile Electrodes	Natural samples compared with Fe-doped synthetic single crystal TiO₂.	246
27	Models for the Photoelectrolytic Decomposition of Water at Semiconducting Oxide Anodes	Although title is general, theoretical study focuses on the TiO₂-electrolyte interface and the effect of surface states.	247
28	Photosynthetic Production of H₂ and H₂O₂ on Semiconducting Oxide Grains in Aqueous Solutions	Hydrogen peroxide formation observed in TiO₂ powder suspensions as in Ref. 243 for TiO₂ films.	248

Table 5. Continuation.

Entry number	Title of article	Comments	Reference
29	Influence of pH on the Potential Dependence of the Efficiency of Water Photo-oxidation at n-TiO_2 Electrodes	Quantum efficiency for water photooxidation is shown to be pH-dependent.	249
30	Photocatalytic Water Decomposition and Water-Gas Shift Reactions over NaOH-Coated, Platinized TiO_2	As in Entry 24 (Ref. 244) by the same research group.	250
31	Photosensitized Dissociation of Water using Dispersed Suspensions of n-Type Semiconductors	Focus of study on TiO_2 and $SrTiO_3$ using EDTA as an electron donor and Fe^{3+} as acceptor for tests of water reduction and oxidation activity respectively (c.f. Figs. 1b and 1c).	251
32	Photocatalytic Hydrogen Evolution from an Aqueous Hydrazine Solution	Pt-TiO_2 photocatalyst used and both H_2 and N_2 evolution observed.	252
33	Conditions for Photochemical Water Cleavage. Aqueous Pt/TiO_2 (Anatase) Dispersions under Ultraviolet Light	As in Entries 24 and 25 (Refs. 244, 245) photocatalyst dispersions studied.	253
34	Colloidal Semiconductors in Systems for the Sacrificial Photolysis of Water. 1. Preparation of a Pt/TiO_2 Catalyst by Heterocoagulation and its Physical Characterization	--	254

6.2 Studies on the Mechanistic Aspects of Processes at the TiO_2-Solution Interface

Also contained in the compilation in Table 5 are some early studies oriented toward the *mechanistic* aspects of the photoelectrochemical oxidation of water (and other compounds) at the n-TiO_2-electrolyte interface, as exemplified by Entries 23 and 29 (Refs. 243 and 249 respectively). More recent and representative studies of this genre include Refs. 258–289.

6.3 Visible Light Sensitization of TiO_2

Rather problematic with TiO_2 in terms of the attainable process efficiency is its rather wide band gap (3.0–3.2 eV). Consequently, only a small fraction (~ 5%) of the overall solar spectrum can be harnessed by this material. Thus, the early work (as in Table 5, Entries 16 and 17, Refs. 236 and 237 respectively) has also included attempts at extending the light response of TiO_2 from the UV to the visible range; see, for example, Refs. 260, 290–296. Reviews of these works are available, see Refs. 20, 32, 297 and 298. For reasons mentioned earlier, we exclude for our discussion, studies oriented toward chemical modification of the TiO_2 surface with a dye. As summarized elsewhere,[297,298] transition metal dopants also modify the interfacial charge transfer and electron-hole recombination behavior of the TiO_2 host. Whether a given dopant exerts a positive or negative effect depends on the particular metal.[297,298]

It must be noted that most studies on metal-doped TiO_2 are oriented toward the photo-oxidation of environmental pollutants (e.g., 4-nitrophenol,[299] 4-chlorophenol[300]) rather than toward the photoelectrolysis of water (but see Section 6.4 of this Chapter). Other aspects of metal doping include the effect of UV radiation of Ag-doped TiO_2 specimens,[301,302] plasma treatment,[303] and thermal treatment.[304] Metal doping by ion implantation of TiO_2 has been discussed.[91,305]

Non-metallic elements such as fluorine, carbon, nitrogen and sulfur have been incorporated into TiO_2. Table 6 contains a compilation of representative studies on this topic. As with the trend noted earlier with metal dopants, only two of the studies in Table 6 are oriented toward water photosplitting or OER.[213,315] Other than the desired optical response, non-metallic dopants also exert electronic effects on the host behavior as with the metal dopants (see above). Thus F-doping is observed to cause a reduction in the e^--h^+ recombination rate[317] while N-doping at high levels has the opposite effect and serves to suppress the photocatalytic activity of the TiO_2 host.[311] Conflicting views exist on non-metal doping, particularly with respect to the mechanistic aspects, as discussed in Ref. 311.

6.4 Recent Work on TiO_2 on Photosplitting of Water or on the Oxygen Evolution Reaction

Table 7 contains a compilation of studies that have appeared since 1985. Several points are worthy of note here. The vast majority of the entries feature studies on TiO_2 powders rather than on electrodes in a photoelectrochemical cell configuration. In this light, the new studies can be regarded as offshoots inspired by the earlier (pre-1985) studies on co-functional photocatalysts and the cyclic cleavage of water.[4,30] Second, many of the new studies address two key issues with the earlier systems:

1. non-stoichiometric evolution of H_2 and O_2, and
2. poor performance stemming from back reactions and electron-hole recombination processes.

With reference to the first point, very little O_2 evolution was observed in many cases in studies on TiO_2 powder suspensions with reports[245,320,324] of stoichiometric H_2 and O_2 evolution (i.e., in the expected 2:1 ratio) being the exceptions rather than the rule. Initially, this discrepancy was attributed by the community to the photo-induced adsorption of the (evolved) O_2 on the TiO_2 surface.

The remarkable effect of a NaOH *dessicant* coating on the TiO_2 surface on the efficiency of water photosplitting appears to have radically changed this thinking (see Ref. 92 and references therein). The new results support the deleterious role that Pt islands on the TiO_2 play in promoting the reverse reaction, $2\,H_2+O_2 \rightarrow 2\,H_2O$. Interestingly, the irradiation geometry also appears to exert an effect on the extent of back reactions.[335] Adsorption of CO on Pt, for example, was also found to inhibit the reverse reaction.[336] Subsequent studies on the role of Na_2CO_3 addition (Ref. 93

Table 6. Representative studies on doping of TiO$_2$ with non-metallic elements.

Entry number	Title of article	Comments	Reference
1	Visible-Light Photocatalysis in Nitrogen-Doped Titanium Oxides.	Both films and powders considered. Substitutional doping with nitrogen shown to bring about band gap narrowing and also high photocatalytic activity with visible light. Experimental data supported with first-principles calculations.	306
2	Formation of TiO$_{2-x}$F$_x$ Compounds in Fluorine-Implanted TiO$_2$.	Fluorine substituted for oxygen sites in the oxide by ion implantation.	307
3	Band Gap Narrowing of Titanium Dioxide by Sulfur Doping.	Oxidative annealing of TiS$_2$ used. Ab initio calculations also reveal mixing of S 3p states with the valence bond to bring about band gap narrowing.	308
4	Efficient Photochemical Water Splitting by a Chemically Modified n-TiO$_2$.	Combustion of Ti metal in a natural gas flame done to substitute carbon for some of the lattice oxygen sites. The photocatalysis performance data have been questioned (see Refs. 214-216).	213
5	Daylight Photocatalysis by Carbon-Modified Titanium Dioxide.	Titanium tetrachloride precursor hydrolyzed with nitrogen bases to yield (surprisingly) C-doped (instead of N-doped) TiO$_2$. Study oriented toward environmental remediation applicability.	309
6	Carbon-Doped Anatase TiO$_2$ Powders as a Visible-Light Sensitive Photocatalyst	Oxidative annealing of TiC used to afford yellow doped powders. Study focus as in Entry 5.	310
7	Nitrogen-Concentration Dependence on Photocatalytic Activity of Ti$_{2-x}$N$_x$ Powders.	Samples prepared by annealing anatase TiO$_2$ under NH$_3$ flow at 550–600 ºC.	311
8	Visible Light-Induced Degradation of Methylene Blue on S-doped TiO$_2$.	As in Entry 3 (Ref. 308) by the same research group.	312
9	Visible-Light Induced Hydrophilicity on Nitrogen-Substituted Titanium Dioxide Films.	Degree of hydrophilicity correlated with the extent of substitution of nitrogen at oxygen sites.	313
10	Spectral Photoresponses of Carbon-Doped TiO$_2$ Film Electrodes.	Raman spectra used to identify disordered carbon in the flame-formed samples in addition to lower nonstoichiometric titanium oxides identified by X-ray diffraction.	314
11	Photoelectrochemical Study of Nitrogen-Doped Titanium Dioxide for Water Oxidation	One of the few studies probing the influence of doping on OER.	315
12	Metal Ion and N Co-doped TiO$_2$ as a Visible-Light Photocatalyst	Co-doped samples prepared by polymerized complex or sol-gel method. Doped N species found to reside at interstitial lattice positions in the host.	316

Table 7. Representative studies that have appeared since 1985 on the photosplitting of water using TiO_2.

Entry Number	Brief outline of study	Reference(s)
1	Ferroelectric substrates (poled $LiNbO_3$) were used to support TiO_2 films. After platinization of TiO_2, water splitting was examined in both liquid and gas phases under Xe arc lamp illumination.	318
2	Both reduced and Pt-modified powder samples were studied in distilled water and in aqueous solutions of HCl, H_2SO_4, HNO_3 and NaOH. Water photodecomposition proceeds moderately in distilled water and in NaOH but is strongly suppressed in acidic aqueous media. The NaOH coating effect mimics that found by other workers earlier (see Ref. 320 and text).	319
3	Sodium carbonate addition to a Pt/TiO_2 suspension in water effective in promoting stoichiometric photodecomposition of water.	321, 322
4	Demonstration of solar H_2 and O_2 production on NiO_x/TiO_2 co-catalyst with Na_2CO_3 or NaOH addition.	323, 324
5	A photoelectrolyzer designed with a TiO_2 photoanode and a membrane of sulfonated polytetrafluoroethylene as the electrolyte. A quantum efficiency of 0.8 was reported.	325
6	Photochemical splitting of water achieved by combining two photocatalytic reactions on suspended TiO_2 particles; namely, the reduction of water to H_2 using bromide ions and the oxidation of water using Fe(III) species. High efficiency also observed for the photoassisted OER on TiO_2 in the presence of Fe(III) ions.	326, 327
7	Pt- and other catalyst supported TiO_2 (P-25) particles studied. Only the HER was observed and stoichiometry H_2 and O_2 formation was not found. Mechanistic reasons proposed have been challenged by other authors (see text).	328
8	HER observed in semiconductor septum cells using TiO_2 or TiO_2-In_2O_3 composites.	106, 107
9	Pure rutile TiO_2 phase isolated from commercial samples containing both rutile and anatase by dissolution in HF. The resultant samples studied for their efficacy in driving the photoassisted OER in the presence of Fe(III) species as electron acceptor (see Entry 6 above).	329
10	A Z-scheme system mimicking the plant photosynthesis model developed with Pt-loaded TiO_2 for HER and rutile TiO_2 for OER. A IO_3^-/I^- shuttle was used as redox mediator.	330
11	Co-doping of TiO_2 with Sb and Cr found to evolve O_2 from an aqueous $AgNO_3$ solution under visible light irradiation.	331
12	HER observed from a mixed water-acetonitrile medium containing iodide electron donor and dye-sensitized Pt/TiO_2 photocatalysts under visible ight irradiation.	332
13	Back-reactions (i.e., O_2 reduction and H_2 oxidation) studied on both TiO_2 or Cr and Sb co-doped TiO_2 samples (see Entry 11 above).	333
14	TiO_2 nanotube arrays prepared by anodization of Ti foil in a F^--containing electrolyte. Pd-modified photocatalyst samples show an efficiency of 4.8% based on photocurrent data for the OER.	210, 210a-c
15	TiO_2 co-doped with Ni and Ta (or Nb) show visible light activity for the OER in aq. $AgNO_3$ and HER in aqueous methanol solution.	334

and Entries 3 and 4 in Table 7) underline the importance of inhibiting back reactions on catalyst-modified TiO_2 samples. By the same token, unusual valence states (Ti^{5+}) that have been proposed[328] to explain the non-stoichiometric gas evolution have been challenged by other authors.[337]

Other factors influencing the yield of H_2 and O_2 in irradiated TiO_2 suspensions include the nature of the co-catalyst (see, for example, Entry 4 in Table 7), the crystal form of TiO_2, particle size of TiO_2, temperature and ambient pressure.[92] The reader is referred to Ref. 92 for further details. Other interesting mechanistic aspects of the water photosplitting process on the TiO_2 surface such as hydrogen atom spillover have also been discussed.[338]

An interesting aspect of the new work on TiO_2, namely that of combining two photosystems (in a Z-scheme) mimicking plant photosynthesis (see Entries 6 and 10 in Table 7) also has its roots in early work in this field (see, for example, Entry 2 in Table 1). Further elaboration of this strategy is contained in Ref. 96.

Finally, some of the studies considered in Table 7 (Entries 11 and 13) buck the trend mentioned earlier that few of the studies on transition-metal doped TiO_2 are oriented toward the water-photosplitting application. These new studies exploit the visible-light sensitization of the doped host material as well as the improved electronic characteristics observed in some cases (particularly the co-doped instance) to enhance the efficiency of the water photosplitting process.

In sum, TiO_2 continues to be a veritable workhorse of the photocatalysis and photoelectrolysis communities. However, this material to date has not yielded systems for evolving H_2 and O_2 at the 10% benchmark efficiency level. Studies on TiO_2 oriented toward visible light sensitization and efficiency enhancement will undoubtedly continue, at an unabated rate, in the foreseeable future. This is because of the extensive and growing market that already exists for this commodity chemical in a variety of *other* application areas and because of its excellent chemical attributes such as inertness and stability.

6.5 Other Binary Oxides

Table 8 contains a compilation of studies on other binary oxides that have been examined for their applicability to drive the photoelectrolysis of water. As cited earlier, general reviews are available on many of the oxides listed in Table 8.[32,62,65] Other than TiO_2, Fe_2O_3 and WO_3 are two of the most widely studied among the binary oxide semiconductors, and studies on these oxides have continued to appear right up to the time of the writing of this Chapter.

Tungsten oxide shares many of the same attributes with TiO_2 in terms of chemical inertness and exceptional photoelectrochemical and chemical stability in aqueous media over a very wide pH range. However, its flat-band potential (V_{fb}) lies positive of that of TiO_2 (anatase) such that spontaneous generation of H_2 by the photogenerated electrons in WO_3 is not possible. This location of V_{fb} has been invoked[347] for the very high IPCE values observed for the photoinduced OER in terms of the rather slow back electron transfer leading to O_2 reduction. A variety of dopants (e.g., F, Mg, Cu) have been tested for WO_3[341,344,350] and Pt-modified samples have been deployed in a Z-scheme configuration.[349] Electron acceptors such as Ag^+[343] and IO_3^-[349]

Table 8. Binary oxides (other than TiO_2) that have been considered for the photoelectrolysis of water.

Entry number	Oxide semiconductor	Energy band gap, eV	Comments	Reference(s)
1	WO_3	2.5–2.8	This material has been used as single crystals, thin films, powders and in mesoporous/ nanostructured form. Both virgin and doped samples studied.	339–350
2	Fe_2O_3	2.0–2.2	As in Entry 1 above.	208, 351–368
3	ZnO	3.37	Unstable under irradiation and OER/HER conditions.	248
4	SnO_2	3.5	Sb-doped single crystal samples used. Stable H_2 and O_2 evolution observed at Pt cathode and SnO_2 photoanode respectively.	369, 370
5	NiO	3.47	A p-type semiconductor with indirect gap optical transition.	371, 372
6	CdO	~ 2.3	A n-type semiconductor. Interestingly, RuO_2-modified samples reduced the yield of O_2 under irradiation.	373
7	PdO	~ 0.8	A p-type semiconductor. Not stable under irradiation in the HER regime.	374
8	Cu_2O	2.0–2.2	Claims of water splitting in powder suspensions challenged by others (see text).	375, 376
9	CuO	1.7	Not photoelectrochemically stable.	353, 377
10	Bi_2O_3	2.8	Both doped and catalytically modified samples studied.	353, 378-380

species have ben used to study the O_2 evolution characteristics of the WO_3 photocatalyst under visible light irradiation. As pointed out very early in the history of study of this material,[339,381] the lower E_g value of WO_3 (relative to TiO_2) results in a more substantial utilization of the solar spectrum. This combined with the advances in nanostructured oxide materials will likely sustain interest in WO_3 from a photoelectrolysis perspective.

The combination of a rather low E_g value, good photoelectrochemical stability and chemical inertness coupled with the abundance of iron on our planet makes Fe_2O_3 an attractive candidate for the photoelectrolysis of water. Thus it is hardly surprising that this material continues to be intensively studied from this perspective. As with TiO_2 and WO_3, Fe_2O_3 (particularly the α-modification) has been examined in single crystal form, as thin films prepared by CVD,[351,353] pyrolytic conversion of iron[354] and spray pyrolysis,[362,364–367] or as sintered pellets from powders.[355–360] A variety of dopants have been deployed to modify the host[356–359,361,364–367] and remarkably, p-type semiconductor behavior has been reported[356,358,359,365] in addition to the more commonly occurring n-type material. The main handicap with Fe_2O_3 is its rather poor electronic and charge transport characteristics regardless of the method of preparation of the material. Specifically, facile e^--h^+ recombination, trapping of electrons at defect sites and the poor mobility of holes conspire to result in very low efficiencies for water oxidation. Attempts to circumvent these problems by using unique photoanode configurations (e.g., nanorod arrays[363]) or compositional tuning (e.g., minimizing sub-stoichiometric phases such as Fe_3O_4, c.f. Refs. 360, 367) are continuing and will undoubtedly contribute to further examinations of this promising material in the future.

By way of contrast, none of the other binary oxides listed in Table 8 appear to hold much promise. While ZnO has enjoyed extensive popularity in the photochemistry community (even comparable to TiO_2 in the early days prior to ~ 1980), it is rather unstable (at least in the forms synthesized up till now) under illumination and in the OER and HER regimes. This problem besets most of the other candidates in Table 8 with the exception of SnO_2 (whose E_g is too high) and possibly Bi_2O_3. The report[375] of photocatalytic water splitting on Cu_2O powder suspensions (with stability in excess of ~ 1900 h!) has been greeted with skepticism by others[376] who have also pointed out that the Cu_2O band-edges are unlikely to bracket the H^+/H_2 and O_2/H_2O redox levels as required (see Figure 1a and earlier discussion in Section 3 of this Chapter). Our own studies[382] on electrodeposited samples of this oxide have utilized a Ni/NiO protective layer, catalyst modification (with e.g., Pt) to drive the HER and the use of optimized electron donors in the anode compartment in a twin-compartment photoelectrochemical cell (Figure 6).[382] Under these conditions, spontaneous HER was observed under visible light irradiation of the p-Cu_2O photocathode. Photoinduced transfer of electrons from p-Cu_2O to an electron acceptor such as methyl viologen was also demonstrated via in situ spectroscopic monitoring of the blue cation radicals.[382] However, the photocurrents generated are only in the μA level necessitating further improvements before assessments of practical viability of Cu_2O for solar H_2 photogeneration.

It is worth noting that some oxides have *too low* a band gap for optimal solar energy conversion. Palladium oxide in Table 8 exemplifies this trend as does PbO_2.[353] On the other hand, PbO has an E_g value around 2.8 eV.[353] Other oxides such as CoO and Cr_2O_3 (both p-type semiconductors) have been very briefly examined early on in the evolution of this field.[353]

In closing this Section, comparative studies on binary oxide semiconductors are available[62,65,353,383] including one study[383] where the electron affinities of several metal oxides (used as anodes in photoelectrolysis cells) were calculated from the atomic electronegativity values of the constituent elements. These electron affinity estimates were correlated with the V_{fb} values measured for the same oxides in aqueous media.[383]

6.6 Perovskite Titanates and Related Oxides

Perovskites have the general formula, ABX_3, with $SrTiO_3$ being a prototype. They contain a framework structure containing corner-sharing TiO_6 octahedra with the A cation in twelve-coordinate interstices.[384,385] Several hundred oxides have this struc

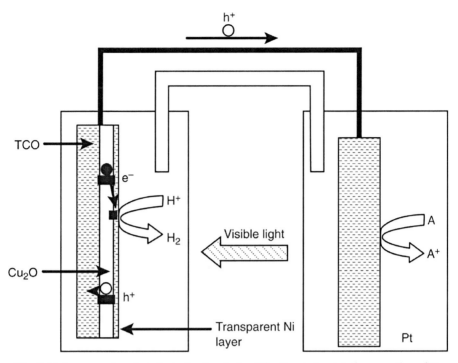

Fig. 6. Twin-compartment photoelectrochemical cell for the photocatalytic generation of H_2 from water using electrodeposited p-Cu_2O (from Ref. 382). TCO is the transparent conducting oxide substrate and A is an electron donor in the anode compartment.

ture. Table 9 lists the studies that have appeared on $SrTiO_3$ with photoelectrolysis of water as a primary objective. As well as the cubic structure exemplified by $SrTiO_3$, a variety of distorted, non-cubic structures occurs in which the framework of TiO_6 octahedra may be twisted. Thus, $BaTiO_3$ is tetragonal at room temperature. Both $SrTiO_3$ and $BaTiO_3$ have energy band gaps around 3.2 eV. With Fe and F doping, the E_g of $BaTiO_3$ has been shrunk from 3.2 eV to ~ 2.8 eV.[413] Relative to $SrTiO_3$, studies on $BaTiO_3$ from a photoelectrolysis perspective are much more sparse.[413–415]

Titanates with tunnel structures have been examined for photoelectrolysis applications.[94] Thus, barium tetratitanate ($BaTi_4O_9$) has a twin-type tunnel structure in which the TiO_6 octahedra are not oriented parallel to one another creating a pentagonal prism space. Alkaline metal hexatitanates ($M_2Ti_6O_{13}$; M = Na, K, Rb) are Wadsley-Andersson type structures in which TiO_6 octahedra share an edge at one level in linear groups of three, giving a tunnel structure with rectangular space. The reader should consult Refs. 91 and 94 for reviews of water photolysis studies using these types of oxides. These materials have been used in powder form in suspensions usually modified with a co-catalyst such as RuO_2.[416–425]

More complex perovskites exist containing two different cations which may occupy either the A or B sites and many of these also have a layered structure. Two

Table 9. Studies on the use of $SrTiO_3$ anodes or powders for the photoelectrolysis of water.

Entry number	Title of paper	Comments	Reference
1	Photoelectrochemical Reactions at $SrTiO_3$ Single Crystal Electrodes.	Cell found to work efficiently even without a pH gradient in the anode and cathode compartments.	386
2	Strontium Titanate Photoelectrodes. Efficient Photoassisted Electrolysis of Water at Zero Applied Potential.	As above but the water photosplitting driven by light only with no external bias. Photoanode stability also confirmed as in the evolution of H_2 and O_2 in the correct 2:1 stoichiometric ratio.	287
3	Photoelectrolysis of Water in Cells with $SrTiO_3$ Anodes.	Maximum quantum efficiency at zero bias (10% at $h\nu = 3.8$ eV) found to be ~an order of magnitude higher than TiO_2.	388
4	Photoeffects on Semiconductor Ceramic Electrodes.	Photoresponse of $SrTiO_3$ found to be better than that of $BaTiO_3$. Unlike the use of single crystals in the above studies (Entries 1–3), polycrystalline electrodes with large area were used.	389
5	Surface Photovoltage Experiments on $SrTiO_3$ Electrodes.	The role of surface states in mediating charge transfer between electrode and electrolyte elucidated.	390
6	Photocatalytic and Photoelectrochemical Hydrogen Production on Strontium Titanate Single Crystals.	Both metal-free and platinized samples studied in aqueous alkaline electrolytes or in the presence of NaOH-coated crystals.	391
7	Photocatalytic Decomposition of Water Vapour on an $NiO-SrTiO_3$ Catalyst.	A series of studies begun with this particular study which uses powdered photocatalyst. See Entries below.	392
8	Visible Light Induced Photocurrents in $SrTiO_3-LaCrO_3$ Single-Crystalline Electrodes.	Co-doping of La and Cr shifts photoresponse down to 560 nm and strong absorption in the visible range ascribed to $Cr^{3+} \rightarrow Ti^{4+}$ charge transfer.	393
9	The Sensitization of $SrTiO_3$ Photoanodes for Visible Light Irradiation.	As in Entry 8 but using the perovskites $LaVO_3$, Sr_2CrNbO_6 and $SrNiNb_2O_9$ as dopants.	395
10	The Colouration of Titanates by Transition Metal Ions in View of Solar Energy Applications.	——	395
11	Evidence of Photodissociation of Water Vapor on Reduced $SrTiO_3(III)$ Surfaces in a High Vacuum Environment.	First report of photodecomposition of water adsorbed from the gas phase in high vacuum conditions on metal-free, reduced single crystals.	396
12	Oxygen Evolution Improvement at a Cr-Doped $SrTiO_3$ Photoanode by a Ru-Oxide Coating.	——	397

Table 9. Continuation.

Entry number	Title of paper	Comments	Reference
13	Electrochemical Conversion and Storage of Solar Energy	A doped n-SrTiO$_3$ single crystal was combined with a proton-conducting solid electrolyte and a metal hydride allowing for storage of the evolved H$_2$.	398
14	Water Photolysis by UV Irradiation of Rhodium Loaded Strontium Titanate Catalysts. Relation Between Catalytic Activity and Nature of the Deposit from Combined Photolysis and ESCA Studies.	Powdered catalysts used and the water photolysis efficiency is found to have a strong pH dependence.	399
15	Photocatalytic Decomposition of Liquid Water on a NiO-SrTiO$_3$ Catalyst.	As in Entry 7 but for liquid water. Effect of NaOH film (see Entry 6) reproduced for NiO-SrTiO$_3$ powder.	400
16	Study of the Photocatalytic Decomposition of Water Vapour over a NiO-SrTiO$_3$ Catalyst.	Mechanistic aspects probed by using a closed gas circulation system and IR spectroscopy (see Entries 7 and 15).	401
17	Photoelectrolysis of Water under Visible Light with Doped SrTiO$_3$ Electrodes.	Sintered samples used with a variety of dopants (Ru, V, Cr, Ce, Co, Rh).	402
18	Mediation by Surface States of the Electroreduction of Photogenerated H$_2$O$_2$ and O$_2$ on n-SrTiO$_3$ in a Photoelectrochemical Cell.	Back reactions probed and the role of surface states elucidated.	403
19	Photocatalytic Decomposition of Water into H$_2$ and O$_2$ over NiO-SrTiO$_3$ Powder. 1. Structure of the Catalyst.	Nickel metal also found at the interface of NiO and SrTiO$_3$. See also Entries 7, 15 and 16.	404
20	Mechanism of Photocatalytic Decomposition of Water into H$_2$ and O$_2$ over NiO-SrTiO$_3$.	HER found to occur on the NiO cocatalyst surface while OER takes place on SrTiO$_3$. See also Entries 7, 15, 16 and 19.	405
21	Water Photolysis over Metallized SrTiO$_3$ Catalysts.	Promoting effect of NaOH not so pronounced as for TiO$_2$.	406
22	Luminescence Spectra from n-TiO$_2$ and n-SrTiO$_3$ Semiconductor Electrodes and Those Doped with Transition-Metal Oxides As Related with Intermediates of the Photooxidation Reaction of Water.	Mechanistic aspects clarified using photo- and electroluminescence measurements.	407
23	Photoinduced Surface Reactions on TiO$_2$ and SrTiO$_3$ Films: Photocatalytic Oxidation and Photoinduced Hydrophilicity.	—	283
24	Stoichiometric Water Splitting into H$_2$ and O$_2$ using a Mixture of Two Different Photocatalysts and an IO$_3$/I$^-$ Shuttle Redox Mediator under Visible Light Irradiation.	A Z-scheme used using a mixture of Pt-WO$_3$ and Pt-SrTiO$_3$ photocatalysts. The latter was co-doped with Cr and Ta.	408

Table 9. Continuation.

Entry number	Title of paper	Comments	Reference
25	Visible-Light-Response and Photocatalytic Activities of TiO_2 and $SrTiO_3$ Photocatalysts Co-doped with Antimony and Chromium.	The band gap of $SrTiO_3$ shrunk to 2.4 eV by co-doping.	409
26	A New Photocatalytic Water Splitting System under Visible Light Irradiation Mimicking a Z-Scheme Mechanism in Photosyn-thesis.	See Entry 23 above.	349
27	Construction of Z-Scheme Type Heterogeneous Photocatalysis Systems for Water Splitting into H_2 and O_2 under Visible Light Irradiation.	A Pt-$SrTiO_3$ doped with Rh is combined with a $BiVO_4$ photocatalyst.	410
28	Electrochemical Approach to Evaluate the Mechanism of Photocatalytic Water Splitting on Oxide Photocatalysts.	Cr or Sb co-doped $SrTiO_3$ samples studied amongst others (cf. Table 7, entry 13)	333
29	H_2 Evolution from a Aqueous Methanol Solution on $SrTiO_3$ Photocatalysts Co-doped with Chromium and Tantalum Ions under Visible Light Irradiation	—	411
30	Photocatalytic Activities of Noble Metal Ion Doped $SrTiO_3$ under Visible Light Irradiation	Mn-, Ru-, Rh- and Ir-doped powder samples studied.	412
31	Nickel and Either Tantalum or Niobium-Co-doped TiO_2 and $SrTiO_3$ Photocatalysts with Visible-Light Response for H_2 or O_2 Evolution from Aqueous Solutions	Co-doping found to afford higher activity for HER compared with Ni alone.	334

main classes of such oxides showing interlamellar activity have been explored for water photolysis:

1. the Dion-Jacobson series of the general formula, $AM_{n-1}B_nO_{3n+1}$ (e.g., $KCa_2Ti_3O_{10}$), and
2. the Ruddlesden-Popper series of general formula, $A_2M_{n-1}B_nO_{3n+1}$ (e.g., $K_2La_2Ti_3O_{10}$).[95]

Corresponding niobates also exist as discussed below. Noble metal co-catalysts (e.g., Pt) are loaded onto these photocatalysts by photocatalytic deposition from H_2PtCl_6 (see above). Since the oxide sheets have a net negative charge (that is balanced by the alkali cations), the $PtCl_6^{2-}$ anions are not intercalated in the host lattice.[95] Instead, the Pt sites are formed on the external surfaces of the layered perovskite powder.

Table 10. Other ternary oxides with the general formula, ABO_3,[a] that have been examined from a water photoelectrolysis perspective.

Entry number	Oxide	Energy band gap eV	Comments	References(s)
1	$FeTiO_3$[b]	2.16	Unstable with leaching of iron observed during photoelectrolysis.	427
2	$YFeO_3$	2.58	N-type semiconductor with an indirect optical transition.	428
3	$LuRhO_3$	~ 2.2	Distorted perovskite structure with p-type semiconductor behavior.	429
4	$BaSnO_3$	~ 3.0	Estimated to be stable toward photoanodic decomposition over a 0.4–14 pH range.	65
5	$CaTiO_3$	~ 3.6	—	65
6	$KNbO_3$	~ 3.1	See next Section.	65
7	$Ba_{0.8}Ca_{0.2}TiO_3$	~ 3.25	—	65
8	$KTaO_3$	~ 3.5	Optical to chemical conversion efficiency of ~ 6% reported. See next Section.	430
9	$CdSnO_3$	1.77	Band-edges not suitably aligned for HER or OER.	431
10	$LaRhO_3$	1.35	See above.	431
11	$NiTiO_3$[c]	~ 1.6	N-type semiconductor crystallizing in the illmenite structure.	432–434
12	$LaMnO_3$	~ 1.1	A p-type semiconductor.	435,436

[a]Not all the oxides in this compilation have the perovskite structure.
[b]Other iron titanates: Fe_2TiO_4 ($E_g = 2.12$ eV) and Fe_2TiO_5 ($E_g = 2.18$ eV) also examined.
[c]Band gap estimated for the transition from the mid-gap Ni^{2+} (3 d^8) level to the CB. Compound can be regarded as NiO *doped* TiO_2.

In many of these cases with layered oxides, the H^+-exchanged photocatalysts show higher activity toward the HER—a trend rationalized by the easy accessibility of the interlayer space to electron donor species such as methanol.[95] Other aspects such as Ni-loading and pillaring of the interlayer spaces have been discussed.[95] Another type of layered perovskites have been studied with the generic composition, $A_nB_nO_{3n+2}$ (n = 4, 5; A = Ca, Sr, La; B = Nb, Ti).[426] Unlike the (100)-oriented structures discussed above, the perovskite slabs in these oxides are oriented parallel to the (110) direction. Thus compounds such as $La_2Ti_2O_7$ and $La_4CaTi_5O_{17}$ were examined in terms of their efficacy toward water splitting under UV irradiation.[426]

In closing this Section, a variety of other ternary oxides (besides the $SrTiO_3$ prototype) have been examined over the years. Table 10 contains a representative listing of these compounds.

6.7 Tantalates and Niobates

We have seen in the preceding Section that oxides with MO_6 octahedra can form perovskite structures and this trend applies to some tantatates and niobates as well.

The perovskite, $KTaO_3$, as well as its Nb-incorporated cousin, $KTa_{0.77}Nb_{0.23}O_3$ were studied early on (1976) in the history of photoelectrolysis of water (see Entry 8 in Table 10).[370,420] Niobium oxides were also considered in early studies aimed at shrinking the large band gaps to values responsive to the visible range of the solar spectrum.[437] Thus in compounds of the type ANb_2O_6 (with A = Ni or Co) we have a conduction band built from d levels of a highly-charged, closed-shell transition metal ion (Nb^{5+}) while the highest filled valence band is also cation-derived from the d levels of either Ni^{2+} or Co^{2+}.[437] Thus the main optical transition should be of Ni^{2+} (or Co^{2+}) $\rightarrow Nb^{5+}$ charge-transfer type in the visible region. The ANb_2O_6 oxide has the columbite structure with a $Fe^{2+} \rightarrow Nb^{5+}$ transition featured by a 2.08 eV gap.[438] Families of Bi_2MNbO_7 (M = Al, Ga, In), $A_2B_2O_7$, $InMO_4$ (M = Nb, Ta) compounds all contain the same octahedral TaO_6 or NbO_6 structural units.[439]

The parent oxide in these cases can be regarded generically as M_2O_5 (M = Nb or Ta). Table 11 contains a listing of the water photosplitting studies that have appeared on M_2O_5, $ATaO_3$ and more complex tantalates and niobates. Layered perovskite type niobates have the generic formula $A[B_{n-1}Nb_nO_{3n+1}]$ with A = K, Rb, Cs and B = Ca, Sr, Na, Pb, etc. For example, with values of n = 2 and 3, we can derive the structures $A_2M_2O_7$ and $AB_2Nb_3O_{10}$ in Table 11 respectively (Entries 3 and 9). Another series of perovskites has the generic formula: $A_nM_nO_{3n+2}$ with A = Ca, Sr, La and M being either Nb, Ta or Ti. Of course, the simplest compound in this series has the AMO_3 composition as exemplified by $SrTiO_3$ or $KTaO_3$ (see above).

The layered oxides featured in this Section and the preceding one have ion-exchange characteristics imparted by the net negative charge residing on the layered sheets. Thus they can assimilate positively charged ions (such as K^+) in the interlamellar spaces. Interestingly, some of these materials (e.g., $K_4Nb_6O_{17}$) have *two* types of interlayer spaces (I and II) which appear alternately.[69] The space "I" is easily hydrated even in air while "II" is hydrated only in a highly humid environment. It is presumed that the NiO co-catalyst exists only in "I" such that the HER occurs mainly in this interlayer space. On the other hand, the OER is thought to occur in the interlayer space, II.[69]

In general, oxides containing early transition metal cations with d^0 electronic configuration such as Ti^{4+}, Nb^{5+} or Ta^{5+} have wide band gaps (> 3.0 eV). In fact Ta_2O_5 has a very high E_g value of ~ 4.0 eV. Thus, these materials do not perform well under visible light irradiation, and in practical scenarios, would only absorb a small fraction of the solar spectrum. As with TiO_2 and the vast majority of the oxides considered earlier, the ternary (and multinary) oxides, namely the titanates, tantalates and niobates suffer from this same handicap. On the other hand, the materials with smaller E_g values listed in Table 10 have other problems related to stability, interfacial energetics, poor charge transfer characteristics, etc.

6.8 Miscellaneous Multinary Oxides

In this *catch-all* Section, we mainly discuss the spinel structures with the generic formulas, AB_2O_4 and A_2MO_4. The unit cell of the spinel structure is a large cube, eight times ($2 \times 2 \times 2$) the size of a typical face-centered cube.[385] We also discuss the delafossite-type structure ABO_2 in which the A cation is in linear coordination

Table 11. Studies on tantalate and niobate photocatalysts for the splitting of water.[a]

Entry number	Compound formula	A cation(s)	B cation(s)	Comments	References
1	Ta_2O_5	–	–	Both crystallized and mesoporous samples studied and in one case, (Ref. 441), NiO co-catalyst was used.	440, 441
2	$ATaO_3$	Li, Na, K	–	Excess alkali cation enhances catalytic activity. Co-catalysts not found to be essential although NiO was also used in addition in some studies.	442–446
3	$A_2M_2O_7$ [b]	Sr	–	Have layered perovskite structure. Samples with both Ta and Nb also studied. Strontium niobate compound is ferroelectric at room temperature. In contrast, the tantalum analog is paramagnetic.	426, 447–449
4	ANb_2O_6	Ni, Co, Zn	–	See text.	438, 450
5	ATa_2O_6	Mg, Ba, Sr	–	Orthorhombic structure used with NiO co-catalyst to enhance photocatalyst activity.	442, 451
6	A_2BNbO_6 [c]	Sr	Fe	–	452
7	$A_3BNb_2O_9$	Sr	Fe	–	452
8	$A_4Nb_6O_{17}$	K, Rb	–	Perhaps the most studied of the ni-obates. NiO co-catalyst used in some cases as was aqueous methanol solution. Composites with CdS also studied.	453–457
9	$AB_2Nb_3O_{10}$	K, Rb, Cs	Ca, Sr, Pb	Layered perovskite structure.	458–460
10	$A_2B_2Ti_{3-x}Nb_xO_{10}$	K, Rb, Cs	La	Partial substitution of Ti with Nb leads to a decrease in the negative charge density of the perovskite sheets.	461
11	$A_3Ta_3Si_2O_{13}$	K	–	Pillared structure with TaO_6 pillars linked by Si_2O_7 ditetrahedral units.	462
12	$A_2BTa_5O_{15}$	K	Ln	Used with NiO co-catalyst. The Pr and Sm compounds show high activity.	463
13	$ATaO_4$	In	–	Crystallizes in the monoclinic wolfra-mite-type structure, like the $FeNbO_4$ compound (see text).	464
14	$A_2Nb_4O_{11}$	Cs	–	Structure consists of NbO_6 and NbO_4 octahedra.	465

[a]Also see Refs. 64, 69, 71, 74, 75 and 77–79.
[b]Belongs to the series A_nMnO_{3n+2} with A = Ca, Sr, La and M = Nb or Ti. The $Sr_2Nb_2O_7$ structure (Entry 3), for example, is the reduced formula of $Sr_4Nb_4O_{14}$ with n = 4 above.
[c]The $Sr_{1.9}Fe_{1.1}NbO_6$ compound was also studied here.

and the B cation is in octahedral coordination with oxygen. One way to visualize this structure is parallel arrangement of sheets with edge-shared BO_6 octahedral with the A cations occupying the interlayer regions of space. Finally, complex oxides containing V and W are also considered. Table 12 contains a listing of these oxides.

Table 12. Miscellaneous multinary oxides for the photodecomposition of water.

Entry number	Oxide semicon- ductor(s)	Energy band gap(s)[a], eV	Comments	Refer- ence
1	Cd_2SnO_4, $CdIn_2O_4$ and Cd_2GeO_4	2.12 (indirect), 2.23 (forbidden) and 3.15 (indirect)	Found to be unsuitable as electrodes in photoelectrolysis cells.	466
2	$ZnFe_2O_4$	?	HER observed by visible light irradiation of H_2S solution.	467
3	$BiVO_4$	2.3	Ag^+ used as electron scavenger and photocatalytic OER observed.	468
4	$Bi_2W_2O_9$, Bi_2WO_6 and Bi_3TiNbO_9	3.0, 2.8 and 3.1	Structure consists of perovskite slabs interleaved with Bi_2O_2 layers.	469
5	$AgVO_3$, $Ag_4V_2O_7$ and Ag_3VO_4	?	Only Ag_3VO_4 evolves O_2 in aqueous $AgNO_3$ solution (with Ag^+ as electron acceptor) under visible light irradiation.	470
6	$ACrO_4$ (A=Sr or Ba)	2.44 and 2.63	The Sr compound shows much lower activity than the Ba counterpart for HER in aqueous methanol.	471
7	$CuMnO_2$	1.23	Photocatalytic HER observed in H_2S medium.	472
8	$PbWO_4$	?	Has tetragonal structure. Used with RuO_2 co-catalyst for water photosplitting with a Hg-Xe lamp as radiation source.	473
9	$CuFeO_2$	?	Photocatalytic water splitting observed under visible light irradiation.	71 (cf.Refs. 474,475)

[a]Values for E_g are listed in the order of appearance of the corresponding oxide compound in column 2.

It is interesting to note that some of the newer studies (e.g., Entry 4, Table 12) are rooted in early investigations dating back to 1981. Thus, Bi_2WO_6 (as well as $Bi_4Ti_3O_{12}$) were examined[438] within the context of shrinking E_g values of oxide semiconductors. Both these compounds have Bi_2O_2 layers, the former with WO_4 layers (comprised of corner-shared WO_6 octahedra) and the latter with double perovskite layers of composition $Bi_2Ti_3O_{10}$. These structures are distorted from pure tetragonal symmetry.

7 Nitrides, Oxynitrides and Oxysulfides

We have seen that introduction of nitrogen into the TiO_2 lattice has a favorable effect in terms of sensitizing it to the visible range of the electromagnetic spectrum (Table 6). The line between doping and new phase formation is one of degree and the studies on nitridation of a given parent oxide exemplify this point. Thus the band gap of Ta_2O_5 shrinks from ~ 4.0 eV to ~ 2.1 eV by nitriding it in a NH_3 atmosphere to yield

Ta_3N_5.[476] This material evolves H_2 and O_2 under visible irradiation ($\lambda < 600$ nm) in the presence of sacrificial electron acceptors such as Ag^+ and a co-catalyst for HER such as Pt.[476] Control of solution pH was found to be critical for suppressing the photoanodic corrosion of the photocatalyst which is signaled by N_2 evolution. The shrinking of the optical band gap was attributed to a conduction band derived from Ta 5d orbitals and a higher-lying valence band derived from N 2p orbitals than the counterpart built from O 2p orbitals.[476]

Partial nitridation affords TaON which is also found to be active for water oxidation or reduction under visible light irradiation (420 nm $\leq \lambda \leq$ 500 nm).[477–480] These studies employed aqueous methanol, a Ru co-catalyst (for HER) or a Ag^+ electron acceptor for OER.[477,478] The band gap of TaON was estimated to be 2.5 eV.

Two other oxynitrides, namely $LaTiO_2N$ and $Y_2Ta_2O_5N_2$,[483] have been reported to be effective for evolving H_2 or O_2 from H_2O under visible light irradiation. The former is derived from $La_2Ti_2O_7$ (see Entry 3, Table 11 for the Nb or Ta analog) by nitriding in NH_3 atmosphere.[481,482] The two structures $La_2Ti_2O_7$ and $LaTiO_2N$ have been compared;[482] the oxynitride has the same structure as a perovskite of the ABO_3 genre and is composed of a TiO_xN_y octahedral skeletal structure ($x + y = 6$). The band gap of $LaTiO_2N$ is estimated to be ~ 2.1 eV. $Y_2Ta_2O_5N_2$ is obtained by nitriding $YTaO_4$ power, which in turn is synthesized via a solid-state reaction between Y_2O_3 and Ta_2O_5.[483] While the $YTaO_4$ has a band gap of 3.8 eV, it shrinks to ~ 2.2 eV on nitridation to $Y_2Ta_2O_5N_2$.[483] Water reduction and oxidation were observed under visible light irradiation for $Y_2Ta_2O_5N_2$ modified with a Pt-Ru co-catalyst.[483]

The final compound in this series is Ge_3N_4 which was formed by nitridation of GeO_2 powder under atmosphere NH_3 flow at elevated temperatures.[484] The band gap of β-Ge_3N_4 is estimated to be ~ 3.8–3.9 eV; the RuO_2-loaded material was found to result in overall water splitting under irradiation from a high-pressure Hg lamp.[484]

It must be borne in mind that in all the cases above where the band gap of the semiconductor was effectively shrunk to values in the 2.0–2.5 eV range, *overall water splitting* (that is, evolution of H_2 and O_2 with no *sacrificial reagents*) was not observed under visible light irradiation. This contrasts with the β-Ge_3N_4 case where, however, UV radiation had to be used.

Other than N, sulfur is another non-metallic element that has worked for sensitizing TiO_2 to visible light (Table 6). Thus oxysulfides have been examined as potential photocatalysts for water splitting. Two sets of studies on $Sm_2Ti_2S_2O_5$ and $Ln_2Ti_2S_2O_5$ (Ln = Pr, Nd, Sm, Gd, Tb, Dy, Ho and Er) have been reported from a photocatalytic water splitting perspective.[484,485] The samarium compound was prepared by heating a mixture of Sm_2S_3, Sm_2O_3 and TiO_2 in sealed tubes at elevated temperatures.[484] This compound has the same structure as the Ruddlesden-Popper type layered perovskites (see Section 6.7 this Chapter). This material, with a band gap of ~ 2 eV, evolves H_2 in the presence of Ag^+ and O_2 in the presence of Na_2S, Na_2SO_3 or methanol under visible light irradiation ($\lambda \leq 650$ nm).[485]

The other lanthanide compounds in this series were prepared either by a similar method as above or alternately by a polymerized complex method using $Ti(OiPr)_4$ and $Ln(NO_3) \cdot 6H_2O$ to yield the lanthanoid titanate precursor which was subsequently sulfided in a flowing H_2S atmosphere.[485] The $Sm_2Ti_2S_2O_5$ compound was found to have the highest activity amongst all the homologues tested. The band gaps vary from a low value of 1.94 eV (Er compound) to 2.13 eV (for the Sm compound).

8 Metal Chalcogenide Semiconductors

8.1 Cadmium Sulfide

Like TiO_2 in the case of oxide semiconductors, CdS is the prototype of compound semiconductors containing a chalcogen (S, Se or Te). Next to TiO_2, CdS is perhaps the most extensively studied from a photocatalytic water splitting perspective, particularly in powdered form in aqueous dispersions. Table 13 contains a chronological listing of representative studies that have appeared on CdS dating back to ~ 1960— well before when the first studies on TiO_2 had begun to appear. The energy band gap of CdS is 2.42 eV and being lower than that of TiO_2, this material can harness a more sizeable fraction of the solar spectrum. On the other hand, CdS is prone to photocorrosion (see Refs. 89, 151–153, 155, 179, 180, 509) unlike TiO_2. Given this, the claims in Ref. 493 (see Entry 8 in Table 13) are rather surprising although the kinetics of hole transfer to RuO_2 in ultrafine CdS particles may be fast enough so that the photogenerated holes are intercepted (by electron donors or co-catalysts) before they attack the CdS lattice itself. Thus CdS has been more extensively used to *photocatalytically* evolve H_2 from water in the presence of an electron donor such as cysteine, EDTA, sulfide or sulfite species (see Table 13). In particular, H_2 generation from an aqueous medium containing H_2S has particular appeal because H_2S is produced in large quantities as an undesirable by-product in coal- and petroleum-related industries.[496] Similarly, sulfite is a pollutant and thus photooxidation of sulfite (to sulfate) and co-generation of H_2 using CdS has a value-added benefit. Therefore, it is not surprising that this approach has been vigorously pursued since 1985 in several studies.[454,511–519]

Other aspects of studies on CdS since 1985, from a water photosplitting perspective, include visible light cleavage of water on CdS photoanodes coated with a composite polymer containing either RuO_2 and $Ru(bpy)_3^{2+}$ [520,521] or a conducting polypyrrole-polybithiophene co-polymer containing dispersed RuO_2,[155] water splitting using polypyrrole-coated CdS photoanodes,[89,151–153] visible light photoassisted HER from methanol-water solutions containing mixtures of composite $CdS-SiO_2$ particles and a platinized wide band gap semiconductor such as TiO_2, ZnO, SnO_2 or WO_3,[522] HER caused by intraparticle electron transfer from CdS to $K_4Nb_6O_{17}$,[454] and the effect of deposited cobalt phthalocyanine on the photocatalytic HER from mixed suspensions of CdS and Cu_xS.[523] Studies[524] on solid solutions of CdS and ZnS[525] will be discussed in a later Section. We finally close this Section noting interesting effects of heat treatment with KCl on the efficacy of CdS powders to evolve H_2 from water.[515]

Table 13. Representative studies on CdS up to ~1985 on applications broadly related to the photosplitting of water.

Entry number	Title of article	Comments	Reference
1	Becquerel Photovoltaic Effect in Binary Compounds.	Appears to be the first study on the mechanism of the photovoltaic effect on a CdS/electrolyte interface.	486
2	Photosensitized Electrolytic Oxidation of Iodide Ions on Cadmium Sulfide Single Crystal Electrode.	The stability of CdS probe under irradiation in an electrolyte containing iodide species. Rotating ring disk voltammetry was used as the methodology.	487
3	Suppression of Surface Dissolution of CdS Photoanode by Reducing Agents.	As above but a variety of electron donors used. The use of sulfite for electrode stabilization demonstrated.	488
4	Photochemical Diodes.	A CdS single crystal platelet used in conjunction with Pt to drive the HER from sulfide solutions.	489
5	Superoxide Generation in the Photolysis of Aqueous Cadmium Sulfide Dispersions. Detection by Spin Trapping.	First of many studies examining the behavior of powder dispersions. Electron transfer by irradiation of CdS shown to occur to methyl viologen. The corresponding radicals observed only when an electron donor (EDTA) is simultaneously present.	490
6	Effect of Platinization of the Photoproperties of CdS Pigments in Dispersion. Determination of H_2 Evolution, O_2 Uptake, and Electron Spin Resonance Spectroscopy	As above but platinized and virgin samples compared. Photoinduced HER also reported.	491
7	Photochemical Hydrogen Production using Cadmium Sulfide Suspensions in Aerated Water	Cysteine or EDTA used as electron donor and platinized samples used as in Entry 5. The H_2 yield only marginally reduced in the presence of O_2.	492
8	Visible Light Induced Water Cleavage in CdS Dispersions Loaded with Pt and RuO_2, Hole Scavenging by RuO_2.	Dual-function Pt and RuO_2 loaded onto CdS for catalyzing HER and OER respectively. Stoichiometric evolution of gases noted with no degradation of CdS after 60 h of irradiation.	493
9	H_2 Production Photosensitized by Aqueous Semiconductor Dispersions.	As in Entry 6 but a fuller study by the same group.	494
10	Photoelectrochemical Cells with Polycrystalline Cadmium Sulfide as Photoanodes	Light induced HER driven with reductants (sulfide, EDTA) in a photoelectro-chemical cell containing three compartments. Also see Refs. 496 and 497 by same group.	495
11	The Effect of Sputtered RuO_2 Overlayers on the Photoelectro-chemical Behavior of CdS Electrodes	Instead of RuO_2 co-catalyst islands, the use of overlayers examined to protect CdS against photocorrosion. See also Entry 29 in Table 2 and Ref. 498.	180
12	Visible Light-Induced Formation of Hydrogen and Thiosulfate from Aqueous Sulfide/Sulfite Solutions in CdS Suspensions.	Addition of sulfite shown to enhance the cleavage of H_2S into H_2 and S. See also Refs. 495–497 and Ref. 500 for follow-up work.	499

Table 13. Continuation.

Entry number	Title of article	Comments	Reference
13	Light-Induced Generation of H_2 at CdS-Monograin Membranes.	A polyurethane film is embedded with CdS particles to compartmentalize HER and OER sites. See also text (Section 2).	101
14	Photocatalytic Hydrogen Production from Solutions of Sulfite using Platinized Cadmium Sulfide Powder	See Entry 12 above.	501
15	Magnetic Field Effects on Photo-sensitized Electron Transfer Reactions in the Presence of TiO_2 and CdS Loaded Particles.	No effect of magnetic field is seen up to 4000 G.	502
16	Visible Light Induced Oxygen Evolution in Aqueous CdS Suspensions.	$PtCl_6^{2-}$ is used as scavenger for conduction band electrons. The co-catalytic activity of Rh_2O_3 found to be superior to RuO_2 in promoting OER in alkaline conditions. No OER observed without redox co-catalyst.	503
17	Visible Light Induced Hydrogen Production from In Situ Generated Colloidal Rhodium-Coated Cadmium Sulfide in Surfactant Vesicles.	The first of a series of studies (see also Refs. 505–507) exploring the use of vesicles and reverse micelles for photo-induced charge separation and HER.	504
18	Photochemical Hydrogen Production with Cadmium Sulfide Suspensions.	Platinized powders used and HER studied in various irradiated solutions (S^{2-}, SO_3^{2-}, $H_2PO_2^-$). Photoetching the CdS microcrystals shown to significantly improve the HER rate.	508
19	Photocorrosion by Oxygen Uptake in Aqueous Cadmium Sulfide Suspensions.	Photo-uptake of O_2 in aqueous CdS suspensions loaded with RuO_2 or Rh_2O_3 shown to lead to photocorrosion of CdS to $CdSO_4$.	509
20	Hydrogen Production through Microheterogeneous Photocatalysis of Hydrogen Sulfide Cleavage. The Thiosulfate Cycle.	Thiosulfate was efficiently generated in irradiated CdS dispersions containing sulfite and bisulfide (or sulfide) ions. The study focuses on CdS CB and VB processes involving electrons, holes and thiosulfate.	510

8.2 Other Metal Chalcogenides

The other compounds that have been examined are compiled in Table 14. These materials have been studied both as photoanodes (Entries 2–4) and in powder form (Entries 1, 5–10). As with their oxide counterparts, the chalcogenide family is also rich in solid-solution chemistries and Entries 6, 7, 9 and 10 in Table 14 exemplify this trend. These materials are further discussed in Section 12 of this Chapter.

Interest in d-band semiconductors derives from the fact that holes in d-bands are not equivalent to broken chemical bonds and thus do not constitute a primary pathway for photocorrosion as in the case of semiconductors (e.g., CdS) with p-orbital derived valence bands.[33,541,542] Thus MoS_2,[541,542] WS_2,[541,542] PtS_2,[543] RuS_2[33,532–534]

Table 14. Metal chalcogenides (other than CdS) that have been examined as photocatalysts for the decomposition of water.

Entry number	Metal chalcogenide(s)	Band gap energy eV	Comments	Reference(s)
1	n-ZnS	~ 3.5–3.8^a	Ni-, Pb-, halogen- and Cu-doped samples also studied.	526–534
2	n-CdSe	1.75	Used in bipolar configuration with CoS. Polymer-coated samples also examined.	155, 205, 206
3	n-RuS$_2$	1.2–1.85	–	33, 532–534
4	n-MSe$_2$ (M = Mo, W)	~1.0	Used in conjunction with p-InP photocathodes in p-n photoelectrolysis cells.	204
5	NaInS$_2$	2.3	Material consists of InS$_2^-$ layers. Evolves H$_2$ from a sulfite medium under visible light irradiation.	535
6	AgInZn$_7$S$_9$	2.3	Solid solution of AgInS$_2$ and ZnS. See Section 12 of this Chapter.	536
7	(AgIn)$_x$Zn$_{2(1-x)}$S$_2$	2.3 (x = 0.22)	As in Entry 6. See Section 12 of this Chapter.	537
8	Cu(In,Ga)(Se,S)$_2$	1.6–2.0	Made from electrodeposited Cu(In,Ca)Se$_2$ precursors. Thin films characterized but HER efficacy not evaluated yet.	538
9	ZnS-CuInS$_2$-AgInS$_2$	2.4	As in Entry 6 but solid solution also contains the copper compound. See Section 12 of this Chapter.	539
10	(CuIn)$_x$Zn$_{2(1-x)}$S$_2$	2.3 (x = 0.09)	Solid solution between CuInS$_2$ and ZnS. See Section 12 of this Chapter.	540

^{a}Depends on the crystallographic phase; the α phase has an E_g value of 3.8 eV and the β phase has an E_g value of 3.6 eV.

and FeS$_2$ [544,545] are not photooxidized to molecular sulfur under irradiation in contact with water. However, in the case of MoS$_2$, WS$_2$ and FeS$_2$, SO$_4^{2-}$does ultimately form (see also Entry 19, Table 13) presumably by attack of the primary oxidation products of water on the crystal-bound sulfur. On the other hand, RuS$_2$ appears to be stable and evolves O$_2$ (or Cl$_2$) from aqueous electrolytes under visible or near-IR radiation.[533] WS$_2$, MoS$_2$ and PtS$_2$ have a layer-type structure while FeS$_2$ and RuS$_2$ belong to the pyrite family.[33]

9 Group III-V Compound Semiconductors

We have already seen (see Table 4) that p-InP photocathodes are capable of evolving H$_2$ from HCl or HClO$_4$ electrolytes with very high efficiency.[66,199,201] Photocathodes made from p-GaInP₂ (a solid solution of GaP and InP, see Section 12 of this Chapter) biased with a GaAs p-n junction have also evolved H_2 with high efficiency.[126] As with their chalcogenide semiconductor counterparts, Group III-V semiconductors, in n-type form, undergo photoanodic corrosion instead of evolving O_2 under illumination in aqueous media. On the other hand, these materials are relatively more stable against *cathodic* photocorrosion and the photogenerated minority carriers (electrons) on the p-type semiconductor surface can be used to reduce water to H_2, particularly in the presence of a co-catalyst such as Pt or Ru.[6] The reader is referred to chapters in Ref. 6 that provide reviews of work up to ~ 1988 on the HER on irradiated GaS and InP surfaces. Table 15 provides a chronological listing of selected studies up to ~ 1985 on Group III-V compound semiconductors that have been examined from a water photosplitting and H_2 generation perspective. The E_g values for GaAs, GaP and InP are 1.43 eV, 2.25 eV and 1.30 eV respectively.

As with the n-TiO₂ and n-SrTiO₃ counterparts discussed earlier in Section 6.2 of this Chapter (see also Ref. 407), luminescence probes have proven to be very useful for unraveling the mechanistic details of the cathodic processes both at n-type (e.g., n-GaAs)[556] and p-type (e.g., p-InP)[557,558] Group III-V semiconductor surfaces. Finally, these semiconductors share another trend with those discussed earlier (metal chalcogenides) in that the majority of the studies since ~ 1990 have been directed at solid solutions (alloys of GaP and InP, GaAs and InAs etc.). These newer studies will be addressed in Section 12 of this Chapter.

In summary, Group III-V semiconductors have several positive features that make them attractive for water photosplitting applications. The combination of high carrier mobility and an optimal band gap (particularly for many of the alloys, see below) coupled with reasonable photoelectrochemical stability for the p-type material under HER conditions, should inspire continuing scrutiny of Group III-V semiconductors. The control of surface chemistry is also particularly crucial to avoid problems with surface recombination. For example, the studies on p-InP photocathode surfaces have shown that a (*controlled*) ultra-thin interfacial oxide layer is critical for minimizing carrier recombination at the surface.[66,199,201,554]

10 Germanium and Silicon

Given the very high level of technological infrastructure that already exists for these elemental semiconductors because of microelectronics applications, it is not surprising that both these materials were examined early on in the evolution of the field of photoelectrolysis of water. As mentioned in an introductory paragraph, cathodic reduction of the Ge surface is accompanied by H_2 evolution.[58,559] However, we are not aware of studies under irradiation of Ge electrodes from a HER or OER perspective. The extreme instability of this semiconductor in aqueous media coupled with its low band gap (E_g = 0.66 eV) make it rather unattractive for water photosplitting applications.

Table 15. Representative studies up to ~ 1985 on Group III-V compound semiconductors from a water photosplitting perspective.[a]

Entry number	Semiconductor(s)	Comments	Reference(s)
1	GaP (n- and p-type)	Catalytic effect of electrodeposited metals studied.	143
2	p-GaP	Combined with either n-TiO$_2$ or n-SrTiO$_3$ in a twin-photosystem configuration.	546
3	p-GaS, p-GaP, p-InP	The current-potential behavior in acidic and alkaline media measured and modeled.	221, 222
4	p-GaP	Combined with a Pt anode. A maximum energy conversion efficiency of 0.1% achieved with a bias potential of 1.3 V.	547
5	InP (both n- and p-type)	Applicability of electrode to the photoelectrolysis of water explored.	548
6	GaAs (both n- and p-type)	The HER reinvestigated. Two reduction steps identified and both processes assigned to conduction band electrons.	549
7	p-GaP, p-GaAs, p-InP	Reasons for the very low efficiency for HER studied.	550
8	p-GaP	Combined with n-Fe$_2$O$_3$ in a monolithic configuration.	130
9	p-InP	Surface modification results in very high efficiencies for HER. See also Entries 1 and 3 in Table 4.	66, 199, 201
10	p-GaP	Study addresses the large gap between the photocurrent onset potential and the flatband potential.	551
11	p-GaP, p-GaAs	Large overpotential with respect to V_{fb} required for HER addressed. See also Entries 6 and 9 above.	552
12	p-InP	HER reported to be the main photocathodic process in 1-M NaOH but photocorrosion found to compete with HER in 0.5-M H$_2$SO$_4$.	553
13	p-InP	Conflicting views on effect of metal islands on the photocathode surface.	554, 555

[a] Also see Ref. 6.

In contrast, Si, particularly in p-type form, has been examined for its H$_2$ evolution efficacy under irradiation in several studies dating back to ~ 1976. Thus, although H$_2$ evolution was observed on heavily doped p-Si photocathodes in salt water, the efficiency was found to be poor.[560] Interestingly, this was attributed to the presence of a surface oxide layer (see above). Subsequent studies have all focused on catalytic modification of p-Si surfaces so that the HER rate is enhanced; a variety of metal catalysts have been examined in this regard.[143,181-184,561,562] An early study also considered n-type Si as a photoanode with a protective SnO$_2$ layer to prevent it from undergoing corrosion.[563] Ruthenium oxide layers have also been studied,[564] this time on p-Si surfaces.

One fundamental problem with Si from a water splitting perspective is its low band gap (1.1 eV). Thus while HER can be driven on a p-Si surface, a *spontaneous* water decomposition device cannot be constructed. One way around it is to seek a *photocatalysis* alternative where, instead of water, aqueous ethanol is used as an electron donor.[565] Thus the photogenerated holes are utilized to oxidize ethanol to CO_2 with simultaneous evolution of H_2. Powdered n-Si photocatalysts were used in this study along with polypyrrole and platinized Ag co-catalyst.[565] Finally, the stability of n-Si in aqueous media has been claimed to have been remarkably improved by B, Al or In doping along with surface modification with Pt or Pd islands.[566]

Tandem type hydrogenated amorphous Si (a-Si) electrodes having an [n-i-p-n-i-p] structure and a similar tandem a-Si electrode having [n-i-p-n-i-p] layers deposited on p-type crystalline Si showed cathodic photocurrents accompanied by the HER.[567] These electrodes, when connected to a RuO_2 counterelectrode (where OER occurred) caused sustained water splitting without external bias with solar-to-chemical conversion efficiencies of 1.98% and 2.93%.[567]

Studies on the use of Si for splitting solvents other than water (e.g., HBr, HI)[131,203,567] will be discussed in a subsequent Section of this Chapter.

11 Silver Halides

The analog photography community is well-versed with the fact that silver halides are tunable band gap semiconductors. Thus the photoactivity of AgCl can be tuned from the UV into the visible light region by a process known as self-sensitization, which is due to the formation of Ag clusters during the photoreaction. The formation of these clusters introduces new levels within the forbidden gap that can now be populated by visible light (Fig. 7).

The photocatalytic oxidation of water to O_2 on thin nanostructured AgCl layers has been reported.[568] In subsequent work by the same group, AgCl photoanodes have been combined with either p-$GaInP_2$[569,570] or an amorphous silicon solar cell[571] in the cathodic part of an electrochemical cell to split water. Modification of AgCl with gold colloids was found to enhance H_2 and O_2 production.[571,572]

12 Semiconductor Alloys and Mixed Oxides

12.1 Semiconductor Composites

The distinction between the two classes of materials considered in this Section pertains to the presence or absence of mixing at the molecular level. Thus in alloys, solid solutions of two or more semiconductors are formed where the lattice sites are interspersed with the alloy components. Semiconductor alloys, unlike their metallic counterparts, have a much more recent history and their development driver has been mainly optoelectronic (e.g., solid-state laser) applications. In mixed semiconductor composites, on the other hand, the semiconductor particles are in *electronic* contact but the composite components do not undergo mixing at the molecular level.

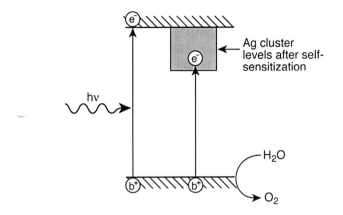

Fig. 7. Energy level scheme for the photoassisted OER from water using AgCl layers after initial self-sensitization to generate Ag clusters.

In the case of alloys, we will consider, in turn, metal oxides, metal chalcogenides and, finally, Group III-V semiconductors. We have seen earlier (Sections 6.3 and 7) how non-metallic elements such as F, N and S can alloy with the metal oxide lattice, these species occupying anion sites within the host framework. The corresponding oxyfluoride, oxynitride and oxysulfide compounds are thus generated (Section 7).

Other solid-solutions involving oxide semiconductors that have been examined include TiO_2-MnO_2,[80] ZnO-CdO,[80] TiO_2-MO_2 (M = Nb, Ta),[573] TiO_2-In_2O_3,[107] TiO_2-V_2O_5,[106] and Fe_2O_3-Nb_2O_5.[80] Tungsten-based mixed-metal oxides, $W_nO_mM_x$ (M = Ni, Co, Cu, Zn, Pt, Ru, Rh, Pd and Ag) have been prepared using electrosynthesis and high-throughput (combinatorial) screening,[574] but it is not clear how many of these compounds are true alloys (rather than mixtures). An interesting oxide alloy with lamellar structure, $In_2O_3(ZnO)_m$, has been reported[575] with photocatalytic activity for HER from an aqueous methanol and OER from an aqueous $AgNO_3$ solution. This alloy consists of layers of wurtzite-type ZnO slabs interspersed with InO_3 lamella; the band gaps of $In_2O_3(ZnO)_3$ and $In_2O_3(ZnO)_9$ are 2.6 eV and 2.7 eV, respectively.[575]

We have seen several examples of solid solutions or alloys involving metal chalcogenides (see Table 4, Entries 6–10). Other widely studied systems include CdS_xSe_{1-x} and $Cd_xZn_{1-x}S$, involving, respectively, substitution in the anion and cation sub-lattices. The latter has been especially examined from a water photosplitting perspective (see for example, Refs. 524 and 576).

Amongst the Group III-V semiconductor alloys, $Al_xGa_{1-x}As$ and $Ga_xIn_{1-x}P$ have been most widely studied. In particular, the alloy composites $Al_{0.4}Ga_{0.6}As$ and $Ga_{0.5}In_{0.5}P$ (or equivalently, $GaInP_2$) have band gap values of 1.9 eV—close to the ideal value in terms of photoelectrolysis applications (see above). Further, both these alloys are lattice-matched to GaAs, allowing for epitaxial growth on this substrate. However, the growth of high-quality, oxygen-free AlGaAs and the fabrication of a high conductance cell interconnect have plagued this alloy material.[577] It must be noted that tandem, monolithic GaInP/GaAs solar cells have yielded very high efficiencies in the 27.3%–29.5% range.[577,578]

The photoelectrochemical stability of p-GaInP$_2$ has been studied in three different electrolytes of varying pH.[579] Tandem cells consisting of a GaInP$_2$ homojunction grown epitaxially on a GaAs homojunction with a GaAs tunnel-diode interconnect, were utilized to photoelectrochemically split water.[126–128] Subsequent work by the same group modified this hybrid electrode structure with an additional top layer of p-GaInP$_2$.[128] To quantify the efficiency gains from a photoelectrolysis system, an integrated photovoltaic cell/electrolysis system was deployed by this group using a GaInP$_2$/GaAs multi-junction cell.[125] Finally, a less expensive alternative for fabricating GaInP$_2$/GaAs junctions was explored using a combined close-spaced vapor transport/liquid-phase epitaxy; arrays of mesas of GaInP$_2$/GaAs were selectively grown on Si substrates.[580]

Much less work has appeared on AlGaAs alloys. A bipolar electrode configuration of Al$_{0.15}$Ga$_{0.85}$As (E_g = 1.6 eV) and Si was used in conjunction with OER and HER co-catalysts, RuO$_2$ and Pt, respectively, to drive water photosplitting at 18.3% conversion efficiency.[209] Clearly, among all the photoelectrode materials discussed up till now, the Group III-V compounds, namely, InP and the alloyed materials, have yielded the most impressive results.

The incentive for using mixed semiconductors derives from the possibility of securing interparticle electron transfer and thus mitigate carrier recombination. For example, the conduction band of WO$_3$ lies at a lower energy (relative to the vacuum reference level) than TiO$_2$.[581,582] Thus, in a TiO$_2$-WO$_3$ composite, the photogenerated electrons in TiO$_2$ are driven to WO$_3$ before they have an opportunity to recombine with the holes in the TiO$_2$ particle. Other examples illustrative of this approach were discussed earlier in this Chapter and include CdS-TiO$_2$[583] and CdS-K$_4$Nb$_6$O$_{17}$.[454] Other examples of mixed semiconductors include TiO$_2$-LaCrO$_3$,[200] CdS-LaCrO$_3$,[436] Fe$_2$O$_3$-TiO$_2$,[584] and Cu$_2$O-TiO$_2$.[585] However, not all these composites have been examined from a water photosplitting perspective. Note that a *bilayer* configuration of the two semiconductors is not fundamentally different (at least from an electron transfer perspective) than a suspension containing mixed semiconductor particles (composites) in electronic contact.

Closely related is the so-called photochemical diode,[489] consisting of either a metal/ semiconductor Schottky barrier or a p-n junction, which generates the voltage needed on illumination, to split water. Photochemical diodes are discussed along with other twin-photosystem configurations in the next Section.

13 Photochemical Diodes and Twin-Photosystem Configurations for Water Splitting

As mentioned earlier, photochemical diodes[489] can be either of the Schottky type, involving a metal and a semiconductor, or a p-n junction type, involving two semiconductors (which can be the same, i.e., a homojunction or different, a heterojunction). Only the latter type is considered in this Section involving two irradiated semiconductor/ electrolyte interfaces. Thus n-TiO$_2$ and p-GaP crystal wafers were bonded together (through the rear Ohmic contacts) with conductive Ag epoxy cement.[489] The resultant heterotype p-n photochemical diode was suspended in an acidic aqueous medium and irradiated with simulated sunlight. Evolution of H_2 and O_2 was noted, albeit at a very slow rate.[489]

This type of device has been contrasted[489] with a series connection of a photovoltaic p-n junction solar cell and a water electrolyzer. Unlike the latter which is a majority carrier system (i.e., the n-side of the junction is the cathode and the p-side becomes the anode), in a photochemical diode, minority carriers (holes for the n-type and electrons for the p-type) are injected into the electrolyte. This distinction translates to certain advantages in terms of the overall energetics of the solar energy conversion system (see Ref. 489).

Since this original work in 1977, another study has appeared combining p-GaP and n-Fe$_2$O$_3$.[130] Co-catalysts (RuO$_2$ on the n-Fe$_2$O$_3$ surface and Pt on the p-GaP surface) several to enhance H_2 and O_2 evolution from seawater.[130] In a broad sense, arrays of p-n diodes that are embedded in glass and used for the splitting of HBr,[131] and the semiconductor/redox electrolyte/ semiconductor photoelectrode configuration,[586] may be regarded as extensions of the photochemical diode approach. These devices are discussed in the next Section.

The p-n photoelectrolysis approach,[60] on the other hand, simply combines a n-type semiconductor photoanode and a p-type semiconductor photocathode in an electrolysis cell (Fig. 2c). The pros and cons of this twin-photosystem approach (which mimicks plant photosynthesis) were enumerated earlier in this Chapter (see Section 2). Table 16 provides a compilation of the semiconductor photocathode and photoanode combinations that have been examined. Reference 67 may also be consulted in this regard for combinations involving n-WSe$_2$, n-MoSe$_2$, n-WS$_2$, n-TiO$_2$, p-InP, p-GaP and p-Si semiconductor electrodes.

14 Other Miscellaneous Approaches and Hydrogen Generation from Media Other than Water

Many of the examples of configurations discussed earlier included bipolar, monolithic structures, involving also in some cases a p-n junction biasing a top photocathode on which the HER occurred (see for example, Refs. 125-128). In another study, heavy doping with acceptor impurities introduced p$^+$ regions on a n-Si wafer and these p-n junctions in turn (modified with either Pt or Pd islands) were used to drive the photoelectrolysis of HI to H_2 and I_2.[203] An interesting hybrid configuration involves a AgCl photoanode operated in conjunction with a Pt cathode.[571] The latter, however, was biased with an amorphous Si solar cell; the AgCl photoanode and the a-Si:H solar cell were simultaneously illuminated.[571] This particular device additionally had a chemical bias from a pH gradient between the cell compartments amounting to 0.21 V. Finally, a triple junction p-i-n a-Si has been used to bias a Pt cathode in an integrated (monolithic) photovoltaic cell-electrolysis system.[125] Other studies combining amorphous Si and photoelectrochemical devices for water splitting have been described.[212,589]

In another interesting variant drawn from early work,[586] a semiconductor/redox electrolyte/semiconductor (SES) configuration was deployed as a photoanode. Thus this SES structure consisted of single crystal wafers of n-CdS and n-TiO$_2$ separated

Table 16. Photoelectrolysis cells using n-type semiconductor photoanodes and p-type semi-conductor photocathodes.

Entry number	Photoanode	Photocathode	Comments	Reference
1	n-TiO_2	p-GaP	Either 1 N H_2SO_4 or 1-N NaOH was employed as electrolyte. The photo-voltage was 0.58 V for the acid and 0.40 V for the base. Deterioration of the cell performance noted.	227
2	n-GaP	p-GaP	Photocorrosion of the anode surface noted.	60
3	n-TiO_2	p-GaP	Water splitting noted without need for external bias. See also Entry 1 on the same system.	60
4	n-TiO_2	p-CdTe	–	546
5	n-TiO_2	p-GaP	–	546
6	n-$SrTiO_3$	p-CdTe	–	546
7	n-$SrTiO_3$	p-GaP	Best performance among the four combinations tested. (See Entries 4-6 above.)	546
8	n-WSe_2	p-WSe_2	Combination used to split HI into H_2 and I_2.	587
9	n-MSe_2 (M = Mo or W)	p-InP	Combination used to split HBr or HI with high efficiency. See also Entry 6 in Table 4.	204
10	n-GaAs	p-InP	–	588

by a thin layer of NaOH, sodium sulfide and sulfur. The inside wall of TiO_2 was coated with Pd to mediate electron transfer between n-TiO_2 and the sulfide-polysulfide redox electrolyte. It was shown[586] that the SES photoanode operated in conjunction with a Pt counterelectrode and 1-M NaOH electrolyte could evolve H_2 and O_2 without external bias. The OER occurs on the n-TiO_2 surface and HER occurs on the Pt counterelectrode surface.

Technologically, perhaps the most highly developed hybrid system involves the use of coupled p-n Si junctions to drive the photoelectrolysis of HBr.[131,590] In this approach, small (0.2 mm diameter) Si spheres (a waste product from the microelectronics industry) are embedded in glass and backed by a conductive matrix to form arrays in contact with the HBr solution. Thus n-Si on p-Si surfaces form photocathodes and the reverse p-Si on n-Si surfaces become photoanodes. The two inverted junctions are electronically coupled and the Si microsphere surfaces are also coated with an ultrathin noble metal layer (acting both as a co-catalyst and a corrosion-protective coating). Each of the junctions produce a photovoltage of ~ 0.55 V so that the total open-circuit voltage of a coupled junction pair is ~ 1.1 V—sufficient to decompose HBr to H_2 and Br_2. The combination of some of the handicaps with this system (e.g., use of a highly corrosive HBr electrolyte) and an unfavorable energy climate, conspired to effectively douse commercial interest in this system.

We have seen several examples above for H_2 generation from solvents other than water, namely H_2S, HBr and HI. In the case of H_2S, the main incentive derives from the fact (as noted earlier) that it is a waste byproduct. On the other hand, the oxidation half-reactions in HBr and HI are kinetically less challenging than the $4e^-$ oxidation of water to O_2. This fact coupled with the lower decomposition voltages of HBr and HI relative to water, have instigated the evaluation of these alternative solvents to H_2O for H_2 generation.

15 Concluding Remarks

The use of irradiated semiconductor-liquid interfaces for hydrogen generation is now a mature field of research. Indeed, impressive results have been obtained at the laboratory scale over the past three decades and a myriad of new semiconductor materials have been discovered. On the other hand, much needs to be done to improve the H_2 generation efficiencies. The photoelectrolysis process must be engineered and scaled up for routine practical use. In this regard, oxide and chalcogenide semiconductors appear to be particularly promising, especially from a process economics perspective. While interesting chemistry, physics, and materials science discoveries will continue to push this field forward, in the author's crystal ball, two types of R&D will be crucial: the use of combinatorial, high throughput methods for photocatalyst development[368,574] and innovations in reactor/process engineering once efficiencies at the laboratory scale have been optimized at a *routinely-attainable* ~ 10% benchmark. Only then will the long sought after goal of efficiently making H_2 from sunlight and water using inexpensive and stable semiconductors be realized.

Acknowledgments

Partial support from the U. S. Department of Energy (Basic Energy Sciences) for the author's research on photoelectrochemical solar energy conversion is gratefully acknowledged. Professors B. A. Parkinson, J. Augustynski, C. A. Grimes and M. Matsumura kindly provided reprints/ preprints. Dr. C. R. Chenthamarakshan and Ms. Gloria Madden are thanked for assistance in literature research and manuscript preparation.

References

1. V. A. Myamlin and Yu. V. Pleskov, *Electrochemistry of Semiconductors*, (English Edition), Plenum Press, NY, 1967.
2. S. R. Morrison, *The Chemical Physics of Surfaces*, Plenum Press, NY and London, 1977.
3. S. J. Fonash, *Solar Cell Device Physics*, Academic Press, San Diego, 1981.
4. M. Grätzel, editor, *Energy Resources through Photochemistry and Catalysis*, Academic Press, NY, 1983.
5. Yu. V. Pleskov and Yu. Ya. Gurevich, *Semiconductor Photoelectrochemistry*, Consultants Bureau, NY and London, 1986.
6. H. O. Finklea, editor, *Semiconductor Electrodes*, Elsevier, Amsterdam, 1988.
7. R. Memming, *Semiconductor Electrochemistry*, Wiley-VCH, Weinheim, 2001.
8. S. Licht, editor, *Semiconductor Electrodes and Photoelectrochemistry – Encyclopedia of Electrochemistry*, Vol. 6, Wiley-VCH, Weinheim, 2001.
9. V. Balzani, editor, *Electron Transfer in Chemistry*, Vol. 4, Wiley-VCH, Weinheim, 2001.

10. M. Green in *Modern Aspects of Electrochemistry* (J. O' M. Bockris, editor), Vol. 2, p. 343, Butterworths, London, 1959.
11. H. Gerischer in *Advances in Electrochemistry and Electrochemical Engineering* (P. Delahay, editor), Vol. 1, p. 139, Interscience, 1961.
12. H. Gerischer in *Physical Chemistry* (H. Eyring, D. Henderson and W. Jost, editors), Vol. IXA, Academic Press, NY, 1970.
13. A. Hamnett in *Comprehensive Chemical Kinetics*,; Vol. 27, Electrode Kinetics: Reactions, (R. G. Compton, editor), Ch. 2, p. 61, Elsevier, Amsterdam, 1987.
14. L. M. Peter and D. Vanmaelkelbergh in *Advances in Electrochemical Science and Engineering* (R. C. Alkire and D. M. Kolb, editors), Vol. 6, p. 77, Wiley-VCH, Weinheim, 1999.
15. L. M. Peter in *Comprehensive Chemical Kinetics* (R. G. Compton and G. Hancock, editors), p. 223, Elsevier, Amsterdam, 1999.
16. M. D. Archer, *J. Appl. Electrochem.* **5**, 1975, 17.
17. A. J. Nozik, *Ann. Rev. Phys. Chem.* **299**, 1978, 89.
18. L. A. Harris and R. H. Wilson, *Ann. Rev. Mater. Sci.* **8**, 1978, 99.
19. K. Rajeshwar, P. Singh and J. DuBow, *Electrochim. Acta* **23**, 1978, 1117.
20. H. R. Maruska and A. K. Ghosh, *Solar Energy* **20**, 1978, 1443.
21. A. J. Bard, *J. Photochem.* **10**, 1979, 50.
22. M. Wrighton, *Acc. Chem. Res.* **12**, 1979, 303.
23. R. H. Wilson, *CRC Crit. Rev. Mater. Sci.* **10**, 1980, 1.
24. A. J. Bard, *Science* **207**, 1980, 139.
25. M. A. Butler and D. S. Ginley, *J. Mater. Sci.* **15**, 1980, 1.
26. J. B. Goodenough, A. Hamnett, M. P. Dare-Edwards, G. Campet and R. D. Wright, *Surf. Sci.* **101**, 1980, 531.
27. R. Memming, *Electrochim. Acta* **25**, 1980, 77.
28. H. Gerischer, *Surf. Sci.* **101**, 1980, 518.
29. A. Heller, *Acc. Chem. Res.* **14**, 1981, 154.
30. M. Grätzel, *Acc. Chem. Res.* **14**, 1981, 376.
31. B. Parkinson, *Solar Cells* **6**, 1982, 177.
31a B. Parkinson, *Acc. Chem. Res.* **17**, 1984, 431.
32. K. Rajeshwar, *J. Appl. Electrochem.* **15**, 1985, 1.
33. H. Tributsch, *J. Photochem.* **29**, 1985, 89.
34. N. S. Lewis, *Acc. Chem. Res.* **23**, 1990, 176.
35. L. M. Peter, *Chem. Rev.* **90**, 1990, 753.
36. N. S. Lewis, *Ann. Rev. Phys. Chem.* **42**, 1991, 543.
37. C. A. Koval and J. N. Howard, *Chem. Rev.* **92**, 1992, 411.
38. A. Kumar, W. C. A. Wilisch and N. S. Lewis, *Crit. Rev. Solid State Mater. Sci.* **18**, 1993, 327.
39. M. X. Tan, P. E. Laibinis, S. T. Nguyen, J. M. Kesselman, C. E. Stanton and N. S. Lewis, *Prog. Inorg. Chem.* **41**, 1994, 21.
40. A. J. Bard and M. A. Fox, *Acc. Chem. Res.* **28**, 1995, 141.
41. A. J. Nozik and R. Memming, *J. Phys. Chem.* **100**, 1996, 13061.
42. N. S. Lewis, *J. Phys. Chem.* **102**, 1998, 4843.
43. N. S. Lewis, *J. Electroanal. Chem.* **508**, 2001, 1.
44. S. Licht, *J. Phys. Chem.* **105**, 2001, 6281.
45. N. S. Lewis, *Nature* **414**, 2001, 589.
46. M. Grätzel, *Nature* **414**, 2001, 338.
47. W. H. Brattain and C. G. B. Garrett, *Bell Syst. Tech. J.* **34**, 1955, 129.
48. J. F. Dewald, *Bell Syst. Tech. J.* **39**, 1960, 615.
49. D. R. Turner, *J. Electrochem. Soc.* **103**, 1961, 252.

50. J. F. Dewald, *J. Phys. Chem. Solids* **14**, 1961, 155.
51. P. J. Boddy and W. Brattain, *J. Electrochem. Soc.* **109**, 1962, 1053.
52. P. J. Boddy, *J. Electrochem. Soc.* **115**, 1968, 199.
53. A. Fujishima and K. Honda, *Bull. Chem. Soc. Jpn.* **44**, 1971, 1148.
54. A. Fujishima and K. Honda, *Nature* **238**, 1972, 37.
55. A. Fujishima, K. Kohayakawa and K. Honda, *Bull. Chem. Soc. Jpn.* **48**, 1975, 1041.
56. A. Fujishima, K. Kohayakawa and K. Honda, *J. Electrochem. Soc.* **122**, 1975, 1487.
57. T. Watanabe, A. Fujishima and K. Honda in Ref. 4, p. 359.
58. For example, R. Memming and G. Neumann, *J. Electroanal. Chem.* **21**, 1969, 295. See also references threin.
59. S. N. Paleocrassas, *Solar Energy* **16**, 1974, 45.
60. A. J. Nozik, *Appl. Phys. Lett.* **29**, 1976, 150.
61. H. Tributsch, *Z. Naturforsch.* **32a**, 1977, 972.
62. H. H. Kung, H. S. Jarrett, A. W. Sleight and A. Feretti, *J. Appl. Phys.* **48**, 1977, 2463.
63. M. Tomkiewicz and H. Fay, *Appl. Phys.* **18**, 1979, 1.
64. R. D. Rauh, J. M. Buzby, T. F. Reise and S. A. Alkaitis, *J. Phys. Chem.* **83**, 1979, 2221.
65. D. E. Scaife, *Solar Energy* **25**, 1980, 41.
66. A. Heller, *Science* **223**, 1984, 1141.
67. L. Fornarini, A. J. Nozik and B. A. Parkinson, *J. Phys. Chem.* **88**, 1984, 3238.
68. I. Willner and B. Steinberger-Willner, *Int. J. Hydrogen Energy* **13**, 1988, 593.
69. K. Domen, A. Kudo, A. Tanaka and T. Onishi, *Catalysis Today* **8**, 1990, 77.
70. J. R. Bolton, *Solar Energy* **57**, 1996, 37.
71. T. Takata, A. Tanaka, M. Hara, J. N. Kondo and K. Domen, *Catalysis Today* **44**, 1998, 17.
72. G. Dimitrov, *Bulgarian Chem. Ind.* **69**, 1998, 102.
73. K. Domen, J. N. Kondo, M. Hara and T. Takata, *Bull. Chem. Soc. Jpn.* **73**, 2000, 1307.
74. A. Kudo, *J. Ceramic Soc. Jpn.* **109**, 2001, S81.
75. K. Domen, M. Hara, J. N. Kondo, T. Takata, A. Kudo, H. Kobayashi and Y. Inoue, *Korean J. Chem. Eng.* **18**, 2001, 862.
76. T. Bak, J. Nowotny, M. Rekas and C. C. Sorrell, *Int. J. Hydrogen Energy* **27**, 2002, 991.
77. A. Kudo, *Catalysis Surveys from Asia* **7**, 2003, 31.
78. H. Kato and A. Kudo, *Catalysis Today* **78**, 2003, 561.
79. A. Kudo, H. Kato and I. Tsuji, *Chem. Lett.* **33**, 2004, 1534.
80. V. M. Aroutiounian, V. M. Arakelyan and G. E. Shahnazaryan, *Solar Energy* **78**, 2005, 581.
81. A. Kudo, guest editor, *Appl. Catalysis B: Environmental* **46**, 2003, 703.
82. H. O. Finklea in Ref. 6, Chapter 2, p. 43.
83. K. W. Frese, Jr. in Ref. 6, Chapter 8, p. 373.
84. L. F. Schneemeyer, A. Heller and B. Miller in Ref. 6, Chapter 9, p. 411.
85. B. Scrosati in Ref. 6, Chapter 10, p. 457.
86. J. Kiwi in Ref. 4, Chapter 9, p. 297.
87. T. Sakata and T. Kawai in Ref. 4, Chapter 10, p. 331.
88. A. Heller in Ref. 4, Chapter 12, p. 385.
89. A. J. Frank in Ref. 4, Chapter 14, p. 467.
90. M. Kaneko and I. Okura, editors, *Photocatalysis,* Kodansha/Springer, Berlin (2002).
91. M. Anpo in Ref. 90, Chapter 10, p. 175.
92. S. Sato in Ref. 90, Chapter 13, p. 223.
93. H. Arakawa and K. Sayama in Ref. 90, Chapter 14, p. 235.
94. Y. Inoue in Ref. 90, Chapter 15, p. 249.
95. K. Domen in Ref. 90, Chapter 16, p. 261.
96. T. Ohno and M. Matsumura in Ref. 90, Chapter 17, p. 279.
97. K. Kalyanasundaram, *Solar Cells* **15**, 1985, 93.

98. R. A. Smith, *Semiconductors*, Cambridge At the University Press, Cambridge, England, 1964.
99. P. A. Cox, *The Electronic Structure and Chemistry of Solids*, Oxford University Press, Oxford, England, 1987.
100. A. B. Ellis, M. J. Geselbracht, B. J. Johnson, G. C. Lisensky and W. R. Robinson, *Teaching General Chemistry – A Materials Science Companion*, Chapter 7, p. 187, American Chemical Society, Washington, D. C., 1993.
101. D. Meissner, R. Memming and B. Kastening, *Chem. Phys. Lett.* **96**, 1983, 34.
102. K. Jackowska and H. T. Tien, *Solar Cells* **23**, 1988, 147.
103. H. T. Tien and J.-W. Chen, *Int. J. Hydrogen Energy* **15**, 1990, 563.
104. S. C. Kondapaneni, D. Singh and O. N. Srivastava, *J. Phys. Chem.* **96**, 1992, 8094.
105. H. T. Tien and A. L. Ottova, *Trends in Photochem. & Photobiol.* **6**, 1999, 153. See also references therein.
106. R. K. Karn, M. Misra and O. N. Srivastava, *Int. J. Hydrogen Energy* **25**, 2000, 407.
107. O. N. Srivastava, R. K. Karn and M. Misra, *Int. J. Hydrogen Energy* **25**, 2000, 495.
108. M. Hara and T. E. Mallouk, *Chem. Commun.* 2000, 1903.
109. A. J. Bard and T. E. Mallouk, *J. Phys. Chem.* **97**, 1993, 7127.
110. Review: M. Yagi and M. Kaneko, *Chem. Rev.* **101**, 2001, 21.
111. J. Kiwi, E. Borgarello, E. Pelizzetti, M. Visca and M. Grätzel, *Angew. Chem. Int. Ed. Engl.* **19**, 1980, 646.
112. E. Borgarello, J. Kiwi, E. Pelizzetti, M. Visca and M. Grätzel, *Nature* **289**, 1981, 158.
113. P. Pichat, J-M. Hermann, J. Disdier, H. Bourbon and M-N. Mozzanega, *Nov. J. Chim.* **5**, 1981, 627.
114. Y. Degani and I. Willner, *J. Chem. Soc., Chem. Commun.* 1983, 710.
115. A. J. Frank, I. Willner, Z. Goren and Y. Degani, *J. Am. Chem. Soc.* **109**, 1987, 3568.
116. P. Maruthamuthu, K. Gurunathan, E. Subramanian and M. V. C. Sastri, *Int. J. Hydrogen Energy* **19**, 1994, 889.
117. K. Gurunathan, P. Maruthamuthu and M. V. C. Sastri, *Int. J. Hydrogen Energy* **22**, 1997, 57.
118. H. Shiroishi, M. Nukaga, S. Yamashita and M. Kaneko, *Chem. Lett.* 2002, 488.
119. M. Grätzel, K. Kalyanasundaram and J. Kiwi, *Struct. Bonding* **49**, 1982, 37.
120. E. Amouyal, *Solar Energy Mater. Solar Cells* **38**, 1995, 249.
121. J-M. Lehn, J-P. Sauvage and R. Ziessel, *Nov. J. Chim.* **4**, 1980, 355.
122. J-M. Lehn, J-P. Sauvage and R. Ziessel, *Nov. J. Chim.* **5**, 1981, 291.
123. J-M. Lehn, J-P. Sauvage, R. Ziessel and L. Hilaire, *Isr. J. Chem.* **22**, 1982, 168.
124. E. Bilgen, *Energy Conversion and Management* **42**, 2001, 1047.
125. O. Khaselev, A. Bansal and J. A. Turner, *Int. J. Hydrogen Energy* **26**, 2001, 127.
126. J. A. Turner, *Science* **280**, 1998, 425.
127. S. S. Kocha, D. Montgomery, M. W. Peterson and J. A. Turner, *Solar Energy Mater. Solar Cells* **52**, 1998, 389.
128. X. Gao, S. S. Kocha, A. J. Frank and J. A. Turner, *Int. J. Hydrogen Energy* **24**, 1999, 319.
129. P. K. Shukla, R. K. Karn, A. K. Singh and O. N. Srivastava, *Int. J. Hydrogen Energy* **27**, 2002, 135.
130. H. Mettee, J. W. Otvos and M. Calvin, *Solar Energy Mater.* **4**, 1981, 443.
131. J. S. Kilby, J. W. Lathrop and W. A. Porter, U. S. Patent 4 021 323 (1977); U. S. Patent 4 100 051 (1978); U. S. Patent 4 136 436 (1979).
132. K. Rajeshwar and J. B. Ibanez, *J. Chem. Educ.* **72**, 1995, 1044.
133. B. Kraeutler and A. J. Bard, *J. Am. Chem. Soc.* **100**, 1978, 4317.
134. B. Kraeutler and A. J. Bard, *J. Am. Chem. Soc.* **100**, 1978, 5985.
135. M. S. Wrighton, *J. Chem. Educ.* **60**, 1983, 877.

136. M. J. Natan and M. S. Wrighton, *Prog. Inorg. Chem.* **37**, 1989, 391.
   136a. M. S. Wrighton, *Science* **231**, 1986, 32.
137. H. Gerischer, *J. Electroanal. Chem.* **82**, 1977, 133.
138. M. A. K. Lodhi, *Energy Convers. Mgmt.* **38**, 1997, 1881.
139. Special issue on energy *Science* **285**, 1999.
140. J. A. Turner, *Science* **285**, 1999, 687.
141. J. A. Turner, M. C. Williams and K. Rajeshwar, *The Electrochemical Society Interface* **13**, 2004, 24.
142. Y. Nakato, K. Abe and H. Tsubomura, *Ber. Bunsen.-Ges. Phys. Chem.* **80**, 1976, 1002.
143. Y. Nakato, S. Tonomura and H. Tsubomura, *Ber. Bunsen.-Ges. Phys. Chem.* **80**, 1976, 1289.
144. Y. Nakato, K. Ueda and H. Tsubomura, *J. Phys. Chem.* **90**, 1986, 5495.
145. S. Menezes, A. Heller and B. Miller, *J. Electrochem. Soc.* **127**, 1980, 1268.
146. T. Skotheim, I. Lundstrom and J. Prejza, *J. Electrochem. Soc.* **128**, 1981, 1625.
147. T. Skotheim, L-G. Petersson, O. Ingamas and I. Lundstrom, *J. Electrochem. Soc.* **129**, 1982, 1737.
148. T. Skotheim, I. Lundstrom, A. E. Delahoy, F. J. Kampas and P. E. Vanier, *Appl. Phys. Lett.* **40**, 1982, 281.
149. R. Noufi, A. J. Frank and A. J. Nozik, *J. Am. Chem. Soc.* **103**, 1981, 1849.
150. R. Noufi, D. Tench and L. F. Warren, *J. Electrochem. Soc.* **127**, 1980, 2709.
151. A. J. Frank and K. Honda, *J. Phys. Chem.* **86**, 1982, 1933.
152. A. J. Frank and K. Honda, *J. Electroanal. Chem.* **150**, 1983, 673.
153. K. Honda and A. J. Frank, *J. Phys. Chem.* **88**, 1984, 5577.
154. A. Gupta and A. S. N. Murthy, *Solar Energy Mater. Solar Cells* **28**, 1992, 113.
   154a. R. A. Simon, A. J. Ricco and M. S. Wrighton, *J. Am. Chem. Soc.* **104**, 1982, 2031.
   154b. R. A. Simon and M. S. Wrighton, *Appl. Phys. Lett.* **44**, 1984, 930.
155. D. Gningue, G. Horowitz, J. Roncali and F. Garnier, *J. Electroanal. Chem.* **269**, 1989, 337.
156. F-R. F. Fan, R. L. Wheeler, A. J. Bard and R. N. Noufi, *J. Electrochem. Soc.* **128**, 1981, 2042.
157. R. E. Malpas and B. Rushby, *J. Electroanal. Chem.* **157**, 1983, 387.
158. R. E. Malpas, F. R. Mayers and A. G. Osborne, *J. Electroanal. Chem.* **153**, 1983, 97.
159. P. A. Kohl, S. N. Frank and A. J. Bard, *J. Electrochem. Soc.* **124**, 1977, 225.
160. M. Tomkiewicz and J. M. Woodall, *J. Electrochem. Soc.* **124**, 1977, 1436.
161. M. A. Russak and J. Reichman, *J. Electrochem. Soc.* **129**, 1982, 542.
162. A. Kampmann, V. Marcu and H-H. Strehblow, *J. Electroanal. Chem.* **280**, 1990, 91.
163. V. Marcu and H-H. Strehblow, *J. Electrochem. Soc.* **138**, 1991, 758.
164. F-R. F. Fan, G. A. Hope and A. J. Bard, *J. Electrochem. Soc.* **129**, 1982, 1647.
165. G. A. Hope, F-R. F. Fan and A. J. Bard, *J. Electrochem. Soc.* **130**, 1983, 1488.
166. P. Leempoel, M. Castro-Acuna, F-R. F. Fan and A. J. Bard, *J. Phys. Chem.* **86**, 1982, 1396.
167. A. Haran, D. H. Waldeck, R. Naaman, E. Moons and D. Cahen, *Science* **263**, 1994, 948.
168. C. W. Sheen, J. X. Shi, J. Martensson, A. N. Parikh and D. L. Allara, *J. Am. Chem. Soc.* **114**, 1992, 1514.
169. Y. Gu, Z. Lin, R. A. Butera, V. S. Smentkowski and D. H. Waldeck, *Langmuir* **11**, 1995, 1849.
170. Y. Gu and D. H. Waldeck, *J. Phys. Chem.* **100**, 1996, 9573.
171. Y. Gu and D. H. Waldeck, *J. Phys. Chem. B* **102**, 1998, 9015.
172. G. Hodes, L. Thompson, J. DuBow and K. Rajeshwar, *J. Am. Chem. Soc.* **105**, 1983, 324.
173. L. Thompson, J. DuBow and K. Rajeshwar, *J. Electrochem. Soc.* **129**, 1982, 1934.
174. S.-I. Ho and K. Rajeshwar, *J. Electrochem. Soc.* **134**, 1987, 2491.

175. F. Decker, M. Fracastoro-Decker, S. Badawy, K. Doblhofer and H. Gerischer, *J. Electrochem. Soc.* **130**, 1983, 2173.
176. F. Decker, J. Melsheimer and H. Gerischer, *Isr. J. Chem.* **22**, 1982, 195.
177. D. Belanger, J. P. Dodelet and B. A. Lombos, *J. Electrochem. Soc.* **133**, 1986, 1113.
178. J. A. Switzer, *J. Electrochem. Soc.* **133**, 1986, 723.
179. K. Rajeshwar and M. Kaneko, *J. Phys. Chem.* **89**, 1985, 3587.
180. W. Gissler, A. J. McEvoy and M. Grätzel, *J. Electrochem. Soc.* **129**, 1982, 1733.
181. D. C. Bookbinder, J. A. Bruce, R. N. Dominey, N. S. Lewis and M. S. Wrighton, *Proc. Natl. Acad. Sci.* **77**, 1980, 6280.
182. R. N. Dominey, N. S. Lewis, J. A. Bruce, D. C. Bookbinder and M. S. Wrighton, *J. Am. Chem. Soc.* **104**, 1982, 467.
183. J. A. Bruce, T. Murahashi and M. S. Wrighton, *J. Phys. Chem.* **86**, 1982, 1552.
184. D. J. Harrison and M. S. Wrighton, *J. Phys. Chem.* **88**, 1984, 3932.
185. H. D. Abruña and A. J. Bard, *J. Am. Chem. Soc.* **103**, 1981, 6898.
186. F-R. F. Fan, B. Reichman and A. J. Bard, *J. Am. Chem. Soc.* **102**, 1980, 1488.
187. M. D. Rosenblum and N. S. Lewis, *J. Phys. Chem.* **88**, 1984, 3103.
188. K. Rajeshwar, M. Kaneko and A. Yamada, *J. Electrochem. Soc.* **130**, 1983, 38.
189. K. Rajeshwar, M. Kaneko, A. Yamada and R. N. Noufi, *J. Phys. Chem.* **89**, 1985, 806.
190. C. P. Kubiak, L. F. Schneemeyer and M. S. Wrighton, *J. Am. Chem. Soc.* **102**, 1980, 6898.
191. L. F. Schneemeyer and B. Miller, *J. Electrochem. Soc.* **129**, 1982, 1977.
192. M. D. Archer and J. R. Bolton, *J. Phys. Chem.* **94**, 1990, 8028.
193. J. O'M. Bockris and O. J. Murphy, *Appl. Phys. Commun.* **2**, 1982-1983, 203.
194. M. Zafir, M. Ulman, Y. Zuckerman and M. Halmann, *J. Electroanal. Chem.* **159**, 1983, 373.
195. J. R. Bolton, S. J. Strickler and J. S. Connolly, *Nature*, **316**, 1985, 495.
196. C. H. Henry, *J. Appl. Phys.* **51**, 1980, 4494.
197. M. F. Weber and M. J. Dignam, *J. Electrochem. Soc.* **131**, 1984, 1258.
198. H. Gerischer in *Solar Energy Conversion* (B. O. Seraphin, editor), p. 115, Springer, Berlin (1979).
199. E. Aharon-Shalom and A. Heller, *J. Electrochem. Soc.* **129**, 1982, 2865.
200. V. Guruswamy and J. O'M. Bockris, *Solar Energy Materials* **1**, 1979, 441.
201. A. Heller and R. G. Vadimsky, *Phys. Rev. Lett.* **46**, 1981, 1153.
202. C. Leygraf, M. Hendewerk and G. A. Somorjai, *Proc. Natl. Acad. Sci. USA* **79**, 1982, 5739.
203. Y. Nakato, M. Yashimura, M. Hiramoto, A. Tsumura, T. Murahashi and H. Tsubomura, *Bull. Chem. Soc. Jpn.* **57**, 1984, 355.
204. C. Levy-Clement, A. Heller, W. A. Bonner and B. A. Parkinson, *J. Electrochem. Soc.* **129**, 1982, 1701.
205. E. S. Smotkin, S. Cervera-March, A. J. Bard, A. Campion, M. A. Fox, T. Mallouk, S. E. Webber and J. M. White, *J. Phys. Chem.* **91**, 1987, 6.
206. S. Cervera-March, E. S. Smotkin, A. J. Bard, A. Campion, M. A. Fox, T. Mallouk, S. E. Webber and J. M. White, *J. Electrochem. Soc.* **135**, 1988, 567.
207. S. Cervera-March and E. S. Smotkin, *Int. J. Hydrogen Energy* **16**, 1991, 243.
208. S. U. M. Khan and J. Akikusa, *J. Phys. Chem. B* **103**, 1999, 7184.
209. S. Licht, B. Wang, S. Mukerji, T. Soga, M. Umeno and H. Tributsch, *J. Phys. Chem. B* **104**, 2000, 8920.
210. G. K. Mor, K. Shankar, O. K. Varghese and C. A. Grimes, *J. Mater. Res.* **19**, 2004, 2989. (a) G. K. Mor, K. Shankar, M. Paulose, O. K. Varghese and C. A. Grimes, *Nano Lett.* **5**, 2005, 191. (b) O. K. Varghese, M. Paulose, K. Shankar, G. K. Mor and C. A. Grimes, *J.*

Nanosci. Nanotechnol. **5**, 2005, 1158. (c) K. Shankar, M. Paulose, G. K. Mor, O. K. Varghese and C. A. Grimes, *J. Phys. D: Appl. Phys.* **38**, 2005, 3543.

211. E. L. Miller, B. Marsen, D. Paluselli and R. Rocheleau, *Electrochem. Solid-State Lett.* **8**, 2005, A247.

212. Y. Yamada, N. Matsuki, T. Ohmori, M. Mametsuka, M. Kondo, A. Matsuda and E. Suzuki, *Int. J. Hydrogen Energy* **28**, 2003, 1167.

213. S. U. M. Khan, M. Al-Shahry and W. B. Ingler, Jr., *Science* **297**, 2002, 2243.

214. A. Fujishima, *Science* **301**, 2003, 1673a.

215. C. Hägglund, M. Grätzel and B. Kasemo, *Science* **301**, 2003, 1673b.

216. K. S. Lackner, *Science* **301**, 2003, 1673c.

217. A. Kumar, P. G. Santangelo and N. S. Lewis, *J. Phys. Chem.* **96**, 1992, 834.

218. B. A. Gregg and A. J. Nozik, *J. Phys. Chem.* **97**, 1993, 13441.

219. H. Gerischer in *Solar Power and Fuels* (J. R. Bolton, editor), Chapter 4, p. 77, Academic Press (1977).

220. F. Williams and A. J. Nozik, *Nature* **271**, 1978, 137.

221. J. O' M. Bockris and K. Uosaki, *Adv. Chem. Ser.* 1976, 33.

222. J. O' M. Bockris and K. Uosaki, *J. Electrochem. Soc.* **125**, 1978, 223.

223. D. C. Card and H. C. Card, *Solar Energy* **28**, 1982, 451.

224. T. Ohnishi, Y. Nakato and H. Tsubomura, *Ber. Bunsen.-Ges. Phys. Chem.* **89**, 1975, 523.

225. A. J. Nozik, *Nature* **257**, 1975, 383.

226. J. G. Mavroides, D. I. Tcherner, J. A. Kafalas and D. F. Kolesar, *Mater. Res. Bull.* **10**, 1975, 1023.

227. H. Yoneyama, H. Sakamoto and H. Tamura, *Electrochim. Acta* **20**, 1975, 341.

228. M. S. Wrighton, D. S. Ginley, P. T. Wolczanski, A. B. Ellis, D. L. Morse and A. Linz, *Proc. Nat. Acad. Sci. USA* **72**, 1975, 1518.

229. K. L. Hardee and A. J. Bard, *J. Electrochem. Soc.* **122**, 1975, 739.

230. J. O' M. Bockris and K. Uosaki, *Energy* **1**, 1976, 143.

231. H. Morisaki, T. Watanabe, M. Iwase and K. Yazawa, *Appl. Phys. Lett.* **29**, 1976, 338.

232. W. Gissler, P. L. Lensi and S. Pizzini, *J. Appl. Electrochem.* **6**, 1976, 9.

233. J. H. Carey and B. G. Oliver, *Nature* **259**, 1976, 554.

234. J. F. Houlihan, D. P. Madacsi and E. J. Walsh, *Mater. Res. Bull.* **11**, 1976, 1191.

235. J-L. Desplat, *J. Appl. Phys.* **47**, 1976, 5102.

236. J. Augustynski, J. Hinden and C. Stalder, *J. Electrochem. Soc.* **124**, 1977, 1063.

237. A. K. Ghosh and H. P. Maruska, *J. Electrochem. Soc.* **124**, 1977, 1516.

238. K. Nobe, G. L. Bauerle and M. Brown, *J. Appl. Electrochem.* **7**, 1977, 379.

239. H. Tamura, H. Yoneyama, C. Iwakura, H. Sakamoto and S. Murakami, *J. Electroanal. Chem.* **80**, 1977, 357.

240. H. Morisaki, M. Hariya and K. Yazawa, *Appl. Phys. Lett.* **30**, 1977, 7.

241. H. Tamura, H. Yoneyama, C. Iwakawa and T. Murai, *Bull. Chem. Soc. Jpn.* **50**, 1977, 753.

242. D. Haneman and P. Holmes, *Solar Energy Materials* **1**, 1979, 233.

243. P. Clechet, C. Martelet, J. R. Martin and R. Oliver, *Electrochim. Acta* **24**, 1979, 457.

244. S. Sato and J. M. White, *Chem. Phys. Lett.* **72**, 1980, 83.

245. T. Kawai and T. Sakata, *Chem. Phys. Lett.* **72**, 1980, 87. See also references therein.

246. J. F. Julião, F. Decker and M. Abramovich, *J. Electrochem. Soc.* **127**, 1980, 2264.

247. J. M. Kowalski, K. H. Johnson and H. L. Tuller, *J. Electrochem. Soc.* **127**, 1980, 1969.

248. M. V. Rao, K. Rajeshwar, V. R. Pai Verneker and J. DuBow, *J. Phys. Chem.* **84**, 1980, 1987.

249. P. Salvador, *J. Electrochem. Soc.* **128**, 1981, 1895.

250. S. Sato and J. M. White, *J. Catal.* **69**, 1981, 128.

251. A. Mills and G. Porter, *J. Chem. Soc., Faraday Trans. I* **78**, 1982, 3659.

252. Y. Oosawa, *J. Chem. Soc., Chem. Commun.* 1982, 221.
253. J. Kiwi and M. Grätzel, *J. Phys. Chem.* **88**, 1984, 1302. See also references therein.
254. D. N. Furlong, D. Wells and W. H. F. Sasse, *J. Phys. Chem.* **89**, 1985, 626.
255. For example: T. A. Egerton and C. J. King, *J. Oil. Col. Chem. Assoc.* **62**, 1979, 386.
256. J. H. Carey, J. Laurence and H. M. Tosine, *Bull. Environ. Contamination Toxicol.* **16**, 1976, 697.
257. J. H. Carey and B. G. Oliver, *Water Poll. Res. J. Canada* **15**, 1980, 157.
258. T. Kobayashi, H. Yoneyama and H. Tamura, *J. Electroanal. Chem.* **124**, 1981, 179.
259. T. Kobayashi, H. Yoneyama and H. Tamura, *J. Electroanal. Chem.* **122**, 1981, 133.
260. P. Salvador, *Solar Energy Materials* **6**, 1982, 241.
261. Y. Nakato, A. Tsumura and H. Tsubomura, *Chem. Phys. Lett.* **85**, 1982, 387.
262. T. Kobayashi, H. Yoneyama and H. Tamura, *J. Electroanal. Chem.* **138**, 1982, 105.
263. Y. Nakato, A. Tsumura and H. Tsubomura, *J. Phys. Chem.* **87**, 1983, 2402.
264. P. Salvador and F. Decker, *J. Phys. Chem.* **88**, 1984, 6116.
265. Y. Oosawa and M. Grätzel, *J. Chem. Soc., Chem. Commun.* 1984, 1629.
266. G. T. Brown and J. R. Darwent, *J. Phys. Chem.* **88**, 1984, 4955.
267. J. Abrahams, R. S. Davidson and C. L. Morrison, *J. Photochem.* **29**, 1985, 353.
268. M. Ulmann, N. R. de Tacconi and J. Augustynski, *J. Phys. Chem.* **90**, 1986, 6523.
269. G. Munuera, A. R. Gonzalez-Elipe, A. Fernandez, P. Malet and J. P. Espinos, *J. Chem. Soc., Faraday Trans. I* **85**, 1989, 1279.
270. M. W. Peterson, J. A. Turner and A. J. Nozik, *J. Phys. Chem.* **95**, 1991, 221.
271. J. Augustynski, *Electrochim. Acta* **38**, 1993, 43.
272. G. Riegel and J. R. Bolton, *J. Phys. Chem.* **99**, 1995, 4215.
273. Y. Nakato, H. Akanuma, J-i. Shimizu and Y. Magari, *J. Electroanal. Chem.* **396**, 1995, 35.
274. L. Sun and J. R. Bolton, *J. Phys. Chem.* **100**, 1996, 4127.
275. D. W. Bahnemann, M. Hilgendorff and R. Memming, *J. Phys. Chem. B* **101**, 1997, 4265.
276. P. A. Connor, K. D. Dobson and A. J. McQuillan, *Langmuir* **15**, 1999, 2402.
277. T. Hirakawa, Y. Nakaoka, J. Nishino and Y. Nosaka, *J. Phys. Chem. B* **103**, 1999, 4399.
278. G. Boschloo and D. Fitzmaurice, *J. Phys. Chem. B* **103**, 1999, 2228.
279. G. Boschloo and D. Fitzmaurice, *J. Phys. Chem. B* **103**, 1999, 7860.
280. S. H. Szczepankiewicz, A. J. Colussi and M. R. Hoffmann, *J. Phys. Chem. B* **104**, 2000, 9842.
281. K-i. Ishibashi, A. Fujishima, T. Watanabe and K. Hashimoto, *J. Phys. Chem. B* **104**, 2000, 4934.
282. A. Tsujiko, T. Kisumi, Y. Magari, K. Murakoshi and Y. Nakato, *J. Phys. Chem. B* **104**, 2000, 4873.
283. M. Miyauchi, A. Nakajima, A. Fujishima, K. Hashimoto and T. Watanabe, *Chem. Mater.* **12**, 2000, 3.
284. A. Yamakata, T. Ishibashi and H. Onishi, *J. Phys. Chem. B* **105**, 2001, 7258.
285. X. Li, C. Chen and J. Zhao, *Langmuir* **17**, 2001, 4118.
286. T. Ohno, K. Sarukawa and M. Matsumura, *J. Phys. Chem. B* **105**, 2001, 2417.
287. S. H. Szczepankiewicz, J. A. Moss and M. R. Hoffmann, *J. Phys. Chem. B* **106**, 2002, 2922.
288. F. Fabregat-Santiago, G. Garcia-Belmonte, J. Bisquert, A. Zaban and P. Salvador, *J. Phys. Chem. B* **106**, 2002, 334.
289. S. Somasundaram, N. Tacconi, C. R. Chenthamarakshan, K. Rajeshwar and N. R. de Tacconi, *J. Electroanal. Chem.* **577**, 2005, 167.
290. C. Stalder and J. Augustynski, *J. Electrochem. Soc.* **126**, 1979, 2007.
291. J. F. Houlihan, D. B. Armitage, T. Hoovler, D. Bonaquist and L. N. Mullay, *Mater. Res. Bull.* 13, 1978, 1205.

292. A. Monnier and J. Augustynski, *J. Electrochem. Soc.* **127**, 1980, 1576.
293. P. Salvador, *Mater. Res. Bull.* **15**, 1980, 1287.
294. Y. Matsumoto, J. Kurimoto, T. Shimizu and E. Sato, *J. Electrochem. Soc.* **128**, 1981, 1090.
295. K. E. Karakitsou and X. E. Verykios, *J. Phys. Chem.* **97**, 1993, 1184. See also references therein.
296. Z. Zhang, C-C. Wang, R. Zakaria and J. Y. Ying, *J. Phys. Chem. B* **102**, 1998, 10871.
297. M. R. Hoffman, S. T. Martin, W. Choi and D. W. Bahnemann, *Chem. Rev.* **95**, 1995, 69.
298. A. L. Linsebigler, G. Lu and J. T. Yates, Jr., *Chem. Rev.* **95**, 1995, 735.
299. A. DiPaola, G. Marci, L. Palmisomo, M. Schiavello, K. Uosaki, S. Ikeda and B. Ohtani, *J. Phys. Chem. B* **106**, 2002, 637.
300. L. Zang, W. Macyk, C. Lange, W. F. Maier, C. Antonius, D. Meissner and H. Kisch, *Chem. Eur. J.* **6**, 2000, 379.
301. E. Stathatos, P. Lianos, P. Falaras and A. Siokou, *Langmuir* **16**, 2000, 2398.
302. E. Stathatos, T. Petrova and P. Lianos, *Langmuir* **17**, 2001, 5025.
303. I. Nakamura, N. Negishi, S. Kutsuna, T. Ihara, S. Sugihara and K. Takeuchi, *J. Mol. Catal. A: Chem.* **161**, 2000, 205.
304. E. Barborini, A. M. Conti, I. N. Kholmanov, P. Piseri, A. Podesta, P. Milani, C. Cepek, O. Sakho, R. Macovez and M. Sncrotti, *Adv. Mater.* (in press).
305. M. Anpo, M. Takeuchi, K. Ikeue and S. Dohshi, *Curr. Opinion Solid State Mater. Sci.* **6**, 2002, 381. See also references therein.
306. R. Asahi, T. Morikawa, T. Ohwaki, K. Aoki and Y. Taga, *Science* **293**, 2001, 269.
307. T. Yamaki, T. Sumita and S. Yamamoto, *J. Mater. Sci. Lett.* **21**, 2002, 33.
308. T. Umebayashi, T. Yamaki, H. Itoh and K. Asai, *Appl. Phys. Lett.* **81**, 2002, 454.
309. S. Sakthivel and H. Kisch, *Angew. Chem. Int. Ed.* **42**, 2003, 4908.
310. H. Irie, Y. Watanabe and K. Hashimoto, *Chem. Lett.* **32**, 2003, 772.
311. H. Irie, Y. Watanabe and K. Hashimoto, *J. Phys. Chem. B* **107**, 2003, 5483.
312. T. Umebayashi, T. Yamaki, S. Tanaka and K. Arai, *Chem. Lett.* **32**, 2003, 330.
313. H. Irie, S. Washizuka, N. Yoshino and K. Hashimoto, *Chem. Commun.* 2003, 1298.
314. K. Noworyta and J. Augustynski, *Electrochem. Solid-State Lett.* **7**, 2004, E31.
315. G. R. Torres, T. Lindgren, J. Lu, C-G. Granqvist and S-E. Lindquist, *J. Phys. Chem. B* **108**, 2004, 5995.
316. Y. Sakatani, H. Ando, K. Okusako, H. Koike, J. Nunoshige, T. Takata, J. N. Kondo, M. Hara and K. Domen, *J. Mater. Sci.* **19**, 2004, 2100.
317. A. Hattori, M. Yamamoto, H. Tada and S. Ito, *Chem. Lett.* 1998, 707.
318. Y. Inoue, M. Okamura and K. Sato, *J. Phys. Chem.* **89**, 1985, 5184.
319. A. Kudo, K. Domen, K. Maruya and T. Onishi, *Bull. Chem. Soc. Jpn.* **61**, 1988, 1535.
320. K. Yamaguchi and S. Sato, *J. Chem. Soc., Faraday Trans. I* **81**, 1985, 1237.
321. K. Sayama and H. Arakawa, *J. Chem. Soc., Chem. Commun.* 1992, 150.
322. K. Sayama and H. Arakawa, *J. Chem. Soc., Faraday Trans.* **93**, 1997, 1647.
323. H. Arakawa and K. Sayama, *Res. Chem. Intermed.* **26**, 2000, 145.
324. A. Kudo, K. Domen, K. Maruya and T. Onishi, *Chem. Phys. Lett.* **133**, 1989, 517.
325. Yu. V. Pleskov and M. D. Krotova, *Electrochim. Acta* **38**, 1993, 107.
326. T. Ohno, D. Haga, K. Fujihara, K. Kaizaki and M. Matsumura, *J. Phys. Chem. B* **101**, 1997, 6415.
327. K. Fujihara, T. Ohno and M. Matsumura, *J. Chem. Soc., Faraday Trans.* **94**, 1998, 3705.
328. T. Abe, E. Suzuki, K. Nagoshi, K. Miyashita and M. Kaneko, *J. Phys. Chem. B* **103**, 1999, 1119.
329. T. Ohno, K. Sarukawa and M. Matsumura, *J. Phys. Chem. B* **105**, 2001, 2417.
330. R. Abe, K. Sayama, K. Domen and H. Arakawa, *Chem. Phys. Lett.* **344**, 2001, 339.
331. H. Kato and A. Kudo, *J. Phys. Chem. B* **106**, 2002, 5029.

332. R. Abe, K. Sayama and H. Arakawa, *Chem. Phys. Lett.* **379**, 2003, 230.
333. Y. Matsumoto, U. Unal, N. Tanaka, A. Kudo and H. Kato, *J. Solid State Chem.* **177**, 2004, 4205.
334. R. Riishiro, H. Kato and A. Kudo, *Phys. Chem. Chem. Phys.* **7**, 2005, 2241.
335. S. Tabata, N. Nishida, Y. Masaki and K. Tabata, *Catal. Lett.* **34**, 1995, 245.
336. S. Sato and J. M. White, *J. Am. Chem. Soc.* **102**, 1980, 7206.
337. C. L. Perkins, M. A. Henderson, D. E. McCready and G. S. Herman, *J. Phys. Chem. B* **105**, 2001, 595.
338. S. Sato, *J. Catal.* **92**, 1985, 11.
339. M. A. Butler, R. D. Nasby and R. K. Quinn, *Solid State Commun.* **19**, 1976, 1011.
340. M. A. Butler, *J. Appl. Phys.* **48**, 1977, 1914.
341. C. E. Derrington, W. S. Godek, C. A. Castro and A. Wold, *Inorg. Chem.* **17**, 1978, 977.
342. J. R. Darwent and Mills, *J. Chem. Soc., Faraday Trans.* **278**, 1982, 359.
343. W. Erbs, J. DeSilvestro, E. Borgarello and M. Grätzel , *J. Phys. Chem.* **88**, 1984, 4001.
344. P. Maruthamuthu, M. Ashok Kumar, K. Gurunathan, E. Subramanian and M. V. C. Sastri, *Int. J. Hydrogen Energy* **14**, 1989, 525.
345. C. Santato, M. Ulmann and J. Augustynski, *Adv. Mater.* **13**, 2001, 511.
346. H. Wang, T. Lindgren, J. He, A. Hagfeldt and S-E. Lindquist, *J. Phys. Chem. B* **104**, 2000, 5686.
347. C. Santato, M. Ulmann and J. Augustynski, *J. Phys. Chem. B* **105**, 2001, 936.
348. C. Santato, M. Odziemkowski, M. Ulmann and J. Augustynski, *J. Am. Chem. Soc.* **123**, 2001, 10639.
349. K. Sayama, K. Mukasa, R. Abe, Y. Abe and H. Arakawa, *J. Photochem. Photobiol. A: Chem.* **148**, 2002, 71.
350. D. W. Hwang, J. Kim, T. J. Park and J. S. Lee, *Catal. Lett.* **80**, 2002, 53.
351. K. L. Hardee and A. J. Bard, *J. Electrochem. Soc.* **123**, 1976, 1024.
352. R. K. Quinn, R. D. Nasby and R. J. Baughman, *Mater. Res. Bull.* **11**, 1976, 1011.
353. K. L. Hardee and A. J. Bard, *J. Electrochem. Soc.* **124**, 1977, 215.
354. L-S. R. Yeh and N. Hackerman, *J. Electrochem. Soc.* **124**, 1977, 833.
355. J. H. Kennedy and K. W. Frese, Jr., *J. Electrochem. Soc.* **125**, 1978, 709.
356. C. Leygraf, M. Hendewerk and G. A. Somorjai, *J. Phys. Chem.* **86**, 1982, 4484.
357. R. Shinar and J. H. Kennedy, *Solar Energy Materials* **6**, 1982, 323. See also references therein.
358. C. Leygraf, M. Hendewerk and G. A. Somorjai, *Proc. Natl. Acad. Sci. USA* **79**, 1982, 5739.
359. C. Leygraf, M. Hendewerk and G. A. Somorjai, *J. Solid State Chem.* **48**, 1983, 357.
360. M. P. Dare-Edwards, J. B. Goodenough, A. Hamnett and P. R. Trevellick, *J. Chem. Soc., Faraday Trans. I* **79**, 1983, 2027.
361. K. Gurunathan and P. Maruthamuthu, *Int. J. Hydrogen Energy* **20**, 1995, 287.
362. A. Watanabe and H. Kozuka, *J. Phys. Chem. B* **107**, 2003, 12713.
363. T. Lindgren, H. Wang, N. Beermann, L. Vayssieres, A. Hagfeldt and S-E. Lindquist, *Solar Energy Mater. Solar Cells* **71**, 2002, 231.
364. C. J. Sartoretti, M. Ulmann, B. D. Alexander, J. Augustynski and A. Weiden-Kaff, *Chem. Phys. Lett.* **376**, 2003, 194.
365. W. B. Ingler, Jr. and S. U. M. Khan, *Thin Solid Films* **461**, 2004, 301.
366. W. B. Ingler, Jr., J. P. Baltrus and S. U. M. Khan, *J. Am. Chem. Soc.* **126**, 2004, 10238.
367. C. J. Sartoretti, B. D. Alexander, R. Solarska, I. A. Rutkawska, J. Augustynski and R. Cerny, *J. Phys. Chem. B* **109**, 2005, 13685.
368. M. Woodhouse, G. S. herman, and B. A. Parkinson, *Chem. Mater.* **17**, 2005, 4318. a) A Duet and M. Grätzel, *J. Phys. Chem., B*, **109**, 2005, 17184.

369. M. S. Wrighton, D. L. Morse, A. B. Ellis, D. S. Ginley and H. B. Abrahamson, *J. Am. Chem. Soc.* **98**, 1976, 44.
370. J. M. Bolts and M. S. Wrighton, *J. Phys. Chem.* **80**, 1976, 2641.
371. M. P. Dare-Edwards, J. B. Goodenough, A. Hamnett and N. D. Nicholson, *J. Chem. Soc., Faraday Trans. II* **77**, 1981, 643.
372. F. P. Koffyberg and F. A. Benko, *J. Electrochem. Soc.* **128**, 1981, 2476.
373. A. Harriman, *J. Chem. Soc., Faraday Trans. I* **79**, 1983, 2875.
374. M. P. Dare-Edwards, J. B. Goodenough, A. Hamnett and A. Katty, *Mater. Res. Bull.* **19**, 1984, 435.
375. M. Hara, T. Kondo, M. Komoda, S. Ikeda, K. Shinohara, A. Tanaka, J. N. Kondo and K. Domen, *Chem. Commun.* 1998, 357.
376. P. E. de Jongh, D. Vanmaekelbergh and J. J. Kelly, *Chem. Commun.* 1999, 1069.
377. F. P. Koffyberg and F. A. Benko, *J. Appl. Phys.* **53**, 1982, 1173.
378. P. Maruthamuthu, K. Gurunathan, E. Subramanian and M. V. C. Sastri, *Int. J. Hydrogen Energy* **18**, 1993, 9.
379. P. Maruthamuthu, K. Gurunathan, E. Subramanian and M. V. C. Sastri, *Int. J. Hydrogen Energy* **19**, 1994, 889.
380. K. Gurunathan, *Int. J. Hydrogen Energy* **29**, 2004, 933.
381. G. Hodes, D. Cahen and J. Manassen, *Nature* **260**, 1976, 312.
382. S. Somasundaram, N. R. de Tacconi, C. R. Chenthamarakshan and K. Rajeshwar, *Int. J. Hydrogen Energy*, in press.
383. M. A. Butler and D. S. Ginley, *J. Electrochem. Soc.* **125**, 1978, 228.
384. A. F. Wells, *Structural Inorganic Chemistry*, 3rd Edition, Oxford University Press, Clarendon, 1962.
385. A. R. West, *Basic Solid State Chemistry*, Second Edition, John Wiley & Sons, Chichester, 2000.
386. T. Watanabe, A. Fujishima and K. Honda, *Bull. Chem. Soc. Jpn.* **49**, 1976, 355.
387. M. S. Wrighton, A. B. Ellis, P. T. Wolczanski, D. L. Morse, H. B. Abrahamson and D. S. Ginley, *J. Am. Chem. Soc.* **98**, 1976, 2774.
388. J. G. Mavroides, J. A. Kafalas and D. F. Kolesar, *Appl. Phys. Lett.* **28**, 1976, 241.
389. M. Okuda, K. Yoshida and N. Tanaka, *Japan J. Appl. Phys.* **15**, 1976, 1599.
390. J. G. Mavroides and D. F. Kolesar, *J. Vac. Sci. Technol.* **15**, 1978, 538.
391. F. T. Wagner and G. A. S0morjai, *J. Am. Chem. Soc.* **102**, 1980, 5494.
392. K. Domen, S. Naito, M. Soma, T. Onishi and K. Tamaru, *J. Chem. Soc., Chem. Commun.* 1980, 543.
393. A. Mackor and G. Blasse, *Chem. Phys. Lett.* **77**, 1981, 6.
394. P. H. M. de Korte, L. G. J. de Haart, R. U. E. 't Lam and G. Blasse, *Solid State Commun.* **38**, 1981, 213.
395. G. Blasse, P. H. M. de Korte and A. Mackor, *J. Inorg. Nucl. Chem.* **43**, 1981, 1499.
396. S. Ferrer and G. A. Somorjai, *J. Phys. Chem.* **85**, 1981, 1464.
397. P. Salvador, V. M. Fernandez and C. Gutierrez, *Solar Energy Mater.* **7**, 1982, 323.
398. J. Schoonman, *Ber. Bunsenges. Phys. Chem.* **86**, 1982, 660.
399. J. M. Lehn, J. P. Sauvage, R. Ziessel and L. Hilaire, *Isr. J. Chem.* **22**, 1982, 168.
400. K. Domen, S. Naito, T. Onishi and K. Tamaru, *Chem. Phys. Lett.* **92**, 1982, 433.
401. K. Domen, S. Naito, T. Onishi, K. Tamaru and M. Soma, *J. Phys. Chem.* **86**, 1982, 3657.
402. M. Matsumura, M. Hiramoto and H. Tsubomura, *J. Electrochem. Soc.* **130**, 1983, 326.
403. P. Salvador and C. Gutierrez, *Surf. Sci.* **124**, 1983, 398.
404. K. Domen, A. Kudo, T. Onishi, N. Kosugi and H. Kuroda, *J. Phys. Chem.* **90**, 1986, 292.
405. K. Domen, A. Kudo and T. Onishi, *J. Catal.* **102**, 1986, 92.
406. K. Yamaguchi and S. Sato, *Nov. J. Chem.* **10**, 1986, 217.
407. Y. Nakato, H. Ogawa, K. Morita and H. Tsubomura, *J. Phys. Chem.*, **90**, 1986, 6210.

408. K. Sayama, K. Mukasa, R. Abe, Y. Abe and H. Arakawa, *Chem. Commun.* 2001, 2416.
409. H. Kato and A. Kudo, *J. Phys. Chem. B* **106**, 2002, 5029.
410. H. Kato, M. Hori, R. Konta, Y. Shimodaira and A. Kudo, *Chem. Lett.* **33**, 2004, 1348.
411. T. Ishii, H. Kato and A. Kudo, *J. Photochem. Photobiol. A: Chem.* **163**, 2004, 181.
412. R. Konta, T. Ishii, H. Kato and A. Kudo, *J. Phys. Chem. B* **108**, 2004, 8992.
413. D. M. Schleich, C. Derrington, W. Godek, D. Weisberg and A. Wold, *Mater. Res. Bull.* **12**, 1977, 321.
414. R. D. Nasby and R. K. Quinn, *Mater. Res. Bull.* **11**, 1976, 985.
415. J. H. Kennedy and K. W. Frese, Jr., *J. Electrochem. Soc.* **123**, 1976, 1683.
416. M. Shibata, A. Kudo, A. Tanaka, K. Domen, K. Maruya and T. Onishii, *Chem. Lett.* 1987, 1017.
417. Y. Inoue, T. Kubokawa and K. Sato, *J. Chem. Soc., Chem. Commun.* 1990, 1298.
418. Y. Inoue, T. Kubokawa and K. Sato, *J. Phys. Chem.* **95**, 1991, 4059.
419. Y. Inoue, T. Niiyama, Y. Asai and K. Sato, *J. Chem. Soc., Chem. Commun.* 1992, 579.
420. M. Kohno, S. Oyura and Y. Inoue, *J. Mater. Chem.* **6**, 1996, 1921.
421. T. Takata, Y. Furumi, K. Shinohara, A. Tanaka, M. Hara, J. N. Kondo and K. Domen, *Chem. Mater.* **9**, 1997, 1063.
422. M. Kohno, S. Ogura, K. Sato and Y. Inoue, *J. Chem. Soc., Faraday Trans.* **93**, 1997, 2433.
423. S. Ikeda, M. Hara, J. N. Kondo, K. Domen, H. Takahashi, T. Okuho and M. Kakihana, *J. Mater. Res.* **13**, 1998, 852.
424. S. Ogura, M. Kohno, K. Sato and Y. Inoue, *J. Mater. Chem.* **8**, 1998, 2335.
425. M. Kohno, T. Kaneko, S. Ogura, K. Sato and Y. Inoue, *J. Chem. Soc., Faraday Trans.* **94**, 1998, 89.
426. H. G. Kim, D. W. Hwang, J. Kim, Y. G. Kim and J. S. Lee, *Chem. Commun.* 1999, 1077. See also references therein.
427. D. S. Ginley and M. A. Butler, *J. Appl. Phys.* **48**, 1977, 2019.
428. M. A. Butler, D. S. Ginley and M. Eibschutz, *J. Appl. Phys.* **48**, 1977, 3070.
429. H. S. Jarrett, A. W. Sleight, H. H. Kung and J. L. Gillson, *J. Appl. Phys.* **51**, 1980, 3916.
430. A. B. Ellis, S. W. Kaiser and M. S. Wrighton, *J. Phys. Chem.* **80**, 1976, 1325.
431. H. Yoneyama, T. Ohkubo and H. Tamura, *Bull. Chem. Soc. Jpn.* **54**, 1981, 404.
432. P. Salvador, C. Gutierrez and J. B. Goodenough, *Appl. Phys. Lett.* **40**, 1982, 188.
433. C. Gutierrez and P. Salvador, *J. Electroanal. Chem.* **134**, 1982, 325.
434. P. Salvador, C. Gutierrez and J. B. Goodenough, *J. Appl. Phys.* **53**, 1982, 7003.
435. P. Lianos and J. K. Thomas, *Chem. Phys. Lett.* **125**, 1986, 299.
436. T. Kida, G. Guan, Y. Minami, T. Ma and A. Yoshida, *J. Mater. Chem.* **13**, 2003, 1186.
437. G. Blasse, G. J. Dirksen and P. H. M. de Korte, *Mater. Res. Bull.* **16**, 1981, 991.
438. J. Koenitzer, B. Khazai, J. Hormadaly, R. Kershaw, K. Dwight and A. Wold, *J. Solid State Chem.* **35**, 1980, 128.
439. Z. Zhou, J. Ye and H. Arakawa, *Int. J. Hydrogen Energy* **28**, 2003, 663.
440. K. Sayama and H. Arakawa, *J. Photochem. Photobiol. A: Chem.* **77**, 1994, 243.
441. Y. Takahara, J. N. Kondo, T. Takata, D. Lu and K. Domen, *Chem. Mater.* **13**, 2001, 194.
442. H. Kato and A. Kudo, *Chem. Phys. Lett.* **295**, 1998, 487.
443. H. Kato and A. Kudo, *Catal. Lett.* **58**, 1999, 153.
444. H. Kato and A. Kudo, *J. Phys. Chem. B* **105**, 2001, 4285.
445. H. Kato, K. Asakura and A. Kudo, *J. Am. Chem. Soc.* **125**, 2003, 3082.
446. A. Yamakata, T. Ishibashi, H. Kato, A. Kudo and H. Onishi, *J. Phys. Chem. B* **107**, 2003, 14383.
447. D. W. Hwang, H. G. Kim, J. Kim, K. Y. Cha, Y. G. Kim and J. S. Lee, *J. Catal.* **193**, 2000, 40.
448. A. Kudo, H. Kato and S. Nakagawa, *J. Phys. Chem. B* **104**, 2000, 571.

449. H. Kato and A. Kudo, *J. Photochem. Photobiol. A: Chem.* **145**, 2001, 129.
450. A. Kudo, S. Nakagawa and H. Kato, *Chem. Lett.* 1999, 1197.
451. H. Kato and A. Kudo, *Chem. Lett.* 1999, 1207.
452. N. Hatanaka, T. Kobayashi, H. Yoneyama and H. Tamura, *Electrochim. Acta* **27**, 1982, 1129.
453. K. Domen, A. Kudo, A. Shinozaki, A. Tanaka, K. Maruya and T. Onishi, *J. Chem. Soc., Chem. Commun.* 1986, 356.
454. J. Yoshimura, A. Kudo, A. Tanaka, K. Domen, K. Maruya and T. Onishi, *Chem. Phys. Lett.* **147**, 1988, 401.
455. A. Kudo, A. Tanaka, K. Domen, K. Maruya, K. Aika and T. Onishi, *J. Catal.* **111**, 1988, 67.
456. Y. Il Kim, S. Salim, M. J. Hug and T. E. Mallouk, *J. Am. Chem. Soc.* **113**, 1991, 9561.
457. K. Sayama, A. Tanaka, K. Domen, K. Maruya and T. Onishi, *J. Catal.* **124**, 1990, 541.
458. K. Domen, J. Yoshimura, T. Sekime, A. Tanaka and T. Onishi, *Catal. Lett.* **4**, 1990, 339.
459. J. Yoshimura, Y. Ebina, J. Kondo and K. Domen, *J. Phys. Chem.* **97**, 1993, 1970.
460. Y. W. Liou and C. M. Wang, *J. Electrochem. Soc.* **143**, 1996, 1492.
461. T. Takata, Y. Furumi, K. Shinohara, A. Tanaka, M. Hara, J. N. Kondo and K. Domen, *Chem. Mater.* **9**, 1997, 1063.
462. A. Kudo and H. Kato, *Chem. Lett.* 1997, 867.
463. A. Kudo, H. Okutomi and H. Kato, *Chem. Lett.* 2000, 1212.
464. Z. Zhou, J. Ye, K. Sayama and H. Arakawa, *Nature* **414**, 2001, 625.
465. Y. Miseki, H. Kato and A. Kudo, *Chem. Lett.* **34**, 2005, 54.
466. F. P. Koffyberg and F. A. Benko, *Appl. Phys. Lett.* **37**, 1980, 320.
467. G. Lu and S. Li, *Int. J. Hydrogen Energy* **17**, 1992, 767.
468. A. Kudo, K. Ueda, H. Kato and I. Mikami, *Catal. Lett.* **53**, 1998, 229.
469. A. Kudo and S. Hijii, *Chem. Lett.* 1999, 1103.
470. R. Konto, H. Kato, H. Kobayashi and A. Kudo, *Phys. Chem. Chem. Phys.* **5**, 2003, 3061.
471. J. Yin, Z. Zou and J. Ye, *Chem. Phys. Lett.* **378**, 2003, 24.
472. Y. Bessekhouad, M. Trari and J. P. Donmerc, *Int. J. Hydrogen Energy* **28**, 2003, 43.
473. N. Saito, H. Kadowaki, H. Kobayashi, K. Ikarashi, H. Nishiyama and Y. Inoue, *Chem. Lett.* **33**, 2004, 1452.
474. R. D. Shannon, D. B. Rogers and C. T. Prewitt, *Inorg. Chem.* **10**, 1971, 713.
475. D. B. Rogers, R. D. Shannon, C. T. Prewitt and J. L. Gillson, *Inorg. Chem.* **10**, 1971, 723.
476. G. Hitoki, A. Ishikawa, T. Takata, J. N. Kondo, M. Hara and K. Domen, *Chem. Lett.* 2002, 736.
477. G. Hitoki, T. Takata, J. N. Kondo, M. Hara, H. Kobayashi and K. Domen, *Chem. Commun.* 2002, 1698.
478. M. Hara, J. Nunoshige, T. Takata, J. N. Kondo and K. Domen, *Chem. Commun.* 2003, 3000.
479. M. Hara, T. Takata, J. N. Kondo and K. Domen, *Catal. Today* **90**, 2004, 313.
480. D. Yamasita, T. Takata, M. Hara, J. N. Kondo and K. Domen, *Solid State Ionics* **172**, 2004, 591.
481. A. Kasahara, K. Nukumizu, G. Hitoki, T. Takata, J. N. Kondo, M. Hara, H. Kobayashi and K. Domen, *J. Phys. Chem. A* **106**, 2002, 6750.
482. A. Kasahara, K. Nukumizu, T. Takata, J. N. Kondo, M. Hara, H. Kobayashi and K. Domen, *J. Phys. Chem. B* **107**, 2003, 791.
483. M. Liu, W. You, Z. Lei, G. Zhou, J. Yang, G. Wu, G. Ma, G. Luan, T. Takata, M. Hara, K. Domen and C. Li, *Chem. Commun.* 2004, 2192.
    483a. J. Sato, N. Saito, Y. Yamada, K. Maeda, T. Takata, J. N. Kondo, M. Hara, H. Kobayashi, K. Domen and Y. Inoue, *J. Am. Chem. Soc.* **127**, 2005, 4150.

484. A. Ishikawa, T. Takata, J. N. Kondo, M. Hara, H. Kobayashi and K. Domen, *J. Am. Chem. Soc.* **124**, 2002, 13547.
485. A. Ishikawa, T. Takata, T. Matsumura, J. N. Kondo, M. Hara, H. Kobayashi and K. Domen, *J. Phys. Chem. B* **108**, 2004, 2637.
486. R. Williams, *J. Chem. Phys.* **32**, 1960, 1505.
487. A. Fujishima, E. Sugiyama and K. Honda, *Bull. Chem. Soc. Jpn.* **44**, 1971, 304.
488. T. Inoue, T. Watanabe, A. Fujishima, K. Honda and K. Kohayakawa, *J. Electrochem. Soc.* **124**, 1977, 719.
489. A. J. Nozik, *Appl. Phys. Lett.* **30**, 1977, 567.
490. J. R. Harbour and M. L. Hair, *J. Phys. Chem.* **81**, 1977, 1791.
491. J. R. Harbour, R. Wolkow and M. L. Hair, *J. Phys. Chem.* **85**, 1981, 4026.
492. J. R. Darwent and G. Porter, *J. Chem. Soc., Chem. Commun.* 1981, 144.
493. K. Kalyanasundaram, E. Borgarello and M. Grätzel, *Helv. Chim. Acta* **64**, 1981, 362.
494. J. R. Darwent, *J. Chem. Soc., Faraday Trans. II* **77**, 1981, 1703.
495. M. Neumann-Spallart and K. Kalyanasundaram, *Ber. Bunsenges* **85**, 1981, 1112.
496. E. Borgarello, K. Kalyanasundaram, M. Grätzel and E. Pelizzetti, *Helv. Chim. Acta* **65**, 1982, 243.
497. K. Kalyanasundaram, E. Borgarello, D. Duonghong and M. Grätzel, *Angew. Chem.* **93**, 1981, 1012.
498. A. J. McEvoy and W. Gissler, *J. Appl. Phys.* **53**, 1982, 1251.
499. D. H. M. W. Thewissen, A. H. A. Tinnemans, M. Eeuwhorst-Reinten, K. Timmer and A. Mackor, *Nouv. J. Chim.* **7**, 1983, 191.
500. D. H. M. Thewissen, K. Timmer, E. A. van der Zouwen-Assink, A. H. A. Tinnemans and A. Mackor, *J. Chem. Soc., Chem. Commun.* 1985, 1485.
501. Matsumura, Y. Saho and H. Tsubomura, *J. Phys. Chem.* **87**, 1983, 3807.
502. J. Kiwi, *J. Phys. Chem.* **87**, 1983, 2274.
503. N. M. Dimitrijević, S. Li and M. Grätzel, *J. Am. Chem. Soc.* **106**, 1984, 6565.
504. Y-M. Tricot and J. H. Fendler, *J. Am. Chem. Soc.* **106**, 1984, 2475.
505. Y-M. Tricot and J. H. Fendler, *J. Am. Chem. Soc.* **106**, 1984, 7359.
506. M. Meyer, C. Wallberg, K. Kurihara and J. H. Fendler, *J. Chem. Soc., Chem. Commun.* 1984, 90.
507. R. Rafaeloff, Y-M. Tricot, F. Nome and J. H. Fendler, *J. Phys. Chem.* **89**, 1985, 533.
508. N. Bühler, K. Meier and J-F. Reber, *J. Phys. Chem.* **88**, 1984, 3261.
509. D. Meissner, R. Memming, L. Shuben, S. Yesodharan and M. Grätzel, *Ber. Bunsenges Phys. Chem.* **89**, 1985, 121.
510. E. Borgarello, N. Serpone, M. Grätzel and E. Pelizzetti, *Int. J. Hydrogen Energy* **10**, 1985, 737.
511. Yy. A. Gruzdkov, E. N. Savinov and V. N. Parmon, *Int. J. Hydrogen Energy* **12**, 1987, 393.
512. D. Hayes, F. Grieser and D. N. Furlong, *J. Chem. Soc., Faraday Trans.* **86**, 1990, 3637.
513. M. Matsumura, H. Ohnishi, K. Hanafusa and H. Tsubomura, *Bull. Chem. Soc. Jpn.* **60**, 1987, 2001.
514. L. Borrell, S. Cervera-March, J. Giménez, R. Simarro and J. M. Andújar, *Solar Energy Mater. Solar Cells* **25**, 1992, 25.
515. S. C. March, L. Borrell, J. Giménez, R. Simarro and J. M. Andújar, *Int. J. Hydrogen Energy* **17**, 1992, 683.
516. K. Kobayakawa, T. Miura, A. Suzuki, Y. Sato and A. Fujishima, *Solar Energy Mater. Solar Cells* **30**, 1993, 201.
517. T. Sato, K. Masaki, T. Yoshioka and A. Okuwaki, *J. Chem. Tech. Biotech.* **58**, 1993, 315.
518. C. A. Linkous, T. E. Mingo and N. Z. Muradov, *Int. J. Hydrogen Energy* **19**, 1994, 203.

519. M. Subrahmanyam, V. T. Supriya and P. R. Reddy, *Int. J. Hydrogen Energy* **21**, 1996, 99.

520. M. Kaneko, T. Okada, S. Teratani and K. Taya, *Electrochim. Acta* **32**, 1987, 1405.

521. A. S. Feiner and A. J. McEvoy, *J. Electroanal. Chem.* **308**, 1991, 327.

522. A. Sobczynshi, A. J. Bard, A. Campion, M. A. Fox, T. Mallouk, S. E. Webber and J. M. White, *J. Phys. Chem.* **91**, 1987, 3316.

523. E. N. Savinov, L. Gongxuan and V. N. Parmon, *React. Kinet. Catal. Lett.* **48**, 1992, 553.

524. J-F. Reber and M. Rusek, *J. Phys. Chem.* **90**, 1986, 824.

525. G. C. Morris and R. Vanderveen, *Solar Energy Mater. Solar Cells* **26**, 1992, 217. See also references therein.

526. S. Yanagida, T. Azuma and H. Sakurai, *Chem. Lett.* 1982, 1069.

527. J. Bücheler, N. Zeug and H. Kisch, *Angew. Chem.* **94**, 1982, 792.

528. J-F. Reber and K. Meier, *J. Phys. Chem.* **88**, 1984, 5903.

529. A. Kudo and M. Sekizawa, *Catal. Lett.* **58**, 1999, 241.

530. A. Kudo and M. Sekizawa, *Chem. Commun.* 2000, 1371.

531. I. Tsuji and A. Kudo, *J. Photochem. Photobiol. A: Chem.* **156**, 2003, 249.

532. H. Ezzaouia, R. Heindl, R. Parsons and H. Tributsch, *J. Electroanal. Chem.* **145**, 1983, 279. See also references therein.

533. H.-M. Kühne and H. Tributsch, *Ber. Bunsenges Phys. Chem.* **88**, 1984, 10.

534. H.-M. Kühne and H. Tributsch, *J. Electroanal. Chem.* **201**, 1986, 263. See also references therein.

535. A. Kudo, A. Nagane, I. Tsuji and H. Kato, *Chem. Lett.* 2002, 882.

536. A. Kudo, I. Tsuji and H. Kato, *Chem. Commun.* 2002, 1958.

537. I. Tsuji, H. Kato, H. Kobayashi and A. Kudo, *J. Am. Chem. Soc.* **126**, 2004, 13406.

538. J. E. Leisch, R. N. Bhattacharya, G. Teeter and J. A. Turner, *Solar Energy Mater. Solar Cells* **81**, 2004, 249.

539. I. Tsuji, H. Kato and A. Kudo, *Angew. Chem. Int. Ed.* **44**, 2005, 3565.

540. I. Tsuji, H. Kato, H. Kobayashi and A. Kudo, *J. Phys. Chem. B* **109**, 2005, 7323.

541. H. Tributsch and J. C. Bennett, *J. Electroanal. Chem.* **81**, 1977, 97.

542. H. Tributsch, *Struct. Bonding* **49**, 1982, 127.

543. H. Tributsch and I. Gorochov, *Electrochim. Acta* **27**, 1982, 215.

544. W. Jaegermann and H. Tributsch, *J. Appl. Electrochem.* **13**, 1983, 743.

545. A. Ennami and H. Tributsch, *Solar Cells* **13**, 1984, 197.

546. K. Ohashi, J. McCann and J. O'M. Bockris, *Nature* **266**, 1977, 610.

547. M. Tomkiewicz and J. M. Woodall, *Science* **196**, 1977, 990.

548. A. M. van Wezemael, W. H. Laflére, F. Cardon and W. P. Gomes, *J. Electroanal. Chem.* **87**, 1978, 105.

549. H. Gerischer, N. Müller and O. Haas, *J. Electroanal. Chem.* **119**, 1981, 41.

550. M. P. Dare-Edwards, A. Hamnett and J. B. Goodenough, *J. Electroanal. Chem.* **119**, 1981, 109.

551. G. Horowitz, *Appl. Phys. Lett.* **40**, 1982, 409.

552. J. J. Kelly and R. Memming, *J. Electrochem. Soc.* **129**, 1982, 730.

553. K. Uosaki and H. Kita, *Solar Energy Mater.* **7**, 1983, 421.

554. A. Heller, *J. Phys. Chem.* **89**, 1985, 2962.

555. J. O' M. Bockris and R. C. Kainthla, *J. Phys. Chem.* **89**, 1985, 2963.

556. S. Kaneko, K. Uosaki and H. Kita, *J. Phys. Chem.* **90**, 1986, 6654.

557. A. Etcheberry, J. Vigneron, J. L. Sculfort and J. Gautron, *Appl. Phys. Lett.* **55**, 1989, 145.

558. A. Etcheberry, personal communication, June (2005).

559. R. Memming in *Comprehensive Treatise of Electrochemistry* (B. E. Conway, J. O'M. Bockris and R. White, editors), Vol. 7, Chapter 8, p. 529, Plenum Press, NY and London, 1983. See also references therein.

560. R. M. Candea, M. Kastner, R. Goodman and N. Hickok, *J. Appl. Phys.* **47**, 1976, 2724.
561. M. Szklarczyk and J. O' M. Bockris, *Appl. Phys. Lett.* **42**, 1983, 1035.
562. M. Szklarczyk and J. O' M. Bockris, *J. Phys. Chem.* **88**, 1984, 1808.
563. M. Noda, *Int. J. Hydrogen Energy* **7**, 1982, 311.
564. W. Gissler and A. J. McEvoy, *Solar Energy Mater.* **10**, 1984, 309.
565. Y. Taniguchi, H. Yoneyama and H. Tamura, *Chem. Lett.* 1983, 269.
566. Y. Nakato, A. Tsumura and H. Tsubomura, *Chem. Lett.* 1982, 1071.
567. Y. Sakai, S. Sugahara, M. Matsumura, Y. Nakato and H. Tsubomura, *Can. J. Chem.* **66**, 1988, 1853.
568. M. Lanz, D. Schürch and G. Calzaferri, *J. Photochem. Photobiol. A: Chem.* **120**, 1999, 105.
569. D. Schürch, A. Currao, S. Sarkar, G. Hodes and G. Calzaferri, *J. Phys. Chem. B* **106**, 2002, 12764.
570. D. Schürch and A. Currao, *Chimia* **57**, 2003, 204.
571. A. Currao, V. R. Reddy, M. K. van Veen, R. E. I. Schropp and G. Calzaferri, *Photochem. Photobiol. Sci.* **3**, 2004, 1017.
572. A. Currao, V. R. Reddy and G. Calzaferri, *Chem. Phys. Chem.* **5**, 2004, 720.
573. Y. Murata, S. Fukuta, S. Ishikawa and S. Yokoyama, *Solar Energy Mater. Solar Cells* **62**, 2000, 157.
574. S. H. Baeck, T. F. Jaramillo, C. Brändli and E. W. McFarland, *J. Comb. Chem.* **4**, 2002, 563.
575. A. Kudo and I. Mikami, *Chem. Lett.* 1998, 1027.
576. J. Kobayashi, K. Kitaguchi, H. Tsuiki, A. Ueno and Y. Kotera, *Chem. Lett* 1985, 627.
577. J. M. Olson, S. R. Kurtz, A. E. Kibbler and P. Faine, *Appl. Phys. Lett.* **56**, 1990, 623.
578. K. A. Bertress, S. R. Kurtz, D. J. Friedman, A. E. Kibbler, C. Kramer and J. M. Olson, *Appl. Phys. Lett.* **65**, 1994, 989.
579. O. Khaselev and J. A. Turner, *J. Electrochem. Soc.* **145**, 1998, 3335.
580. M. G. Mauk, A. N. Tata and B. W. Feyock, *J. Cryst. Growth* **225**, 2001, 359.
581. I. Shiyanovskaya and M. Hepel, *J. Electrochem. Soc.* **146**, 1999, 243.
582. N. R. de Tacconi, C. R. Chenthamarakshan, K. Rajeshwar, T. Pauporté and D. Lincot, *Electrochem. Commun.* **5**, 2003, 220.
583. N. Serpone, E. Borgarello and M. Grätzel, *Chem. Commun.* 1984, 342.
584. F-T. Liou, C. Y. Yang and S. N. Levine, *J. Electrochem. Soc.* **129**, 1982, 342.
585. W. Siripala, A. Ivanovskaya, T. F. Jaramillo, S-H. Baeck and E. F. McFarland, *Solar Energy Mater. Solar Cells* **77**, 2003, 229.
586. Y. Nakato, N. Takamori and H. Tsubomura, *Nature* **295**, 1982, 312.
587. F. R. F. Fan and A. J. Bard, *J. Electrochem. Soc.* **128**, 1981, 945.
588. R. C. Kainthla and J. O' M. Bockris, *Int. J. Hydrogen Energy* **13**, 1988, 375.
589. E. L. Miller, R. E. Rocheleau and X. M. Deng, *Int. J. Hydrogen Energy* **28**, 2003, 615.
590. J. R. White, F-R. F. Fan and A. J. Bard, *J. Electrochem. Soc.* **132**, 1985, 544.

8

Photobiological Methods of Renewable Hydrogen Production

Maria L. Ghirardi, Pin Ching Maness, and Michael Seibert

National Renewable Energy Laboratory (NREL), Golden, CO

1 Introduction

Oxygenic photosynthetic microbes such as green algae and cyanobacteria normally absorb sunlight and store the energy in the form of polysaccharides such as starch (in green algae) or glycogen (in cyanobacteria). These storage biomolecules are mobilized, as required, to produce the energy needed to drive microbial metabolism. The conversion of light energy into chemical potential, "the light reactions of photosynthesis", is well described and has been reviewed recently (cf. Ref. 32,132,184, and references therein). Under certain conditions[65,84,85,159,164,168,249,263,265] these microbes can use the energy from sunlight to produce H_2 gas instead of fixing carbon, and thus are at least partially able to sustain growth and cellular repair function although at lower than normal levels for up to several days.[65] However, H_2 production is not the normal function of algal and cyanobacterial photosynthesis. Indeed, the H_2-producing enzymes are not even synthesized under normal growth conditions (with the exception of some cyanobacteria), and their genes are only expressed following exposure to specific environmental conditions.[11,68,102,255] Nevertheless, it is this type of biological function that offers the potential to efficiently generate renewable H_2 in direct light conversion processes.

The photobiological production of H_2 by oxygenic, photosynthetic organisms occurs in four major sequential steps:

1. light absorption and transfer of excitation energy to reaction center (RC) molecules;
2. charge separation by the RCs, electron extraction from water, and charge equilibration between the photosystems;

3. transfer of low potential (high energy) electrons to intermediate acceptors such as ferredoxin or some combination of an organic carrier and ferredoxin; and

4. H_2 gas release associated with nitrogenase or hydrogenase enzymes.

The overall light-conversion efficiency of the process is determined mainly by the absorption spectrum of the organism, the amount of that energy delivered to the H_2-producing enzyme, and the energetic requirements of the enzyme catalyzing the H_2-evolving step. The purpose of Sections 2 and 3 is to show how green algae and cyanobacteria approach each step and how the process might be mobilized to produce useful amounts of H_2. Section 4 will discuss other types of biohydrogen systems, Section 5 will examine the major current scientific issues preventing application of photobiological H_2-production technology, and Section 6 will outline some future directions.

2 Green Algae

2.1 Mechanism of Hydrogen Production

Photosynthesis in green algae takes place in cellular organelles known as chloroplasts, which contain an array of thylakoid membrane vesicles forming both stacked and unstacked regions.[28] The thylakoid membranes bind many of the electron transport components of photosynthesis, and their arrangement in a vesicular structure defines an inside (lumenal space) and an outside (stromal region).[28] The light reactions of photosynthesis in plants, algae, and cyanobacteria are organized according to the so-called Z-scheme[189] depicted schematically in Fig. 1. The antenna chlorophyll (Chl) molecules absorb sunlight and funnel the energy to the RC complexes of two sequential photosystems, called photosystem II (PSII) and photosystem I (PSI). This process drives light-induced, charge-separation reactions in both RCs. The resultant chemical potential generated in PSII is coupled to a series of electron transfer steps that use four photons to generate one O_2 molecule, four protons, and four electrons per two H_2O molecules as indicated in Eq. 1.

The electrons and protons derived from water are normally used to fix CO_2 and produce sugars using the Benson-Calvin Cycle (Fig 1). However, under certain conditions, electrons on the reducing side of PSI can be diverted at the level of ferredoxin (Fd) to another pathway (Fig 1), employing an [FeFe]-hydrogenase (H_2ase) to evolve H_2 gas. During photosynthetic H_2 production, two protons and two electrons are recombined in a reaction catalyzed by the hydrogenase enzyme to yield one H_2 molecule (Eq. 2). The overall balanced reaction, combining Eqs. 1 and 2, uses eight photons (Eq. 3) because two light reactions (both PSII and PSI) are involved. Thus, the overall quantum yield of hydrogenase-catalyzed photosynthetic H_2 production on a per-absorbed photon basis is 25%, however, the maximum theoretical solar conversion efficiency to H_2 on an energy basis is only about 13% (see below).

$$2 H_2O + 4 \text{ photons} = 4 H^+ + 4e^- + O_2 \qquad (1)$$

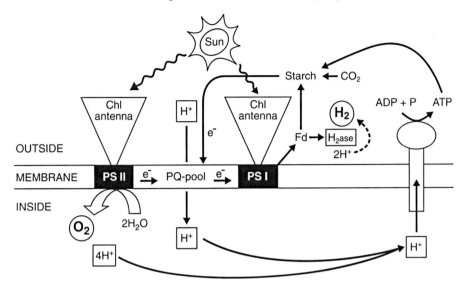

Fig. 1. Abbreviated Z-scheme for the light reactions of oxygenic photosynthesis showing electron transport pathways to carbon fixation and hydrogen production.

$$2 H^+ + 2e^- + 2 \text{ photons} = H_2 \tag{2}$$

$$2 H_2O + 8 \text{ photons} = O_2 + 2H_2 \tag{3}$$

The absorption spectrum of green algae closely resembles that of plants, since they contain the same types of light-absorbing antenna pigments (mainly Chls and carotenoids). As a result, these organisms can absorb at maximum about 43–45% of the incident solar irradiation, with an average energy content of 2.25 electron volts (eV)/photon (between 400 and 700 nm).[28] After light absorbed by the antenna pigments is transferred as excitation energy to one of the two specialized RC Chl-protein complexes, the RCs convert the energy into chemical potential, in the form of a charge-separated state. In PSII, this state corresponds to $P680^+Pheo^-$ (Fig. 2), where $P680^+$ is the oxidized form of a Chl monomer (previously P680 was identified as a dimer) and $Pheo^-$ is the reduced state of the primary electron acceptor, pheophytin. In PSI, the initial charge-separated state is $P700^+A_0^-$ (Fig. 2), where $P700^+$ is the oxidized form of a Chl dimer, and A_0^-, the reduced electron acceptor, is a Chl monomer. All the co-factors involved in charge separation are bound to proteins that comprise the PSII and PSI RC complexes (Figs. 1 and 2).

The amount of energy stored in *each* of these two initial charge-separated states (i.e., their band gaps) is about 1.8 V for PSII and 1.5 V for PSI, and this represents about 71% of the absorbed light energy.[189] Moreover, as shown in Figure 2, the band gap of 1.8 eV associated with PSII straddles the energy bands of both the water oxidation (O_2 evolution) and water reduction (H_2 production) processes, suggesting that PSII itself has the potential to directly oxidize water molecules to O_2 and reduce protons to H_2. However, the initial charge-separated states in PSII and PSI are not

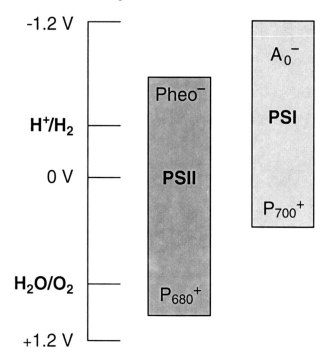

Fig. 2. Energy levels of the charge-separated states of Photosystems I and II (PSI and PSII).

stable, and quickly recombine (half-lives in the nanosecond time domain)[28] unless the negative charges are rapidly transferred down the chains of electron acceptors, and the positive charges filled by electrons from donor molecules. In PSII, $P680^+$ is ultimately re-reduced by electrons extracted from water molecules (redox midpoint potential, $+ 0.815$ eV), and the negative charge at $Pheo^-$ is shuttled via several plastoquinone (PQ) molecules through the cytochrome b6/f protein complex to plastocyanin, where it re-reduces the PSI electron donor, $P700^+$ (Figs. 1 and 2). The negative charge in A_0^- at PSI, on the other hand, is transferred through several FeS centers to a soluble ferredoxin molecule (redox midpoint potential, $- 0.45$ eV). The resulting stable effective charge-separated state, O_2 and Fd^- has a band gap of 1.26 eV, having absorbed 2 x 2.25 eV photons per electron transferred. This yields an energy conversion efficiency of only about 30% on a per absorbed photon basis. It is noteworthy that the loss of efficiency (from 71% to 30%), utilizing two light reactions is a trade-off between high efficiency and high product stability.

The light-catalyzed oxidation of water by the PSII RC complex yields three products, O_2, electrons, and protons (Fig. 1). Oxygen gas diffuses out of the cells while the protons accumulate in the lumenal space, since the latter cannot cross the lipophilic thylakoid membranes. During electron transfer from PSII to PSI, two electrons, coupled with two protons, are transported from the stromal side (outside) of the thylakoid membrane to the lumen (inside). This also contributes to the accumulation of a proton gradient across the membrane. The proton gradient creates a proton-

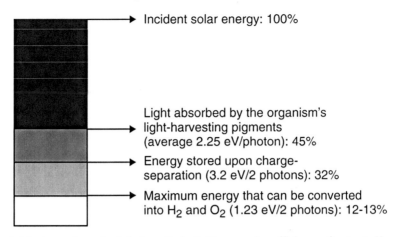

Fig. 3. Maximum theoretical photosynthetic light conversion efficiency of water to H₂ using sunlight. The arrows indicate efficiency losses at various stages after photosynthesis starts, and the remaining efficiency at each stage is indicated as a percentage.

motive force across the thylakoid membrane that drives the synthesis of ATP, a molecule that stores the energy from absorbed sunlight in the form of a stable, high-energy (7 kcal/mol) phosphate bond. The synthesis of ATP is catalyzed by an AT-Pase, a trans-membrane protein complex that links the dissipation of the proton gradient to the phosphorylation of ADP (see Fig. 1).

Although hydrogenase-linked H₂ production does not require ATP utilization, normal aerobic fixation of atmospheric CO_2 does. As will be discussed below, when CO_2 fixation does not occur (as is the case under anaerobic, sulfur-deprived conditions), the accumulation of ATP molecules in the stroma inhibits ATPase function. This results in the non-dissipation of the proton gradient and causes the build-up of the proton-motive force. It has been shown that, under these conditions, photosynthetic electron transport is down-regulated[9,17] and consequently reductant is not available for efficiently producing H₂.[140]

2.2 Hydrogenase-Catalyzed H₂ Production

The last step in the photosynthetic algal H₂ production process occurs with the sequential transfer of two negative charges from reduced ferredoxin to a soluble [FeFe]-hydrogenase,[98,216] which is almost isoenergetic ($\Delta E_h = -0.020$)[11] with ferredoxin. The reduction of protons to H₂ gas, catalyzed by the hydrogenase enzyme, is facilitated by the rapid diffusion of H₂ gas produced in the cells out of the hydrogenase catalytic site and into the extracellular medium,[51] and by the continuous reduction of ferredoxin coupled to the light reactions of photosynthesis. Both drive the process away from equilibrium and increase the rate of H₂ production. As expected, continuous removal of product (by purging with a neutral gas, for instance) also increases the rate of H₂ production by reducing product inhibition.[84,85]

Besides water oxidation, endogenous substrate (such as starch) catabolism can also generate reductants for the algal photosynthetic electron transport chain at the

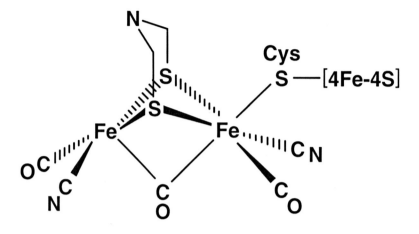

Fig. 4. Schematic diagram of the catalytic H-cluster of an [FeFe]-hydrogenase.

level of the plastoquinone pool[101,131,280] through the NAD(P)H/plastoquinone oxido-reductase pathway (see Fig. 1; arrow from starch to the PQ-pool). The relative contribution of water oxidation and substrate degradation may vary according to the light intensity and other environmental parameters,[137] reflecting the varying amounts of O_2 being evolved by PSII.

Figure 3 summarizes the maximum theoretical efficiencies of each step in the photosynthetic production of H_2 by green algae, and provides the rationale for further developing green algal photobiological systems.

2.3 [FeFe]–hydrogenases

Algal hydrogenases belong to the class of [FeFe]-hydrogenases, characterized by the presence of an H-cluster at their catalytic centers.[2b,255] As shown in Fig. 4, H-clusters are unique structures, consisting of a 4Fe4S center bridged by a cysteine residue to a 2Fe2S cluster. The latter is stabilized by CO and CN ligands,[194] and by an uncharacterized dithiolate bridge between the irons of the 2Fe2S cluster (Y in Fig. 4)[243] Both propanedithiol and a dithiomethylamine structures have been postulated, based on crystallographic data[181] and on theoretical calculations.[182] Inorganic chemistry supports the idea of an N-protonated derivative.[143] The H-cluster structure endows [FeFe]-hydrogenases with the fastest catalytic turnover number of all hydrogenases, but algal enzymes are also extremely sensitive to O_2 inactivation.[2b] The latter occurs by the irreversible binding and oxidation of the distal iron in the 2Fe2S cluster by O_2.[70] Interestingly, the same Fe atom also binds CO, but in a reversible manner.[70] Recent work from our laboratory[199] demonstrated that the assembly of the H-cluster is a unique process, unlike that of the normal assembly process of 4Fe4S centers or that of the metallo-center in [NiFe]-hydrogenases (see Section 3.3). H-cluster assembly requires the function of the S-adenosylmethionine (SAM)-like[235] algal proteins, HydEF and HydG,[199] which have homologous genes in all known organisms that contain [FeFe]-hydrogenases. These proteins act through the intermediate formation of radicals, although the exact reaction mechanism and intermediates in the assembly of FeFe hydrogenases have not as yet been determined. [FeFe]-hydrogenases are also found in strictly anaerobic bacteria such as Clostridia and Desulfovibria,[255,281] thermophiles such as *Thermatoga maritima*,[253] in the hydrogenosomes of protists such as *Trichomonas vaginalis*,[40] and in fungi.[260] In almost all cases, [FeFe]-hydrogenases are H_2-producing enzymes, although H_2-oxidizing [FeFe]-hydrogenases have been described in *Desulfovibrio vulgaris*.[2b] Two algal [FeFe]-hydrogenase genes(and, in many cases, the two corresponding proteins) have been identified in a number of green algae such as *Chlamydomonas reinhardtii*,[68,102] *Scenedesmus obliquus*,[67,269] *Chlorella fusca*,[266] and *Chlamydomonas moewusii*.[267] Moreover, H_2-production activity has been reported in the green alga, *Platymonas subcordiformis*,[86] thirty other genera of green algae, two species of yellow-green algae, and one species of diatom.[31,35,32] Although it is not clear what specific role each hydrogenase plays in these organisms, both enzymes in *C. reinhardtii* produce H_2 *in vitro*, when reduced mediators are used.[51,126]

Besides being inactivated by O_2,[74,99,216] algal [FeFe]-hydrogenases are also inhibited transcriptionally in the presence of O_2.[68,102] Under aerobic conditions, the hydrogenase genes are not transcribed (they are turned OFF), but a short exposure to anaerobic conditions, either in the light or in the dark, induces their transcription (they are turned ON). Research is underway to determine the mechanism of algal gene expression, in order to allow researchers to turn the genes ON or OFF at will. Work in our laboratory has demonstrated two complementary types of [FeFe]-hydrogenase regulatory activity in *C. reinhardtii*.[201] First, regulation of H_2-production activity was shown to occur at the transcriptional level,[200] possibly modulated by the energy state of the cells. Second, H_2-production activity was also shown to depend on the expression of 2 genes, encoding for proteins required for the assembly of an active hydrogenase.[199] The identification of the assembly or maturation genes for [FeFe]-hydrogenases also allowed us to produce active enzyme heterologously in organisms that contain only NiFe-hydrogenases, by co-expressing the hydrogenase structural gene with its maturation genes.[126,199]

3 Cyanobacteria

3.1 Mechanisms of Hydrogen Production

Cyanobacteria, previously known as blue-green algae, appeared in the fossil record about 2.8 to 3.5-billions years ago during the Proterozoic era of the Precambrian period.[81] They are true prokaryotes, belong to the *Bacteria* domain, and as such do not contain a separate organelle where oxygenic photosynthesis takes place. Instead, cyanobacterial photosynthesis takes place on the intra-cytoplasmic membranes (ICM), possibly in direct contact with the cytoplasmic membranes,[134,237] from which the biogenesis of photosynthetic RCs likely originated.[273] Cyanobacteria also carry out oxygenic photosynthesis using two photosystems, in a manner similar to green algae and higher plants. In addition to photosynthesis, respiratory activity also occurs on the ICM[223] and perhaps respiration shares some of its electron transport carriers with the photosynthetic apparatus. The co-mingling of cyanobacterial electron transport components in photosynthesis and respiration poses potential challenging issues for H_2 production not observed with green algae. Many cyanobacteria can also fix atmospheric N_2 and convert it to ammonia. Serving as the evolutionary link between the more primitive bacteria and the more advanced green algae and higher plants, cyanobacteria possess features common to both groups and are found in diverse ecosystems.

The light absorbing pigments of cyanobacteria encompass a larger variety of forms than those found in plants and green algae, except that the hydrogenase is linked to NAD(P)H oxidation, and as such can more efficiently absorb incident light in the middle part of the visible spectrum. Besides Chls and carotenoids, cyanobacteria contain major light-absorbing antenna structures called phycobilisomes. Phycobilisomes are attached to the periplasmic side of the ICM[37] and transfer excitation energy to the PSII RC in an manner similar to that of the normal plant-type antenna complexes.[73,16] Both phycocyanin and allophycocyanin pigments absorb red light maximally in phycobilisomes, while phycoerythrins absorb green light, the region of the solar spectrum not absorbed strongly by green algae or higher plants.[52]

Once absorbed by the antenna pigments, the light energy is converted into chemical energy in the form of electrons, protons and ATP using the Z scheme of photosynthesis (Fig. 1). The actual release of H_2 by cyanobacteria, in the last step of the process, can occur by one of two separate reactions: indirectly via an ATP-dependent, nitrogenase-mediated reaction or directly by a hydrogenase-catalyzed reaction similar to that described above for green algae except that the hydrogenase is linked to NAD(P)H oxidation. Without the ATP-requirement, the hydrogenase system is two to three times more energy efficient (at least theoretically) in terms of light-energy conversion to H_2 than the nitrogenase system. In addition, all N_2-fixing cyanobacteria contain an H_2-oxidizing uptake hydrogenases[228] to recycle or recapture the energy in H_2 to improve the efficiency of N_2 fixation. Obviously, the net H_2 output is diminished in the presence of uptake-hydrogenase activity, the deletion of which leads to enhanced H_2 production. The following Sections will address these issues in more detail.

3.2 Hydrogenase-Catalyzed H_2 Production

A bidirectional or reversible [NiFe]-hydrogenase has been reported in both N_2-fixing and non-N_2-fixing cyanobacteria, yet its presence is not universal.[240-242] Nevertheless, all non-N_2-fixing cyanobacteria tested thus far contain the bidirectional hydrogenase.[228] H_2 evolution in the cyanobacterium, *Anabaena* 7120, occurs when the cells were illuminated with low intensity light.[111] Light enhancement was very short-lived, and the addition of a ferredoxin antagonist lowered the H_2-production activity. This was the first evidence suggesting that light-driven H_2 evolution via a bidirectional hydrogenase could be linked to a photo-reduced ferredoxin, perhaps involving pathways and components similar to those of green algae. However, based on its high affinity for H_2 (K_m = 2.3 µM), the authors[111] proposed that the physiological role of this hydrogenase was likely in H_2 uptake. Furthermore, based on its extreme sensitivity to O_2, being inactivated by levels as low as 0.1% O_2,[112] the bidirectional hydrogenase likely functions only under anaerobic conditions. The actual role of this hydrogenase in cyanobacteria is still being debated. Appel and coworkers[12] proposed that the bidirectional hydrogenase in *Synechocystis* sp. PCC 6803 serves as an electron valve to dissipate low potential electrons to ensure immediate electron transport at the onset of photosynthesis. This was consistent with earlier evidence from Abdel-Basset and Bader,[1] who showed that a burst of H_2 is seen during a dark-to-light transition and that as soon as O_2 accumulated, H_2 was consumed rapidly. This hydrogenase was also rapidly inactivated by O_2.

Despite the lack of activity in the presence of O_2, the inactivation of cyanobacterial [NiFe]-hydrogenases by O_2 is reversible. Exposure of partially purified bidirectional hydrogenase from *Oscillatoria limnetica* to air for 50 days at 20 °C results in its total deactivation; however, it recovers nearly 80% of its original activity upon restoration of anaerobic conditions.[112] Immunoblot studies demonstrated that the *Anabaena variabilis* hydrogenase protein is present under both anaerobic and aerobic conditions, suggesting that regulation of hydrogenase activity occurs by reactivating a pre-existing deactivated form of the protein when anaerobic conditions were restored.[233] The bidirectional hydrogenase is now believed to be a constitutively expressed enzyme commonly found in both N_2-fixing and non-N_2 fixing cyanobacteria, where it is synthesized in an inactive form even under normal O_2-evolving photosynthetic conditions.[19,241] This is a positive feature with significant implications in solar biohydrogen applications since an induction phase is not necessary.

The molecular biology of the biodirectional hydrogenases may provide additional insight as to their physiological function. The hydrogenase structural protein (the catalytic large and small subunits) are encoded by *hox*YH, and the diaphorase component (the flavo-protein subunit of the hydrogenase that interacts with NAD/NADH)[242] is encoded by *hox*EFU. The latter has a high level of homology with the respiratory complex I of *E. coli*.[10,33,34,225,227] The HoxFUYH components of the hydrogenase display high levels of homology with those of *Ralstonia eutropha*, which is a NAD-reducing hydrogenase.[225] Indeed, biochemical studies confirm the presence of a $NAD(P)^+$-dependent hydrogenase activity (both H_2 evolution and H_2 oxidation) in the cyanobacteria, *Anacystis nidulans*[226] and *Synechocystis* sp. PCC 6803.[227] The presence of the diaphorase subunits (HoxEFU) and their homology with those components in respiratory complex I in *E. coli* implies that the cyanobacterial hydrogenase may play a physiological role in consuming NAD(P)H in support of respiration or cyclic electron transport around PSI.[10,227,242] This may explain the lack of H_2 accumulation in cyanobacteria cultured under photosynthetic conditions, in addition to the O_2 sensitivity of the hydrogenase protein. Notably, sustained levels of H_2 production were detected for the first time in a mutant of *Synechocystis* sp. PCC 6803 defective in type I NADPH-dehydrogenase complex (NDH-1) involved in respiratory activity.[53] Consequently, in order to realize the potential of solar H_2 production in cyanobacteria, scientific breakthroughs are essential in at least two areas: the development of a more O_2-tolerant hydrogenase and the uncoupling of H_2 production from its oxidation via the complex-I respiratory pathway.

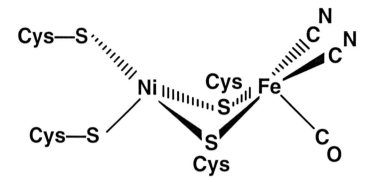

Fig. 5. Schematic diagram of the active site of the [NiFe]-hydrogenase.

3.3 [NiFe]-Hydrogenases

All [NiFe]-hydrogenases contain two subunits. The large catalytic subunit of the bidirectional hydrogenase from cyanobacteria (the *hox*H gene product) has a cysteine motif and a stretch of 26 amino acids near the carboxyl terminus. The latter is typically involved in nickel insertion and binding.[33] [NiFe]-hydrogenases from various microbes are highly conserved and phylogenetically distinct from the [FeFe]-hydrogenases.[255] The large catalytic subunit harbors an unusual binuclear center containing Fe and Ni, bridged via two cysteine residues (Fig. 5).[258,259] Moreover, similar to [FeFe]-hydrogenases, the Fe atom of [NiFe]-hydrogenases is also coordinated to toxic ligands, including one CO and two CN groups.[100] These ligands likely stabilize the Fe ion in a lower spin state or lower redox state than otherwise possible, allowing more favorable binding of H_2 gas or of the hydride ion.[3] The small subunit binds one or more Fe-S clusters, with a role in transferring electrons to or accepting electrons from the catalytic center of the large subunit.[258]

The biosynthesis of the metal center of the [NiFe]-hydrogenase requires the synergistic actions of at least seven accessory proteins. Together, they are involved in the insertion of nickel, and the synthesis and insertion of the CO and CN ligands into the active site. The maturation process (see Fig. 6 for a schematic representation of the process) has been elucidated through a series of mutant studies of the large subunit of the hydrogenase 3 (HycE) in *E. coli*. The auxiliary proteins are encoded by a set of *hyp* (hydrogenase pleiotrophy) genes (*hyp*A,B,C,D,E,F) and by the *hyc*I gene; the latter controls C-terminus processing leading to a mature protein. It was suggested that carbamoyl phosphate (CP) serves as the substrate for the biosynthesis of either the CO or the CN ligand, or both, catalyzed by the HypF enzyme.[191–192] HypF is a carbamoyl-transferase and catalyzes the transfer of a carbamoyl moiety from CP to the HypE protein. This forms a HypE-thiocyanate intermediate, which ultimately serves as the cyanide donor to the metal center of the precursor of HycE (pre-HycE). The donation is assisted by a complex formation between the HypE-thiocyate and HypC and HypD. These latter two proteins provide a platform for supplying the CO ligands to the Fe atom of the active site in pre-HycE.[30,212] The formation of the

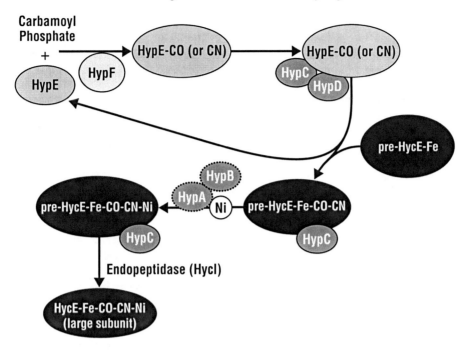

Fig. 6. Schematic representation of the biosynthetic pathway associated with the assembly of the NiFe-hydrogenase catalytic metal center. The precursor to the large subunit is represented by pre-HycE-Fe, and the CO (and/or CN) ligands are provided by the multi-protein complex formed between HypE, HypC and HypD. The source of the ligands to HypE is carbamoyl phosphate, and the insertion reaction is catalyzed by HypF. The Ni atom is inserted into the pre-HycE-Fe-CO-CN-HypC complex in a reaction catalyzed by HypA and HypB. Finally, the HycI endopeptidase processes the C-terminus of the pre-HycE-Fe-CO-CN-HypC complex, yielding the mature NiFe-hydrogenase large subunit.

Fe(CN)$_2$(CO) group precedes nickel insertion into the hydrogenase apoprotein. HypA cooperates with HypB in the storage of and insertion of nickel into pre-HycE.[115,151] A lesion in *hyp*B can be complemented by high nickel ion concentration in the medium.[119,146] HypC is also a chaperone-like protein, forming a complex with preHycE to keep the latter in a conformation more favorable for nickel insertion.[59] Once nickel is incorporated, HypC leaves the complex to allow for the C-terminus processing of pre-HycE, catalyzed by an endopeptidase encoded by *hyc*I.[25,150,217] Nickel insertion is thus a pre-requisite for the C-terminus processing to occur. C-terminal processing, however, is not a universal phenomenon for maturation among all [NiFe]-hydrogenases. The H$_2$-sensing hydrogenase in *R. capsulatus* and the CO-linked hydrogenase from *Rhodospirillum rubrum* both lack the C-terminus extension, therefore no protease is involved in the final maturation process in these cases.[69,256] Figure 6 summarizes all of this information in a schematic diagram.

Based on the assembly/maturation genes in *E. coli* as a model, homologous *hyp*-type of genes have been discovered in other microbes containing [NiFe]-

hydrogenases. In *Ralstonia eutropha*, the gene products of *hyp*A, B, C, D, E, and F are involved in the maturation of its three [NiFe]-hydrogenases: a soluble NAD-reducing hydrogenase, a membrane-bound hydrogenase linked to electron transport, and a H_2-sensing hydrogenase. The last senses ambient H_2, which leads to the expression of the other two energy-yielding hydrogenase proteins.[41,58] An additional gene (*hyp*X) was discovered in *R. eutropha* and *Rhizobium leguminosarum*, whose function is to attach an additional CN ligand to the nickel metal.[29,214] Similarly, *hyp*A-F genes have been found upstream of the [NiFe]-type uptake hydrogenase in *Nostoc* sp. 73102, the assembly of which is carried out putatively by the adjacent *hyp* genes.[96] *Hyp* genes have also been found to cluster near the genes encoding the uptake and H_2-sensing hydrogenases in the purple non-sulfur photosynthetic bacterium, *R. capsulatus*, with putative roles in assembling both [NiFe]-hydrogenases.[256]

3.4 Nitrogenase-Catalyzed H_2 Production

In addition to the bi-directional hydrogenase, many cyanobacteria can produce H_2 via the nitrogenase enzyme, the catalyst responsible for biological N_2-fixation. The nitrogenase gene is only de-repressed in the complete absence, or in the presence of low amounts of fixed nitrogen nutrients such as nitrate or ammonium.[112] The physiological function of nitrogenases is to reduce atmospheric N_2 to ammonium, which is then assimilated into amino acids to support cell growth. In addition to N_2 reduction, nitrogenases also catalyze proton reduction to by-product H_2 as an inherent property of the nitrogenase reaction. The overall process is as follows:

$$N_2 + 8H^+ + 8e^- + 16ATP \rightarrow 2NH_3 + H_2 + 16ADP + 16\ P_i \qquad (4)$$

Hydrogen can still be produced where the normal substrate N_2 is absent, and the enzyme is functional. Under these conditions only 4 moles of ATP are required per mole of H_2 evolved.

Nitrogenase-catalyzed H_2 production has attracted attention in the biohydrogen research field because of its high rate of H_2 production. However, there are some serious issues inherent to this enzyme. Nitrogenases are less efficient energetically compared to the hydrogenases in catalyzing H_2 production due to their requirement for ATP. Since nitrogenases require two ATP molecules for each electron transferred, a minimum of four ATPs are required per mol H_2 produced, provided that N_2 gas is not present.[263] Nitrogenases also exhibit a very low turnover frequency (6.4 s^{-1}) requiring the constant biosynthesis of very large quantities of the enzyme under low fixed nitrogen conditions.[94,112] Cell growth is thus limited in the absence of N_2 gas, leading to lower quantities of total H_2 output due to low cell mass. This will be an obstacle for scale-up of H_2 production in applied systems. Nevertheless, there has been a lot of interest around the world in developing working systems.

Nitrogenases are also extremely labile in an O_2 environment. Cyanobacteria have adopted two strategies to overcome the O_2 sensitivity of the nitrogenase reaction, either temporal or spatial separation of O_2 evolution and H_2 production. Temporal separation has been achieved mainly in non-heterocystous unicellular cyanobacteria such as *Gloeothece* sp. strain ATCC 27152 and *Plectonema boryanum*. Nitrogen fixation occurs only in darkness, during which respiration of the carbohydrates, synthesized and stored during photosynthesis, decreases the ambient O_2 to levels low enough so that nitrogenase function is not inhibited, while at the same time yielding enough ATP to sustain both N_2 fixation and H_2 production.[72,165,211] Under fixed-nitrogen-deprived conditions, spatial separation is accomplished by differentiation of approximately 5–10% of the filamentous cyanobacterial vegetative cells into specialized cells called heterocysts.[80] Lacking PSII, heterocysts perform only PSI activity using carbohydrate provided by the vegetative cells as electron donors. In return, the heterocysts provide the vegetative cells with the amino acid glutamine.[112] Heterocysts develop a thick cell envelope to retard diffusion of O_2 diffusion into the structure, and operate with increased respiratory activities to scavenge O_2.[172] The respiratory process also supplies the heterocyst with excess ATP needed for nitrogenase activity. This spatial separation achieves the anaerobic conditions in a protected environment that allows the nitrogenase to operate even though the organism is in ambient air in the light. Because the nitrogenase enzyme is localized within the heterocysts, strategies to increase the occurrence of heterocyst frequency should subsequently produce more nitrogenase for enhanced H_2 production. Fortunately, the genes responsible for heterocyst differentiation have been identi-fied,[268] so more in-depth understanding of the expression and interactions of these genes should lead to the means to alter the frequency and pattern of heterocyst differentiation. In *Anabaena variabilis*, nitrogen starvation results in the induction of two sets of nitrogenases, one localized in heterocysts, and the other in vegetative cells; however, the latter only functions under anoxic conditions.[245] It is not known yet if this cyanobacterium produces greater amounts of H_2 production due to the presence of the second nitrogenase.

All N_2-fixing cyanobacteria examined thus far contain an H_2-oxidizing uptake hydrogenase for energy recycling.[228] This explains the lack of significant amounts of H_2 accumulation under normal N_2-fixation conditions. In filamentous cyanobacteria, uptake hydrogenase activity is more active in the heterocysts,[62,112] although more recently the uptake hydrogenase gene (*hup*L) has also been detected in the vegetative cells of *Nostoc muscorum*, *Nostoc* sp. PCC 73102, and *Anabaena variabilis*.[18] In order to improve overall H_2 output, modification of the H_2-uptake pathway is essential. Indeed, deletion of the uptake hydrogenase has resulted in a 4- to 6-fold increase in H_2 output in *Anabaena variabilis*.[103] Similarly, deletion of the uptake hydrogenase, but not its bidirectional hydrogenase,[154,155] caused a 4- to 7-fold increase in H_2 production in *Anabaena* sp. PCC7120.

3.5 Nitrogenases

Nitrogenase is composed of two subunits: the MoFe protein (dinitrogenase) and the Fe protein (dinitrogenase reductase).[91,234] The MoFe protein is a tetramer with two pairs of $\alpha\beta$-dimer configurations, encoded by the *nif*D and *nif*K genes.[61,156] Each of the $\alpha\beta$ pairs contain an [8Fe-7S]-metallocluster termed the P-cluster situated at the interface between the α and β subunits, and an iron-molybdenum cofactor (FeMo-co) localized on the α subunit. The Fe protein is a homodimer containing a single bridging [4Fe-4S] cluster at the dimer interface and two Mg-ATP-binding sites. After ATP hydrolysis, the Fe protein undergoes a conformational change and donates electrons to the P-cluster of the MoFe protein. Specifically, the electrons are transferred to the FeMo-co catalytic center, the site of substrate reduction.[210,218] Both the FeMo protein and the Fe protein are extremely sensitive to O_2 inhibition.

Mo-containing nitrogenases are the most common nitrogenases found in nature.[54] However, Bishop and co-workers[26] discovered that in the absence of Mo, *Azotobacter vinelandii* produced an alternative nitrogenase containing vanadium (V) instead of Mo in its active site. The V-containing nitrogenase has since been found in the cyanobacterium, *Anabaena variabilis*.[124,244,245] *Azotobacter vinelandii* is unique in that it also contains a third type of nitrogenase, the Fe-only nitrogenase.[47] Fe-containing nitrogenases have since been observed in the purple non-sulfur photosynthetic bacteria, *Rhodobacter capsulata*,[82] *Rhodospirillum rubrum*,[141] and *Rhodopsu edomonas palustris*.[135] The different metal-containing nitrogenases exhibit differences in reductant distribution between the N_2-reduction and proton-reduction reactions, and, in fact, the alternative nitrogenases produce more H_2 (Eqs. 5 and 6) than does the MoFe enzyme (Eq. 4):

V-nitrogenase: $N_2 + 12H^+ + 12e^- + 16ATP \rightarrow 2NH_3 + 3H_2 + 16ADP + 16 P_i$ (5)

Fe-nitrogenase: $N_2 + 21H^+ + 21e^- + 16ATP \rightarrow 2NH_3 + 7.5H_2 + 16ADP + 16 P_i$ (6)

With the alternative nitrogenases, the cells can be cultured to higher density in N_2 as the sole nitrogen nutrient, while still partitioning a larger portion of the reductant toward H_2 production. This leads to an overall increase in total H_2 output. Thus, in theory, microbes containing the V- and Fe- nitrogenases may have more potential as H_2 producers, provided that the activity of the uptake hydrogenase is inhibited.

4. Other Systems

4.1 Non-Oxygenic Purple, Non-Sulfur Photosynthetic Bacteria

With only one photosystem and the inability to split water, purple, non-sulfur photosynthetic bacteria must carry out anoxygenic photosynthesis using organic carbon compounds, sulfides or thiosulfate as electron donors. These microbes contain several types of bacteriochlorophyll and carotenoid pigments, which absorb light ranging from 400 nm to near 1000 nm. Purple non-sulfur photosynthetic bacteria belong to the family, *Rhodospirillaceae*, and are noted for their diverse modes of carbon and energy metabolism.[118,193,248] Almost all of the photosynthetic bacteria examined thus far contain nitrogenase enzyme for biological N_2 fixation.[93,153,224] Similar to the systems described in Sections 3.4 and 3.5, the nitrogenase reaction in photosynthetic bacteria can also partition part of the reductant toward H_2 production. It was first observed by Ormerod et al. in *Rs. rubrum*[188] and later confirmed in *Rb. capsulatus*[105] that upon feeding carbon compounds from the citric acid cycle, H_2 was produced when the ammonia nutrient was exhausted in the medium. Photosynthetic bacteria have the advantage that they can use many different organic acid donors, including acetic, butyric, lactic, propionic, and formic acids, to support H_2 production in the light. Many of these compounds are dead-end products commonly found in anaerobic dark fermentation processes.[64,20,129,139]

Carbon conversion efficiency is defined as the experimental number of moles of H_2 produced per mole of carbon substrate consumed as a percentage of the theoretical maximum.[129] Based on the organic acid substrate used, the theoretical maximum can be calculated according to the Eq. 7, assuming that all of the substrate is used for H_2 production:

$$C_xH_yO_z + (2x - z) \ H_2O \rightarrow (y/2 + 2x - 2) \ H_2 + x \ CO_2 \qquad (7)$$

Using more easily metabolizable lactate or malate as the substrate, carbon conversion efficiencies as high as 84% have been reported in an uptake-hydrogenase deficient mutant of *Rb. capsulatus*,[187] 80% in *Rs. rubrum*[277] and nearly 100% in *Rb. sphaeroides*.[222] On the other hand, acetate and butyrate are more common organic acid waste byproducts of the *Clostridial*-type fermentation observed in microbes.[43] Carbon conversion efficiencies as high as 73% have been reported in *Rhodopseudomonas* sp. using acetate,[20] and 75% in *Rb. sphaeroides* using butyrate, generated during co-culturing with *Clostridium butyricum*.[166] Other potential carbon substrates are raw algal biomass[116] and lignin-derived aromatic acids. The latter can be converted to H_2 by immobilized *Rhodopseudomonas palustris* with carbon conversion efficiency ranging from 57 to 88%.[66]

Due to the carbon substrate requirement, photosynthetic bacterial systems are most suitable for integration into the second of a two-stage process, following a dark fermentation reaction. In this case the effluent from the first stage fermentation becomes the feedstock for the second-stage photosynthetic process, as described above.

4.2 Mixed Light/Dark Systems

The combination of photosynthetic starch accumulation in the light, followed by starch fermentation under dark, anaerobic conditions has been proposed in the past as an alternative method to produce H_2 biologically. This system, also referred to as indirect biophotolysis,[94] would involve only a single organism such as a green alga[148,164] or a combination of an alga and a photosynthetic bacterium[4,149] or a fermentative[94] bacterium in subsequent reactions. Such a two-stage process has been proposed by the Netherlands Biohydrogen Network where the H_2 is produced from biomass utilizing both stages.[48,49] Research findings from Japan reveal that the inclusion of both fermentative and photosynthetic bacterial stages significantly improves the H_2 molar yield when starch is used as the substrate.[272] Thus, a combined process can significantly improve the overall H_2 molar yield beyond 1–4 moles of H_2 per mole of glucose,[94] which currently limits the H_2 output of dark fermentation alone. The other advantage of this system is that H_2 photoproduction mediated by nitrogenase is not feedback-inhibited even by an atmosphere of 99% H_2.[106] However, one major potential issue lies in the low solar energy conversion efficiency to H_2. Based on the theoretical calculations of Akkermann et al.,[5] the photoconversion efficiency over the utilizable absorption spectrum (400–950 nm) of the photosynthetic bacteria is estimated to be 10%. This calculation is comparable to up to 7% that measured in *Rb. sphaeroides*.[167] It is important to note that these calculations do not account for the energy content of the carbon substrates that are ultimately the products of previous photosynthesis.

The theoretical quantum efficiency of such a system would be rather low, however, if one considers that photosynthetic starch accumulation requires a minimum of 48 quanta/mol glucose and fermentation of the starch-derived glucose presently only yields 2 mol of H_2 per mol glucose. The overall quantum efficiency of the system would be 2 H_2/48 absorbed quanta, or a maximum of 2% on an incident solar quantum basis. The energy conversion efficiencies would be even lower.

The low quantum efficiency is due to the fact that fermentation of starch extracts only 1/6 of the energy of a glucose molecule since, in the absence of O_2, the tricarboxylic acid (TCA) cycle and oxidative phosphorylation pathways are not active. The latter are responsible for extracting the remaining 5/6 of the energy from glucose. Unless these pathways can be activated, the yields of fermentation will not increase past the above value and will, in most cases, be lower than 2%. One potential area for improvement of the H_2 molar yield in dark fermentation is to knock out the NADH-consuming competing pathways. By blocking these pathways, which lead to either ethanol or lactic acid formation, Rachman[205] increased the H_2-molar yield by approximately 50% in *Enterobacter aerogenes*. Presumably the excess NADH could drive H_2 production, catalyzed by the NADH-ferredoxin oxidoreductase/hydrogenase pathway, as has been seen in *Clostridium cellulolyticum*.[88]

Two recent advances are of some note. First, there is an initial report of a co-culture of *Lactobacillus delbrueckii* and an unidentified mold that produces lactate. Lactate is a suitable substrate for photosynthetic bacteria and such a mixed system yielded more than 7 H_2 per glucose.[15] Secondly, a recent development has resulted in the conversion of 1 mol of acetate to nearly 2.9 mol of H_2 with an electrochemically assisted microbial fuel cell device under a dark, anaerobic environment.[144] By applying an over-potential of 250 mV or more, acetate is oxidized at the anode resulting in H_2 production at the cathode. This is the first report of anaerobic conversion of acetate to H_2 in the dark, and represents an innovative way of producing additional H_2 from waste organic acids following the darkness fermentation of sugar. This could potentially result in the production of nearly 12 mol of H_2 from one mol of glucose (minus the small amount of energy in the electrochemical input), and this would make the process very attractive.

4.3 Bio-Inspired Systems

The existence of various cellular regulatory mechanisms limiting biological H_2 photoproduction has and will continue to encourage the search for *in vitro* systems with the capability of performing light absorption/charge-separation function as well as catalytic O_2 and H_2 production. This concept was first demonstrated in systems composed of bacterial [FeFe]-hydrogenase, plant ferredoxin, and plant chloroplasts.[13,22] These initial studies identified limitations in rate of H_2 production due to O_2 inhibition of the reaction[22,190] and to the instability of the hydrogenase preparations.[22,208] The former was circumvented temporarily by adding O_2-scavengers to the system[22,109] and the latter was addressed by attempts to immobilize the enzyme.[24,79,122,127] However, the systems were always too expensive and could not be run efficiently for more than a short period of time.

Other approaches that are currently being pursued, involve *in vitro* arrangements of immobilized biological components that separate O_2 evolution from H_2 production,[264] combinations of biological and chemical components,[36,57,209] and completely artificial systems.[83,89,90,117,163,219,262] All these systems require the following components in common:

(a) efficient light-absorbing pigments;
(b) a coupled, charge-separation component with a band gap larger than 1.3-eV straddling both the O_2 and the H_2 redox bands (see Fig. 2), or a sacrificial electron donor other than water; and
(c) a suitable electron acceptor/catalyst for H_2-generation.

In bio-inspired systems, porphyrins, phthalocyanines, carotenoids, fullerenes and ruthenium bipyridyl complexes have been the most common choices for light-harvesting and charge-separation components (photosensitizers), particularly when coupled to added electron donors and acceptors to stabilize the initial charge-separated state.[170] However, their band gaps are not large enough to link H_2O oxidation to H_2 production. Sacrificial donors other than water must be utilized, instead, although they significantly increase the cost of the process, unless they are reduced waste materials.[21] Similarly, in bio-inspired systems, the H_2-production catalyst is usually a noble metal such as Pt or even a hydrogenase enzyme.[220] However, there is growing interest in synthesizing electrocatalysts, which have hydrogenase activity but do not contain noble metals.[14,145,239,282,283] Recently, a different system for the efficient transduction of light energy into a proton-motive force was successfully coupled to both ATP synthesis[236] and to the transport of Ca^{+2} ions across a artificial, lipophilic membrane.[23] Additionally, systems involving the oxidation of synthetic manganese complexes are under development[113,114,213,238] in hopes of modeling the water oxidation function of PSII as well. More detailed reviews on bio-inspired systems are available elsewhere (cf. Refs. 95 and 170 and references therein).

5   Scientific and Technical Issues

5.1 General

Although the potential for a highly efficient photobiological H_2 system is well recognized,[179] several research challenges[74,76,78,142,161,202] limit commercial application at this time. These include

(a) the lack of continuity of H_2 production in the light due to the sensitivity of the hydrogenase enzyme to O_2, one of the obligatory products of water oxidation;
(b) competition among different metabolic pathways for photosynthetic reductants;
(c) the down-regulation of photosynthetic electron transport in the absence of CO_2 fixation;

(d) the low light-saturation properties of biological photosynthetic organisms; and

(e) the high cost of photobioreactor materials and operation.

Currently, many of these issues are being addressed by groups both in the U.S. and the rest of the world, and some will be discussed in the following Sections.

5.2 Oxygen Sensitivity of [FeFe]-Hydrogenases

The first issue, the sensitivity of the algal hydrogenase enzyme to O_2, is being addressed in the U.S. using both molecular engineering and physiological approaches.[77,231] In order to engineer increased O_2 tolerance into the algal hydrogenase, research has focused on restricting the accessibility of the catalytic site to O_2 gas,[77] although an alternative approach would be to decrease the sensitivity of the catalytic site itself to O_2.[29] Detailed studies of potential gas diffusion pathways within the [FeFe]-hydrogenase protein structure, using molecular dynamics simulations, have been initiated to identify the most probable pathways for O_2 and H_2 gas diffusion.[50,51] These studies are based on the highly homologous *Clostridium pasteurianum* CpI [FeFe]-hydrogenase X-ray structure, since no structures are currently available for the algal hydrogenases. Interestingly, preliminary results have identified multiple pathways for H_2 gas diffusion out of the hydrogenase's catalytic site, while pinpointing only two major pathways for potential O_2 diffusion into the catalytic site.[50,51] One of the O_2 pathways corresponds to the previously described *H_2-channel*, proposed to link the catalytic site of the *Desulfovibrio desulfuricans* [FeFe]-hydrogenase out of the enzyme.[169,181] These results support a rational mutagenesis approach to prevent O_2 from reaching the enzyme's catalytic site by judicious site-directed mutagenesis, since restricting O_2 diffusion through the two protein channels should not significantly affect H_2 diffusion out to the surface of the enzyme.

Other groups have adopted a random mutagenesis approach to solve the O_2-sensitivity problem of the hydrogenase, as well as the screening of existing culture collections for unusual H_2-producing organisms.[278] These approaches will require the development of high-throughput screening assays in order to handle the large number of colonies to be examined. To date, only a limited number of assays are available, and none is appropriate yet for the level of screening that needs to be done. The most common assays require the use of Clark electrodes,[35,74,97,215,261] a gas chromatograph,[9,159,215] or a mass spectrograph[173] to directly detect H_2 gas, following illumination of anaerobically-induced organisms. It is also possible to measure the presence of hydrogenase activity in cell extracts by determining colorimetrically the oxidation of reduced methylviologen in the dark.[63,102] Moreover, staining of non-denatured protein gels has been used to identify bands containing hydrogenase activity.[98] However, the most promising assay involves the sensitization of a patented palladium/tungsten oxide film by H_2 evolved from algal colonies, following illumination.[229,230,232] This assay, although not implemented in a high-throughput mode yet, was successfully used to identify O_2-tolerant mutants of *Chlamydomonas reinhardtii* generated by random chemical mutagenesis[231] and the [FeFe]-hydrogenase assembly genes from an insertional mutagenesis library.[199] It has the advantage that the screen is specific only for H_2, a characteristic not necessarily shared by the colorimetric assays described above.

The second approach to address the algal hydrogenase O_2-sensitivity issue is based on the discovery of a physiological switch, sulfur deprivation, which down-regulates photosynthetic O_2 evolution[270] to the point where the cultures become anaerobic and express an active algal [FeFe]-hydrogenase in the light.[75,159] This process, which has been reviewed recently,[32,77,162,274] entails a complex interaction between fermentative, photosynthetic and respiratory pathways, and has been extensively studied over the last couple of years.[75,78,130,131,160,162] Biochemical and physiological data demonstrated that, although the majority of the reductants for H_2 production are generated by the photosynthetic water-splitting process,[75,131] starch catabolism is required to maintain the anaerobic conditions necessary for H_2-production activity.[130,250,275] Additional studies provided further information on the biophysical and biological parameters required for optimal H_2-production activity. For example, it was shown that sulfur deprivation can be achieved either by resuspension of the algal cultures in a sulfur-deprived medium or by dilution of the same cultures in sulfur-deprived medium,[136] an energy- and cost-savings measure. The effect of

(a) the addition of limiting amounts of sulfate to the cultures during sulfur deprivation;[130,131]
(b) changes in culture cell density and age;[125,130]
(c) exposure of the algal cultures to different light/dark regimes;[250]
(d) different incident light intensities during the H_2-production phase;[92,137] and
(e) changes in the pH of medium[131] were all examined.

An overall economic analysis of the system has also been performed,[8,76] which is guiding the direction of research in areas that will produce cost-saving results. As a result, a new, lower cost, continuous H_2-production system has been conceived, designed and tested,[65] although it is still not a cost-competitive process yet.

In order to address the low cell density issue associated with sulfur-deprived cell suspensions, cell immobilization strategies have been investigated. This resulted in an 18 times increase in the H_2-production rates on a per-cell-culture volume basis and a 6 times longer period of H_2 production by algal cultures immobilized on glass fibers.[138] Additionally, the immobilized cultures have increased productivity (2.5 times) on a per illuminated area basis, compared to cell suspensions.

Finally, a nuclear gene-encoding a chloroplast envelope sulfate permease gene (SulP), which is responsible for delivering sulfate into the chloroplast in *C. reinhardtii* was identified.[44,45] When SulP is inactivated by antisense technology, the resulting transformants have a phenotype similar to organisms that were continuously exposed to low levels of sulfate. Specifically, the cells exhibit limiting rates of O_2 evolution per cell, very low levels of O_2 in the medium, very little growth and reduced *de novo* chloroplast protein biosynthesis.[45] As such, the *sulP* mutant might be more easily manipulated for use in H_2-producing photobioreactors in the future.[46]

5.3 Oxygen Sensitivity of [NiFe]-Hydrogenases

[NiFe]-hydrogenases in general exhibit higher levels of O_2 tolerance, and the inactivation of the [NiFe]-hydrogenase by O_2 is also more reversible than that of [FeFe]-hydrogenases.[123,152,207,258] [NiFe]-hydrogenases from a number of organisms can be prepared under aerobic environments even though as isolated, the enzyme is in an inactive state. Organisms harboring this category of hydrogenase include *Ralstonia eutropha*,[251] *Thiocapsa reoeopersicina*,[207] Allo*chromatium vinosum*,[133] *Methanobacterium thermoautotrophicum*,[38] *Desulfovibrio vulgaris*,[185] and *D. gigas*.[248] The inactive enzyme can be fully reactivated by several means, including incubation in a H_2 atmosphere, addition of reducing agents such as sodium dithionite or NADH, or simply by evacuation to remove O_2 from the gas phase.[2a,204] Although the inhibition by O_2 is reversible, there is no evidence yet as to whether the [NiFe]-hydrogenase is catalytically active in O_2. This is due to the inclusion in the hydrogenase assay of reducing agents such as sodium dithionite and reduced methyl viologen, both of which scavenge O_2 continuously. One solution to circumvent this limitation is to take advantage of the hydrogen-deuterium (H-D) exchange activity displayed by most hydrogenases, which yields HD as the end product.[69,271] The isotopic exchange rate is comparable with the H_2-production rate measured by reduced methyl viologen.[157] Since no electron mediator or reductant is involved in the exchange reaction, O_2 can be present throughout the assay. Using this assay, the H_2-sensing hydrogenase (HupUV) from *Rhodobacter capsulatus* was found insensitive to the simultaneous presence of O_2.[254] Similarly, the CO-linked, H_2-evolving hydrogenase from *Rubrivivax gelatinosus* was catalytically active and retained approximately 80% and 60% of the H-D exchange activity in the presence of 3.3% and 13% O_2, respectively.[152] These findings are promising and provide the first evidence that certain [NiFe]-hydrogenase can function in O_2.

When [NiFe]-hydrogenases are exposed to O_2 or to an oxidative environment, an oxo- or a hydroxo-bridge forms between the Ni and Fe atoms. This reaction renders the metal cluster insensitive to further O_2 inactivation.[259] During activation, the addition of a reductant (H_2 or sodium dithionite) chemically reduces the oxo- or hydroxo-group to fully reconstitute its enzymatic activity.[70,104] Crystal structures of the [NiFe]-hydrogenases from both *D. gigas* and *D. fructosovorans* also reveal a hydrophobic cavity or gas channel extending from the active site to the surface of the proteins.[169,257,259] Amino acids lining this channel are highly conserved and the channel was postulated to serve as a pathway for H_2 to diffuse in and out of the protein, or alternatively as a reservoir for H_2 gas concentration prior to catalysis. Oxygen can potentially migrate through the same passage, and its actual accessibility to the catalytic site may account for different degrees of O_2 tolerance displayed by various hydrogenases. Sequence alignments of amino acid residues forming the putative gas channel reveal that the more O_2-tolerant, H_2-sensing hydrogenases, such as those from *R. capsulatus* and *R. eutropha*, are often lined with bulkier residues near the active site, presumably to block the accessibility of O_2 to the NiFe center. This hypothesis was recently confirmed in *R. eutropha* in which a widening of its putative gas channel via site-directed mutagenesis rendering the H_2-sensing hydrogenase more sensitive to O_2.[42] Moreover, a third CN ligand, encoded by *hypX*, was found to coordinate to the Ni atom of the soluble [NiFe]-hydrogenase from *R. eutropha*.[29] This additional CN ligand confers improved O_2 tolerance allowing this hydrogenase to function during aerobic cell growth, and a HypX-deficient strain loses its ability to oxidize H_2 (linking to NAD reduction) under an aerobic environment. Together these findings and those discussed in Section 5.2 provide support for the validity of and strategies for engineering a more O_2-tolerant hydrogenase for potential biotechnological applications.

5.4 Competition between Different Pathways for Photosynthetic Reductants

A key intermediate electron transport carrier on the reducing side of PSI in green algae and cyanobacteria is the FeS-containing ferredoxin (see Fd in Fig. 1). Ferredoxin is a soluble protein that competitively interacts with an array of different enzymes (cf. Ref. 128 and references therein) including:

(a) ferredoxin-NADPH oxidoreductase (FNR);
(b) ferredoxin-thioredoxin oxidoreductase (FTR);
(c) ferredoxin nitrite and sulfite reductases;
(d) glutamate synthase; and
(e) hydrogenases.

The first pathway is prevalent during normal, aerobic photosynthesis that generates NADPH required for CO_2 fixation (see Fig. 1). The affinity of ferredoxin for FNR in *C. reinhardtii* is very high and on the order of 0.4 μM.[120] The second pathway is responsible for the reduction of thioredoxins *m* and *f* in the chloroplast stroma.[39] Both thioredoxins are involved in the regulation of different sets of CO_2 fixation enzymes and are reduced by the same FTR. The FTR binds to sites on ferredoxin, other than that at which FNR binds, with a K_m on the order of 1.8 μM.[60,108] The third pathway, through either nitrite or sulfite reductase is only activated under specific nutrient conditions such as, respectively, ammonium and sulfate depletion. The later may be important in regulatory processes leading to H_2 production during sulfur-deprivation, but so far no one has examined this possibility. The K_m of ferredoxin for nitrite reductase is 20 μM[108] and that for sulfite reductase is 25 μM,[107] so these enzymes bind to ferredoxin much less efficiently than either FNR or FTR. Glutamate synthase reduces glutamine to glutamate, a reaction that is responsible for the synthesis of glutamate under ammonia-limiting conditions. Its K_m for ferredoxin has been reported to be 2 μM.[108] Finally, the interaction of ferredoxin with *C. reinhardtii* hydrogenase was reported to have a K_m of 10 μM.[215] More recently measurements with ferredoxin report a K_m of 35 μM for *C. reinhardtii* HydA1 [98] and ~31 μM for HydA2.[126] It is clear, then, that a significant amount of reductant could be diverted potentially to other pathways instead of being used for H_2 production via the hydrogenase enzyme. An engineered system for efficient H_2 photoproduction may well have to address competition between hydrogenase and these other pathways.

Based on the homology of the cyanobacterial bidirectional hydrogenase with the NAD$^+$-reducing hydrogenase from *R. eutropha* and the detection of NAD$^+$-reducing activity from H_2 oxidation (described in Section 3.2), cyanobacterial H_2

production likely derives its reducing equivalents from NAD(P)H instead of photo-reduced ferredoxin. Indeed, purified bidirectional hydrogenase from *Synechocystis* sp. PCC6803 showed that both reduced NADH and NDAPH can serve as the imme-diate electron donor, but not ferredoxin, in support of H_2 production.[227] This suggests that any competition for photoreductant is at the level of NAD(P)H, presumably competing with CO_2 fixation. Indeed, with the complex I mutant M55 (deficient in Type I NADPH-dehydrogenase complex) of *Synechocystis* sp. PCC 6803, which has an impaired CO_2-transport pathway, sustained H_2 production was observed in light.[53] NAD(P)H is also a favorite substrate for respiratory complex I, which is largely confined to the ICM membranes in cyanobacteria.[186] Presumably any H_2 produced during the initial phase of photosynthesis can be re-oxidized by the bidirectional hydrogenase to generate NAD(P)H for the respiratory complex I activity. Based on these observations, removing complex I activity seems to be a logical choice for sustained H_2 production, provided that the CO_2 uptake pathway can be modulated.

5.5 Down-Regulation of Electron Transport Rates

Following prolonged dark, anaerobic incubation in algae, the CO_2 fixation pathway is inactivated and H_2 production is induced.[85] As a consequence (see Section 2.1), photosynthetically-generated ATP is not consumed, the proton gradient across the thylakoid membrane is not dissipated (see Fig. 1) and photosynthetic electron trans-port from PSII to PSI is down-regulated.[9,56,140] This down-regulation process is thought to be mediated by the operation of the violoxanthin/zexanthing cycle, which senses the acidification of the lumen and generates a quencher of absorbed light energy.[110,171,183] Alternatively, the acidification of the medium might be a response to lowering of the proton conductivity of the ATP synthase,[121] which occurs under low CO_2 and low O_2 partial pressures.[17] When the ATP synthase proton conductivity decreases, protons accumulate in the lumen and trigger the violoxanthin/zeaxanthin response described above.[132] Finally, it is worth mentioning that other regulatory mechanisms have been proposed to regulate the rate of electron transport through the photosynthetic chain, such as those that sense the redox state of the plastoquinone pool[9,178] and trigger state transitions[7,276] or mechanisms that sense the reduced state of the ferredoxin/thioredoxin system.[247] All of these further contribute to the build-up of the proton gradient.

In order to prevent the observed decrease in the rates of electron transport to the hydrogenase during H_2 production, researchers are working on the design and ex-pression of an artificial proton channel across the thylakoid membrane of *C. rein-hardti*.[140] These channels would physically short-circuit the thylakoid membrane and as such, dissipate the proton gradient during H_2-producing conditions. However, in order to allow for the synthesis of ATP during periods in which CO_2 fixation is re-quired, the artificial channel will have to be placed under the regulation of the hy-drogenase promoter as well, and would have to have a limited life-time in the mem-brane to allow CO_2 fixation to occur during periods when H_2 production is not occurring. The stimulating effect of proton uncouplers on H_2 photoproduction was reported:

(a) in *Chlamydomonas* by Healey,[279] who added CCCP to anaerobically induced cultures, and by Zhang and Melis,[274] who added methylamine to sulfur-deprived cultures, and

(b) in a marine green alga by,[86,87] who added CCCP and reported large increases in rates.

Although the control rates were rather low in the latter case, both experiments provide support for the artificial channel approach in algae. Whereas the addition of CCCP also increases H_2 production in cyanobacteria, the uncoupling of electron transport and photophosphorylation *per se* does not seem to be involved in the mechanism.[284]

5.6 Low-Light Saturation Properties of Photosynthetic Organisms

The mass culture of photosynthetic organisms under high light intensity is inherently limited by the poor photosynthetic light utilization efficiency of the organisms involved. This property of both algae and cyanobacteria is a consequence of the presence of large light-absorbing arrays of pigments,[162] which lead to the saturation of the light reactions of photosynthesis at about 500 μE m^{-2} s^{-1},[197] less than 1/5 of sunlight. As a result, up to 80–90% of the absorbed photons are wasted as fluorescence or heat,[158] and the photosynthesis rates become limited by the rate of electron transport through the electron transport chain. Among the most promising approaches to overcome this issue is the suggested development of microalgal strains with limited numbers of pigment antenna molecules.[158,159,195,206] This type of work is being conducted in the US and Japan, using various mutagenesis techniques to reduce the content of the light harvesting pigments of the photosystems in different microalgae.[174–177,195–198]

In Japan, a truncated-antenna mutant of the cyanobacterium, *Synechocystis* PCC 6714, was generated by chemical mutagenesis and screened for the lack of phycobilisomes.[71] The mutant lacked phycocyanin, a pigment associated with PSII in these organisms, and it exhibited increased photosynthetic productivity, compared to its wild-type strain, saturating at nearly solar intensities when grown in dense cultures.[174,175] The same research group extended its work to a strain of the green alga *Chlamydomonas perigranulata* isolated from the Red Sea. This alga was submitted to UV and chemical mutagenesis, and screened for small light-harvesting antennae. The positive clones were further characterized.[176] One of the isolated mutants had reduced Chl content and increased photosynthetic productivity, when measured on a per area basis.[177] However, neither the cyanobacterial nor the algal mutant lost its ability to acclimate to different light intensities as previously described.[180,221] As a result, the truncated antenna phenotype was variable as a function of the light intensity during growth.[177]

The work in the USA utilized insertional mutagenesis to generate *Chlamydomonas reinhardtii* mutants that had truncated Chl antenna sizes, followed by fluorescence screening for positive clones.[161] Three mutants, with increasingly smaller Chl antennae have been identified. All of the mutants exhibit light-saturation properties characteristics of small-antenna organisms,[158] although the extent of the antenna truncation and the nature of the mutations varied. The first isolated mutant, *cbs3*, had an inactivated chlorophyll *a* oxygenase (CAO) gene, which caused a specific decrease in the antenna size of PSII.[195] The second mutant, *npq2/lor1*, had defects in the carotenoid biosynthesis pathway and lacked the specific xanthophylls that are normally present in the PSII Chl antenna. While its PSII antenna size was truncated, there was no change in the antenna size of PSI.[196] Finally, the third identified mutant, *tla1*, was unable to produce a large Chl antenna size for either PSII or PSI and is believed to have an affected regulatory gene in *Chlamydomonas reinhardtii*.[198] This mutant is unable to acclimate to changes in environmental light intensity.

Although the work from both groups has been encouraging in terms of increasing the overall productivity in mass cultures of microalgae, it remains to be determined whether the mutations have any negative secondary effects on the H_2-production since H_2 production has not as yet been reported from any of these mutants.

This is not the case for photosynthetic bacteria, where a reduced-pigment mutant from *R. sphaeroides* RV was obtain using UV irradiation.[285] Although this mutant contained less bacteriochlorophyll and carotenoid pigment, it grew at rates similar to the wild type at light intensities below 300 W/m^2. On the other hand, it was photoinhibited after prolonged exposure to light. Hydrogen production activity on a dry cell weight basis was lower in the mutant at non-saturating, low light intensities. However, the rates saturated at higher light intensity in the mutant, and were about 50% higher than those of the wild type control at 300 W/m^2. The mutant cultures achieved this higher light conversion efficiency through increased light penetration into the photobioreactors.

5.7 Photobioreactor and System Costs

The major biological challenges that must be surmounted, before a viable photobiological H_2-production system can be considered, have been discussed in the previous Sections. The non-biological challenges will center on the cost of photobioreactor systems and on the identification of suitable materials for photobioreactors that have sufficient long lifetimes under exposure to harsh, outdoor conditions. A number of articles have discussed various aspects of these issues[76,142,161,202] as well as describing functioning photobioreactor systems, e.g., Refs. 6, 203, 246.

Melis[161] reported a preliminary cost study on a 500-L scaled-up modular photobioreactor, and outlined the following cost distribution for the unit based on operating it for a period of 6 months: photobioreactor materials, 45%; mineral nutrients, 39%; labor; 4%; water supply, 4%; land lease, 2%; power, 1%; and miscellaneous expenses, 5%. He reported the actual cost of the materials for the pilot photobioreactor unit as $0.75 m^{-2}, but the material used and the lifetime of the unit was not stated. This cost is much lower that the "often-quoted" estimates of up to $100 per meter square using glass-covered structures. Glass does not appear to be a realistic material to use in a commercial photobioreactor system because of both initial and replacement costs.

A comprehensive study, undertaken by Amos[8] for the US Department of Energy, concluded that the cost of photobioreactors would have to be brought down to $10/m^2 or less to meet the DOE goal of $2.60 per kg of H_2. Ghirardi and Amos[76] suggested that new, UV-resistant, transparent materials that can contain H_2 gas at a cost of about $0.25/m^2$, the current cost of polyethylene, would have to be developed. Furthermore, the Amos[8] study performed a thorough cost-sensitivity analysis, which in addition to the photobioreactor costs, assessed the biological issues needing resolution before a practical H_2-production system based on the sulfur deprivation of algae could be considered. The projections were based on the expected improvements in the biology of the process (i.e., the algal cell concentration must be increased by a factor of ten on a surface area basis, and the specific rate of H_2 production would have to be improved by a similar factor.)[76] The analysis demonstrated that

Table 1. A summary of the currently known genes in the unicellular green alga, including *Chlamydomonas reinhardtii*, which are involved in the process of H_2 production or in the cellular metabolism related to H_2 photoproduction (adapted from Ref. 162.)

Gene name	Symbol	Reference
[FeFe]-hydrogenase	HydA1	67,102,266
[FeFe]-hydrogenase	HydA2	68,269
[FeFe]-hydrogenase assembly genes	HydEF, HydG	199
Chlorophyll antenna size regulatory gene	Tla1	198
Sulfate permease	SulP	44
Isoamylase	Sta7	200

if the technical challenges were met, it would be possible to sell H_2 produced by algae at a selling price as low as $0.57–3.68 (for production reactors costing $1–10/m^2$ but not including H_2 compression and storage costs). This price range partially overlaps the US Department of Energy's target goal and the current cost of producing H_2 by reforming natural gas. Ghirardi and Amos[76] also calculated a selling price of $1.26/kg. Finally, these authors suggested that an aerobic H_2-production system, employing an O_2-resistant hydrogenase, would further improve the economics of H_2 production.

Ways of meeting the goals for improving the biological performance of research organisms are being addressed by a number of laboratories around the world; however, ways of meeting the materials performance and cost goals must also be addressed. The effect of weathering on potential construction materials for closed photobioreactors has received little attention. However, the biohydrogen applications also share the necessity for covering large areas with collectors/reactors at very low cost with other solar technologies. The effects of accelerated and outdoor weathering on the optical properties of polymers has been studied extensively for solar thermal applications.[55] Hydrogen production adds the unique additional requirement for containing hydrogen and excluding oxygen.

There is a paucity of data on the properties of potential reactor materials and the engineering challenges that must be overcome to make photobiological production of hydrogen a commercial reality. As such, the US DOE Hydrogen Program has started to support research to identify potential construction materials and determine the effects of weathering on critical materials properties such as light transmission, H_2 and O_2 permeation, and outdoor durability. Work to date, which builds on work being done for the DOE Solar Program, has identified acrylics (plexiglass), polycarbonates, fluoropolymers, and PET (polyethyleneteraphthalate) as materials having reasonable outdoor lifetimes, based on the measurement of optical properties. Hydrogen and O_2 permeation measurements are being made of examples of these polymer classes;[27] however, the effects of weathering on gas permeation has yet to be determined.

Additional engineering studies will have to be conducted to identify the most efficient physical designs to implement for photobioreactor systems after the biological and materials challenges are solved.

5.8 Genomics Approaches

Advances in the tools of molecular biology over the past forty years are leading to exciting new approaches to the problem of biohydrogen production. In the case of green algae, we are just now beginning to identify specific genes that are associated with the relevant biological pathways. As seen in Table 1, less than ten are known in *C. reinhardtii* out of the possible 15,000 genes thought to be present in this sequenced organism. Fortunately, the sequences of new organisms are being reported weekly, and the world seems to have plenty of sequencing capability to continue exploring new genomes for years to come. However, analogous to the situation in the 1980's and 1990's, when it was recognized that scientists had to convert from the idea that genes could only be sequenced and annotated one gene at a time, we must now think globally in terms of what array of genes in the genome are expressed under different conditions, what proteins are actually synthesized, and what regulatory processes are involved. This will necessitate the application of new global gene expression (microarray) technology, metabolomics, regulomics, and proteomics, among others. Work in this area is just getting under way, but the first advances will probably come utilizing currently sequenced organisms, such as *Synechocystis* PCC6803, *Rhodospseudomonas palustris*, and *C. reinhardtii*. The basic knowledge that comes out of this type of work not only will increase our basic understanding of microbial H_2 metabolism, but it will also facilitate both enzyme and pathway engineering approaches in candidate organisms aimed at improving the efficiency of H_2 production and their ability to function in a commercial environment.

6 Future Directions

Ancient and current biological process that result in either the production of fossil fuels or biomass, are responsible over 90% of the energy that humans have ever used. All of these processes employ photosynthesis, and it is not unrealistic that natural photosynthesis might be adapted to efficiently produce new forms of stored energy for the future use of mankind besides merely the production of biomass. This review has outlined a number of specific biological and systems challenges that must be addressed to advance science and technology down this path.

The challenges that we can identify now may in fact not be the only ones that must be faced, so we must be prepared to address new issues as we solve the current ones. However, the direct or indirect utilization of sunlight are the only sustainable paths (barring new discoveries in the area of nuclear fusion), which can supply the energy needs of our civilization beyond the 21st century. We cannot count on the continued availability of fossil fuels, the appropriateness of their continued use in the future, or even the availability of uranium, which itself is a limited resource, if we have to depend on it for all our energy needs. Cost effective Biohydrogen production from water represents a challenging, but technically attainable goal. Recent discoveries in the physiology, enzymology, biochemistry, and genetics of the biological processes involved are very encouraging and provide reasons for optimism. Biodiscovery of new organisms and enzymes in the wild, such has been highlighted recently by the voyage of the Sorcerer II,[252] also promises the possibility of new biological materials with which to stimulate this nacent area of research.

Acknowledgments

The authors would like to acknowledge the support of the Hydrogen, Fuel Cell, and Infrastructure Technology; Energy Biosciences; and GTL Programs within the US Department of Energy. We would also like to thank everyone in our laboratory for many helpful discussions, and in particular, Dr. Dan Blake, who provided us with critical information on photobioreactor materials.

References

1. Abdel-Basset, R., and Bader, K. P. 1998. Physiological analysis of the hydrogen gas exchange in cyanobacteria. *J. Photochem. Photobiol. B: Biol.* **43**, 146-151.
2. a) Adams, M. W. W., Mortenson, L. E., and Chen, J. S. 1981. Hydrogenase. *Biochim. Biophys. Acta* **594**, 105-176. b) Adams, M.W.W. 1990. The structure and mechanism of iron-hydrogenases. *Biochim. Biophys. Acta* **1020**, 115-145.
3. Adams, M. W. W., and Stiefel, E. I. 1998. Biological hydrogen production: not so elementary. *Science* **282**, 1842-1843.
4. Akano, T., Muiro, Y., Fukatsu, K., Miyasaka, H., Ikuta, Y., Matsumoto, H., Hamasaki, A., Shioji, N., Mizoguchi, T., Yagi, K. and Maeda, I. 1996. Hydrogen production by photosynthetic microorganisms. *Appl. Biochem. Biotechnol.* **57-58**, 677-688.
5. Akkerman, I., Janssen, M., Rocha, J., and Wijffels, R. 2002. Photobiological hydrogen production: photochemical efficiency and bioreactor design. *Int. J. Hydrogen Energy* **27**, 1195-1208.
6. Akkerman, I., Janssen, M., Rocha, J.M.S., Reith, J.H., and Wijffels, R.H. 2003. Photobiological hydrogen production: Photochemical efficiency and bioreactor design. In, Biomethane & Bio-hydrogen, (J. H. Reith, R. H. Wijffels, and H. Barten, eds.), Chapter 6, Dutch Biological Hydrogen Foundation, Petten, The Netherlands, pp. 124-145.
7. Allen, J.F. 1991. Protein phosphorylation in regulation of photosynthesis. *Biochim. Biophys. Acta* **1098**, 275-335.
8. Amos, W. 2004. Updated Cost Analysis of Photobiological Hydrogen Production from *Chlamydomonas reinhardtii* Green Algae. NREL/MP-560-35593. http://www.nrel.gov/docs/fy04osti/35593.pdf.
9. Antal, T.K., Krendeleva, T.E., Laurinavichene, T.V., Makarova, V.V., Ghirardi, M.L., Rubin, A.B., Tsygankov, A.A. and Seibert, M. 2003. The dependence of algal H$_2$ production on

photosystem II and O_2 consumption activities in sulfur-deprived *Chlamydomonas reinhardtii* cells. *Biochim Biophys Acta* **1607**, 153-160.

10. Appel, J., and Schulz, R. 1996. Sequence analysis of an operon of a NAD(P)-reducing nickel hydrogenase from the cyanobacterium *Synechocystis* sp. PCC6803 gives additional evidence for direct coupling of the enzyme to NAD(P)H-dehydrogenase (complex I). *Biochim. Biophys. Acta.* **1298**, 141-147.

11. Appel, J., and Schulz, R. 1998. Hydrogen metabolism in organisms with oxygenic photosynthesis: hydrogenases as important regulatory devices for a proper redox poising? *J. Photochem. Photobiol. B: Biol.* **47**, 1-11.

12. Appel, J., Phunpruch, S., Steimüller, K., and Schulz, R. 2000. The bidirectional hydrogenase of *Synechocystis* sp. PCC6803 works as an electron valve during photosynthesis. *Arch. Microbiol.* **173**, 333-338.

13. Arnon, D.I., Mitsui, A. and Paneque, A. 1961. Photoproduction of hydrogen gas coupled with photosynthetic phosphorylation. *Science* **134**, 1425.

14. Artero, V., and Fontecave, M. 2005. Some general principles for designing electrocatalysts with hydrogenase activity. *Coord. Chem. Rev.* **249**, 1518-1535.

15. Asada, Y., Ishimi, K., Tokumoto, M., Kohno, H. and Tomiyama M. 2005. Hydrogen Production by Co-cultures of Facultative Anaerobes and Photosynthetic Bacteria. *Abstracts, COST Action 841 Workshop*, Porto, Portugal.

16. Ashby, M. K., and Mullineaux, C. W. 1999. Cyanobacterial ycf27 gene products regulate energy transfer from phycobilisomes to photosystems I and II. *FEMS Microbiol. Lett.* **181**, 253-260.

17. Avenson, T., Cruz, J.A. and Kramer, D.M., 2004, Modulation of energy-dependent quenching of excitons in antennae of higher plants. *Proc. Natl. Acad. Sci. USA* **101**, 5530-5535.

18. Axelsson, R., Oxelfelt, F., and Lindblad, P. 1999. Transcriptional regulation of *Nostoc* uptake hydrogenase. *FEMS Microbiol. Lett.* **170**, 77-81.

19. Axelsson, R., and Lindblad, P. 2002. Transcriptional regulation of *Nostoc* hydrogenases: effects of oxygen, hydrogen, and nickel. *Appl. Environ. Microbiol.* **68**, 444-447.

20. Barbosa, M. J., Rocha, J. M. S., Tramper, J., and Wijffels, R. H. 2001. Acetate as a carbon source for hydrogen production by photosynthetic bacteria. *J. Biotech.* **85**, 25-33.

21. Bard, A.J. and Fox, M.A. 1995. Artificial photosynthesis: solar splitting of water to hydrogen and oxygen. *Acc. Chem. Res.* **28**, 141-145.

22. Benemann, J.R., Berenson, J.A., Kaplan, N.O. and Kamen, M.D. 1973. Hydrogen evolution by a chloroplalst-ferredoxin-hydrogenase system. *Proc. Natl. Acad. Sci. USA* **70**, 2317-2320

23. Bennett, I.M., Farfano, H.M.V., Bogani, F., Primak, A., Liddell, P.A., Otero, L., Sereno, L., Silber, J.J., Moore, A.L., Moore, T.A. and Gust, D. 2002. Active transport of Ca^{2+} by an artificial photosynthetic membrane. *Nature* **420**, 398-401.

24. Berenson, J.A. and Benemann, J.R. 1976. Immobilization of hydrogenase and ferredoxins on glass beads. *FEBS Lett.* **76**, 105-107.

25. Binder, U., Maier, T., and Böck, A. 1996. Nickel incorporation into hydrogenase 3 from *Escherichia coli* requires the precursor form of the large subunit. *Arch. Microbiol.* **165**, 69-72.

26. Bishop, P. E., Jarlenski, D. M. L., and Hetherington, D. R. 1980. Evidence for an alternative nitrogen fixation system *in Azotobacter vinelandii*. *Proc. Natl. Acad. Sci. U.S.A.* **77**, 7342-7346.

27. Blake, D.M. and Kennedy, C. 2005. Hydrogen reactor development and design for photofermentation and photolytic processes. Project PD-19, DOE Hydrogen Program Review, May 23-26, 2005, http://www.hydrogen.energy.gov/).

28. Blankenship, R.E. 2002. *Molecular Mechanisms of Photosynthesis*, Blackwell Science, London.

29. Bleijlevens, B., Buhrke, T., van der Linden, E., Friedrich, B. and Albracht, S.P.J. 2004. The auxiliary protein HypX provides oxygen tolerance to the soluble [NiFe]-hydrogenase of *Ralstonia eutropha* H16 by way of a cyanide ligand to nickel. *J. Biol. Chem.* **279**, 46668-46691.

30. Blokesch, M., Albracht, S. P., Matzanke, B. F., Drapal, N. M., Jacobi, A., and Böck, A. 2004. The complex between hydrogenase-maturation proteins HypC and HypD is an intermediate in the supply of cyanide the active site iron of [NiFe]-hydrogenases. *J. Mol. Bio.* **344**, 155-167.

31. Boichenko, V.A. and Hoffman, P. 1994. Photosynthetic hydrogen production in prokaryotes and eukaryotes: occurrence, mechanism, and functions. *Photosynthetica* **30**, 527-552.

32. Boichenko, V.A., Greenbaum, E., and Seibert, M. 2004. Hydrogen production by photosynthetic microorganisms, in *Photoconversion of Solar Energy: Molecular to Global Photosynthesis*, M.D. Archer and J. Barber, eds., Imperial College Press, London, pp. 397-452.

33. Boison, G., Schmitz, O., Mikheeva, L., Shestakov, S., and Bothe, H. 1996. Cloning, molecular analysis and insertional mutagenesis of the bidirectional hydrogenase genes from the cyanobacterium. *FEBS Lett.* **394**, 153-158.

34. Boison, G., Bothe, H., and Schmitz, O. 2000. Transcriptional analysis of hydrogenase genes in the cyanobacteria *Anacystis nidulans* and *Anabaena variabilis* monitored by RT-PCR. *Curr. Microbiol.* **40**: 315-321.

35. Brand, J.J., Wright, J. and Lien, S. 1989. Hydrogen production by eukaryotic algae. *Biotechnol Bioeng.* **33**, 1482-1488.

36. Brune, A., Jeong, G., Liddell, P.A., Sotomura, T., Moore, T.A., Moore, A.L. and Gust, D., 2004, Porphyrin-sensitized nanoparticulate TiO_2 as the photoanode of a hybrid photoelectrochemical biofuel cell. *Langmuir* **20**, 8366-8371.

37. Bryant, D. 1991. Cyanobacterial phycobilisomes: progress toward complete structural and functional analysis via molecular genetics. In: Bogorad L. and Vails IK (eds) Cell Structure Somatic Cell Genetics of Plants, Vol. 7B (The Photosynthetic Apparatus: Molecular Biology and Operation), pp 257-300. Academic Press, New York.

38. Braks, I. J., Hoppert, M., Roge, S., and Mayer, F. 1994. Structural aspects and immunolocalization of the F_{420}-reducing and non-F_{420}-reducing hydrogenases from *Methanobacterium thermoautotrophicum* Marburg. *J. Bacteriol.* **176**, 7677-7687.

39. Buchanan, B.B. 1991. Regulation of CO_2 assimilation in oxygenic photosynthesis: the ferredoxin/thioredoxin system. Perspective on its discovery, present status and future development. *Arch. Biochem. Biophys.* **289**, 1-9.

40. Bui, E.T.N. and Johnson, P.J. 1996. Identification and characterization of [Fe]-hydrogenases in the hydrogenosome of *Trichomonas vaginalis*. *Mol. Biochem. Parasitol.* **76**, 305-310.

41. Buhrke, T., Bleijlevens, B., Albracht, S.P.J., and Friedrich, B. 2001. Involvement of *hyp* gene products in maturation of the H_2-sensing [NiFe] hydrogenase of *Ralstonia eutropha*. *J. Bacteriol.* **183**, 7087-7093.

42. Buhrke, T., Lenz, O., Kraub, N., and Friedrich, B. 2005. Oxygen tolerance of the H_2-sensing [NiFe] hydrogenase from *Ralstonia eutropha* H16 is based on limited access of oxygen to the active site. *J. Biol. Chem.* **280**, 23791-23796.

43. Chen, C.K., and Blaschek, H.P. 1999. Effect of acetate on molecular and physiological aspects of *Clostridium beijerinckii* NCIMB 8052 solvent production and strain degeneration. *Appl. Environ. Microbiol.* **65**, 499-505.

44. Chen H-C., Yokthongwattana K, Newton A.J. and Melis A. 2003. *SulP*, a nuclear gene encoding a putative chloroplast-targeted sulfate permease in *Chlamydomonas reinhardtii*. *Planta* **218**, 98-106.

45. Chen H-C. and Melis A. 2004. Localization and function of SulP, a nuclear-encoded chloroplast sulfate permease in *Chlamydomonas reinhardtii*. *Planta* **220**, 198-210.
46. Chen H-C., Newton A.J. and Melis A. 2005. Role of SulP, a nuclear-encoded chloroplast sulfate permease, in sulfate transport and H_2 evolution in *Chlamydomonas reinhardtii*. *Photosynth. Res.* **84**, 289-296.
47. Chisnell, J.R., Premakumar, R., and Bishop, P.E. 1988. Purification of a second alternative nitrogenase from a *nif*HDK deletion strain of *Azotobacter vinelandii*. *J. Bacteriol.* **170**, 27-33.
48. Claassen, P.A.M., van Groenestijin, J.W., Janssen, A.J.H., van Niel, E.W.J., and Wijffels, R.H. 2000. Feasibility of biological hydrogen production from biomass for utilization in fuel cells. In: Proceedings of the 1st World Conference and Exhibition on biomass for energy and industry, Sevilla, Sapin, 5-9 June 2000.
49. Claassen, P. A. M., de Vrije, T., and Budde, M. A. W. 2004. Biological hydrogen production from sweet sorghum by thermophilic bacteria. In: Proceedings of the 2nd World Conference on Biomass for Energy, Industry and Climate Protection, 10-14 May 2004, Rome, Italy.
50. Cohen, J., Kim, K., Posewitz, M., Ghirardi, M.L., Schulten, K., Seibert, M. and King, P. 2005a. Molecular dynamics and experimental investigation of H_2 and O_2 diffusion in [Fe]-hydrogenase. *Biochem. Soc. Transact.* **33**, 80-82.
51. Cohen, J., Kim, K., King, P., Seibert, M., and Schulten K. 2005. Finding gas diffusion pathways in proteins: O_2 and H_2 gas transport in CpI [FeFe]-hydrogenase and the role of packing defects. *Structure* **13**, 1321-1329.
52. Cornejo, J., and Beale, S.I. 1997. Phycobilin biosynthetic reactions in extracts of cyanobacteria. *Photosynthesis Res.* **51**, 223-230.
53. Cournac, L., Guedeney, G., Peltier, G., and Vignais, P. M. 2004. Sustained photoevolution of molecular hydrogen in a mutant of *Synechocystis* sp. strain PCC 6803 deficient in the type I NADPH-dehydrogenase complex. *J. Bacteriol.* **186**, 1737-1746.
54. Dalton, H., and Mortenson, L. E. 1972. Dinitrogen (N_2) fixation (with a biochemical emphasis). *Bacteriol. Rev.* **36**, 231-260.
55. Davidson, J.H., Mantell, S., and Jorgensen, G. 2003. Status of the development of polymeric solar water heating systemsAdvances in Solar Energy, (D. Yogi Goswami, ed.) Vol. 15, Chapter 6, ASES, Boulder, CO, pp.149-186.
56. De Vitry, C., Oyuang, Y., Finazzi, G., Wollman, F.A. and Kallas, T., 2004, The chloroplast Rieske iron-sulfur protein – at the crossroad of electron transport and signal transduction. *J. Biol. Chem.* **279**, 44621-44627.
57. De la Garza, L., Jeong, G., Liddell, P.A., Sotomura, T., Moore, T.A., Moore, A.L. and Gust, D. 2003. Enzyme-based photoelectrochemical biofuel cell. *J. Phys. Chem. B* **107**, 10252-10260.
58. Dernedde, J., Eitinger, T., Patenge, N., and Friedrich, B. 1996. hyp gene products in *Alcaligenes eutrophus* are part of a hydrogenase-maturation system. *Eur. J. Biochem.* **235**, 351-358.
59. Drapal, N., and Böck, A. 1998. Interaction of the hydrogenase accessory protein HypC with HycE, the large subunit of *Escherichia coli* hydrogenase 3 during enzyme maturation. *Biochemistry* **37**, 2941-2948.
60. Droux, M., Jacquot, J.P., Miginac-Maslow, M., Gadal, P., Huet, J.C., Crawford, N.A., Yee, B.C., and Buchanan, B.B. 1987. Ferredoxin-thioredoxin reductase, an iron-sulfur enzyme linking light to enzyme regulation in oxygenic photosynthesis: purification and properties of the enzyme from C_3, C_4, and cyanobacterial species. *Arch. Biochem. Biophys.* **252**, 426-439.
61. Einsle, O. Tezcan, F. A., Andrade, S. L., Schmid, B., Yoshida, M., Howard, J. B., and Rees, D. C. 2002. Nitrogenase MoFe-protein at 1.16 A resolution: a central ligand in the FeMo-cofactor. *Science* **297**, 1696-1700.

62. Eisbrenner, G., Roos, P., and Bothe, H. 1981. The number of hydrogenases in cyanobacteria. *J. Gen. Microbiol.* **125**, 383-390.
63. Erbes, D.L., King, D. and Gibbs, M. 1979. Inactivation of hydrogenase in cell-free extracts and whole cells of *Chlamydomonas reinhardi* by oxygen. *Plant Physiol.* **63**, 1138-1142.
64. Fascetti, E., D'addario, E. Todini, O., and Robertiello, A. 1998. Photosynthetic hydrogen evolution with volatile organic acids derived from the fermentation of source selected municipal solid wastes. *Int. J. Hydrogen Energy* **23**, 753-760.
65. Fedorov, A.S., Kosourov, S., Ghirardi, M.L. and Seibert, M. 2005. Continuous hydrogen photoproduction by *Chlamydomonas reinhardtii* using a novel two-stage, sulfate-limited chemostat system. *Appl. Biochem. Biotechnol.*, **121-124**, 403-412.
66. Fißler, J., Kohring, G. W., and Giffhorn, F. 1995. Enhanced hydrogen production from aromatic acids by immobilized cells of *Rhodopsdudomonas palustris*. *Arch. Microbial. Biotechnol.* **44**, 43-46.
67. Florin, L., Tsokogou, A. and Happe, T. 2001. A novel type of [Fe]-hydrogenase in the green alga *Scenedesmus obliquus* is linked to the photosynthetic electron transport chain. *J. Biol. Chem.* **276**, 6125-6130.
68. Forestier, M., King, P., Zhang, L., Posewitz, M., Schwarzer, S., Happe, T., Ghirardi, M.L. and Seibert, M. 2003. Expression of two [Fe]-hydrogenases in *Chlamydomonas reinhardtii* under anaerobic conditions. *Eur. J. Biochem.* **270**, 2750-2758.
69. Fox, J. D., Kerby, R. L., Roberts, G. P., and Ludden, P. W. 1996. Characterization of the CO-induced, CO-tolerant hydrogenase from *Rhodospirillum rubrum* and the gene encoding the large subunit of the enzyme. *J. Bacteriol.* **178**, 1515-1524.
70. Frey, M. 2003. Hydrogenases: hydrogen-activating enzymes. *Chembiochem.* **3**, 153-160.
71. Fujita, Y. and Murakami, A. 1987. Regulation of electron transport composition in cyanobacterial photosynthetic system: stoichiometry among PSI and PSII complexes and their light harvesting antenna and $Cyt\ b_6$-f complex. *Plant Cell Physiol.* **28**, 1547-1553.
72. Gallon, J. R. and Chaplin, A. E. 1988. Recent studies on N_2 fixation by nonheterocystous cyanobacteria. In: Bothe F. J., de Bruijin F. J., Newton, W. E. (eds) Nitrogen Fixation: Hundreds Years After. Gustav Fischer, Stuttgart, Germany, p. 183.
73. Gantt, E. 1986. Phycobillisomes. In: Staehelin L. A. and Arntzen, C. J. (eds). Encyclopedia of Plant Physiology. Photosynthesis III, ("Photosynthetic Membranes and Light-Harvesting Systems'), Vol. 19, pp 260-268, Academic Presss, New York.
74. Ghirardi, M.L., Togasaki, R.K. and Seibert, M. 1997. Oxygen sensitivity of algal H_2-production. *Appl. Biochem. Biotechnol.* **63-65**, 141-151.
75. Ghirardi, M.L.,Zhang, L., Lee, J.W., Flynn, T., Seibert, M., Greenbaum, E. and Melis, A. 2000. Sustained photobiological hydrogen gas production upon reversible inactivation of oxygen evolution in the green alga *Chlamydomonas reinhardtii*. *Trends Biotechnol.* **18**, 506-511.
76. Ghirardi, M.L. and Amos, W. 2004. Hydrogen photoproduction by sulfur-deprived green algae – status of the research and potential of the system. *Biocycle*, **45**, 59.
77. Ghirardi, M.L., King, P.W., Posewitz, M.C., Maness, P.C., Fedorov, A., Kim, K., Cohen, J., Schulten K. and Seibert, M. 2005. Approaches to developing biological H_2-photoproducing organisms and processes. *Biochem. Soc. Transact.* **33**, 70-72.
78. Ghirardi, M.L., Kin, P., Kosourov, S., Forestier, M., Zhang, L. and Seibert, M. 2005. Development of algal systems for hydrogen photoproduction – addressing the hydrogenase oxygen-sensitivity problem, in: *Artificial Photosynthesis*, C. Collings, ed., Wiley – VCH Verlag, Weinheim, Germany, pp. 213-227.
79. Gisby, P.E. and Hall, D.O. 1980. Biophotolytic H_2 production using alginate-immobilized chloroplasts, enzymes and synthetic catalysts. *Nature* **287**, 251-253.
80. Golden, J. W., and Yooon, H. S. 2003. Heterocyst development in *Anabaena*. *Curr. Opin. Microbiol.* **6**, 557-563.

81. Golubić, S. 1973. The relationship between blue-green alage and crbonate deposits. pp. 434-472. *In* The biology of blue-green alage, Carr, N. G., and Whitton, B. A. (eds.), Univ. of California Press, CA.

82. Gollan, U., Schneider, K., Müller, A., Schüddekopf, K., and Klipp, W. 1993. Detection of the *in vivo* incorporation of a metal cluster into a protein. The FeMo cofactor is inserted into the FeFe protein of the alternative nitrogenase in *Rhodobacter capsulatus*. *Eur. J. Biochem.* **215**, 25-35.

83. Grätzel, M.. 1983. Energy Resources through Photochemistry and Catalysis, Academic Press, New York.

84. Graves, D.A., Tevault, C.V., and Greenbaum, E. 1989. Control of photosynthetic reductant: the role of light and temperature on sustained hydrogen photoevolution by *Chlamydomonas* sp. in an anoxic, carbon-dioxide-containing atmosphere. *Photochem. Photobiol.* **50**, 571-576.

85. Greenbaum, E. 1988. Energetic efficiency of hydrogen photoevolution by algal water splitting. *Biophys. J.* **54**, 365-368.

86. Guan, Y., Deng, M., Yu, X., and Zhang, W. 2004. Two-stage photo-biological production of hydrogen by marine green alga *Platymonas subcordiformis*. *Biochem. Engin. J.* **19**, 69-73.

87. Guan, Y., Zhang, W., Deng, M., Jin, M., and Yu, X. 2004. Significant enhancement fo photobiological H2 evolution by carbonylcyanide m-chlorophenylhydrazone in the marine green alga *Platymonas subcordiformis*. *Biotechnol. Letts.* **26**, 1031-1035.

88. Guedon, E., Payot, S., Desvaux, M., and Petitdemange, H. 1999. Carbon and electron flow in *Clostridium cellulolyticum* grown in chemostat culture on synthetic medium. *J. Bacteriol.* **181**, 3262-3269.

89. Gust, D., Moore, T.A. and Moore, A.L. 2000. Photochemistry of supramolecular systems containing C60. *J. Photochem. Photobiol. B. Biology* **58**, 63-71.

90. Gust, D., Moore, T.A. and Moore, A.L. 2001. Mimicking photosynthetic solar energy transduction. *Acc. Chem. Res.* **34**, 40-48.

91. Hageman, R. V., and Burris, R. H. 1978. Nitrogenase and nitrogenase reductase associate and dissociate with each catalytic cycle. *Proc. Natl. Acad. Sci. USA* **75**, 2699-2702.

92. Hahn, J.J., Ghirardi, M.L. and Jacoby, W.A. 2004. Effect of process variables on photosynthetic algal hydrogen production. *Biotechnol. Progr.* **20**, 989-991.

93. Hallenbeck, P. C., Meyer, C. M., and Vignais, P. M.1982. Nitrogenase from the photosynthetic *bacterium Rhodopseudomonas capsulata: purification and molecular properties. J. Bacteriol.* **149**, 708-717.

94. Hallenbeck, P.C. and Benemann, J.R. 2002. Biological hydrogen production: photochemical efficiency and bioreactor design. *Int. J. Hydrogen Energy* **27**, 1185-1194.

95. Hammarström, L., Sun, L., Åkermark, B., Styring, S. 2001. A biomimetic approach to artificial photosynthesis: Ru(II)-polypyridine photo-sensitisers linked to tyrosine and manganese electron donors. *Spectrochim. Acta Part A* **37**, 2145-2160.

96. Hansel, A., Axelsson, R. Lindberg, P., Troshina, O. Y., Wünschiers, R., and Lindblad, P. 2001. Cloning and characterization of a hyp gene in the filamentous cyanobacterium *Nostoc* sp. strain 73102. *FEMS Microbiol.* **201**, 59-64.

97. Hanus, F.J., Carter, K.R. and Evans, H.J. 1980. Techniques for measurement of hydrogen evolution by nodules. *Methods Enzymol.* **69**, 731-739.

98. Happe, T. and Naber, J.D. 1993. Isolation, characterization and N-terminal amino acid sequence of hydrogenase from green alga *Chlamydomonas reinhardtii*. *Eur. J. Biochem.* **214**, 475-481.

99. Happe, T., Mosler, B. and Naber, J.D. 1994. Induction, localization and metal content of hydrogenase in the green alga *Chlamydomonas reinhardtii*. *Eur. J. Biochem.* **222**, 769-774.

100. Happe, R. P., Roseboom, W., Pierlk, A. J., Albracht, S. P. J., and Bagley, K. A. 1997. Biological activation of hydrogen. *Nature* **385**, 126.

101. Happe, T., Hemschemeier, A., Winkler, M. and Kaminski, A. 2002. Hydrogenases in green algae: do they save the algae's life and solve our energy problems? *Trends Plant Scie.* **7**, 246-250.

102. Happe, T. and Kaminski, A. 2002. Differential regulation of the [Fe]-hydrogenase during anaerobic adaptation in the green alga *Chlamydomonas reinhardtii. Eur. J. Biochem.* **269**, 1-11.

103. Happe, T., Schűtz, K., and Bőhme, H. 2000. Transcriptional and mutational analysis of the uptake hydrogenase of the filamentous cyanobacterium *Anabaena variabilis* ATCC 29413. *J. Bacteriol.* **182**, 1624-1631.

104. Higuchi, Y., Ogata, H., Miki, K., Yasuoka, N., and Yagi, T. 1999.Removal of the bridging ligand atom at the Ni-Fe active site of [NiFe] hydrogenase upon reduction with H_2, as revealed by X-ray structure analysis at 1.4 Å resolution. *Structure* **7**, 549-556.

105. Hillmer, P. and Gest, H. 1977a. H_2 metabolism in the photosynthetic bacterium *Rhodopseudomonas capsulata*: H_2 production by growing cultures. *J. Bacteriol.* **129**, 724-731.

106. Hillmer, P. and Gest, H. 1977b. H_2 metabolism in the photosynthetic bacterium *Rhodopseudomonas capsulata*: production and utilization of H_2 by resting cells. *J. Bacteriol.* **129**, 732-739.

107. Hirasawa, M., Boyer, J.M., Gray, K.A., Davis, D.J. and Knaff, D. 1987. The interaction of ferredoxin-linked sulfite reductase with ferredoxin. *FEBS Lett* **221**, 343-348.

108. Hirasawa, M., Droux, M., Gray, K.A., Boyer, J.M., Davis, D.J., Buchanan, B.B. and Knaff, D.B. 1988. Ferredoxin-thioredoxin reductase: properties of its complex with ferredoxin. *Biochim. Biophys. Acta* **935**, 1-8.

109. Hoffman, D., Thauer, R. and Trebst, A. 1977. Photosynthetic hydrogen evolution by spinach chloroplasts coupled to a *Clostridium* hydrogenase. *Z. Naturforsch* **32c**, 257-262.

110. Horton, P., Ruban, A.V., Rees, D., Pascal, A.A., Noctor, G. and Young, A.J. 1991. Control of the light-harvesting function of chloropolast membranes by aggregation of the LhcII chlorophyll protein complex. *FEBS Lett.* **292**, 1-4.

111. Houchins, J. P., and Burris, R. H. 1981. Physiological reactions of the reversible hydrogenase from *Anabaena* 7120. *Pl. Physiol.* **68**, 717-721.

112. Houchins, J. P. 1984. The physiology and biochemistry of hydrogen metabolism in cyanobacteria. *Biochim. Biophys. Acta.* **768**, 227-255.

113. Howard, D.L., Tinoco, A.D., Brudvig, G.W., Vrettos, J.S., Allen, B.C. 2005. Catalytic oxygen evolution by a bioinorganic model of the photosytem II core complexes. *J. Chem. Edu.* **82**, 791-794.

114. Huang, P., Hogblom, J., Anderlund, M.F., Sun, L., Magnuson, A. and Styring, S. 2004, Light-induced multistep oxidation of dinuclear manganese complexes for artificial photosynthesis. *J. Inorg. Biochem.* **98**, 733-745.

115. Hube, M., Blokesch, M., and Böck, A. 2002. Network of hydrogenase maturation in *Escherichia coli*: role of accessory proteins HypA and HybF. *J. Bacteriol.* **184**, 3879-3885.

116. Ike, A. Saimura, C., Hirata, K., and Miyamoto, K. 1996. Environmentally friendly production of H_2 incorporating microalgal CO_2 fixation. *J. Marine Biotech.* **4**, 47-51.

117. Imahori, H. and Sakata, Y. 1999. Fullerenes as novel acceptors in photosynthetic electron transfer. *Eur. J. Org. Chem* **1999**, 2445-2457.

118. Imhoff, J. F. 1995. Taxonomy and physiology of phototrophic purple bacteria and green sulfur bacteria. p. 1-15. In Anoxygenic Photosynthetic Bacteria. Blakenship, R. E., Madigan, M. T., and Bauer, C. E. (eds). Kluwer Academic Publishers, The Netherlands.

119. Jacobi, A., Rossmann, R., and Böck, A. 1992. The hyp operon gene products are required for the maturation of catalytically active hydrogenase isoenzymes in *Escherichia coli*. *Arch. Microbiol.* **158**, 444-451.

120. Jacquot, J.P., Stein, M., Suzuki, A., Liottet, S., Sandoz, G. and Miginiac-Maslow, M. 1997. Residue Glu-91 of *Chlamydomonas reinhardtii* ferredoxin is essential for electron transfer to ferredoxin-thioredoxin reductase. *FEBS Lett.* **400**, 293-296.

121. Kanazawa, A. and Kramer, D.M. 2002. *In vivo* modulation of nonphotochemical exciton quenching (NPQ) by regulation of the chloroplast ATP synthase. *Proc. Natl. Acad. Sci. USA* **99**, 12789-12794.

122. Karube, I., Matsunaga, T., Otsuka, T., Kayano, H. and Suzuki, S. 1981. Hydrogen evolution by co-immobilized chloroplasts and *Clostridium butyricum*. *Biochim. Biophys. Acta* **637**, 490-495.

123. Kemner, J., and Zeikus, J. G. 1994. Purification and characterization of membrane-bound hydrogenase from *Methanosarcina barkeri* MS. Arch. Microbiol. **161**, 47-54.

124. Kentemich, T., Danneberg, G., Hundeshagen, B., and Bothe, H. 1988. Evidence for the occurrence of the alternative, vanadium-containing nitrogenase in the cyanobacterium *Anabaena variabilis*. *FEMS Microbiol. Lett.* **51**, 19-24.

125. Kim, J.P., Kang, C.D., Sim, S.J., Kim, M.S., Park T.H., Lee, D., Kim, D., Kim, J.H., Lee, Y.K. and Pak, D. 2005. Cell age optimization for hydrogen production induced by sulfur deprivation using a green alga *Chlamydomonas reinhardtii* UTEX 90. *J. Microbiol. Biotechnol.* **15**, 131-135.

126. King, P.W., Posewitz, M.C., Ghirardi, M. L., and Seibert M. 2006. Functional studies of [FeFe]-hydrogenase maturation in an *Escherichia coli* biosynthetic system, *J. Bacteriol.* **288**, 2163-2172 .

127. Klibanov, A.M., Kaplan, N.O. and Kamen, M.D. 1978. A rationale for stabilization of oxygen-labile enzymes: application to a clostridial hydrogenase. *Proc. Natl. Acad. Sci. USA* **75**, 3640-3643.

128. Knaff, D. B. 1996. Ferredoxin and ferredoxin-dependent enzymes. In: Oxygenic Photosynthesis: The Light Reactions (D.R. Ort and C.F. Yocum, eds.), Kluwer Academic Publishers, pp.333-361.

129. Koku, H., Eroğlu, İ., Gűndűz, U. Yűcel, M., and Tűrker, L. 2002. Aspect of the metabolism of hydrogen production by *Rhodobacter sphaeroides*. *Intl. J. Hydrogen Energy* **27**, 1315-1329.

130. Kosourov, S., Tsygankov, A., Seibert, M. and Ghirardi, M.L. 2002. Sustained hydrogen photoproduction by *Chlamydomonas reinhardtii* – Effects of culture parameters. *Biotechnol. Bioeng.* **78**, 731-740.

131. Kosourov, S, M Seibert and ML Ghirardi. 2003. Effects of extracellular pH on the metabolic pathways in sulfur-deprived, H_2-producing *Chlamydomonas reinhardtii* cultures. *Plant Cell Physiol.* **44**,146-155.

132. Kramer, D.M., Avenson, T.J. and Edwards, G.E. 2004. Dynamic flexibility in the light reactions of photosynthesis governed by both electron and proton transfer reactions. *Trends Plant Sci.* **9**, 349-357.

133. Kurkin, S., George, S. J., Thorneley, R. N. F., and Albracht, S. P. J. 2004. Hydrogen-induced activation of the [NiFe]-hydrogenase from *Allochromatium vinosum* as studied by stopped-flow infrared spectroscopy. *Biochem.* **43**, 6820-6831.

134. Kunkell, D.D. 1982. Thylakoid centers: structures associated with cyanobacterial photosynthetic membrane system. *Arch. Microbiol.* **133**, 97-99.

135. Larimer, F.W., Chain, P., Hauser, L., Lamerdin, J., Malfatti, S., Do, L., Land, M.L., Pelletier, D. A., Beatty, J.T., Lang, A.S., Tabita, F.R., Gibson, J.L., Hanson, T.E., Bobst, C., Torres, J.L. T., Peres, C., Harrison, F., Gibson, J., and Harwood, C. S. 2004. Com-

plete genome sequence of the metabolically versatile photosynthetic bacterium *Rhodopseudomonas palustris*. *Nature Biotechnol.* **22**, 55-61.

136. Laurinavichene, T.V., Tolstygina, I.V., Galiulina, R.R., Ghirardi, M.L., Seibert, M. and Tsygankov, A. 2002. Different methods to deprive *Chlamydomonas reinhardtii* cultures of sulfur for subsequent hydrogen photoproduction. *Internat'l. J. Hydrogen Energy* **27**, 1245-1249.

137. Laurinavichene, T., Tolstyginina, I. and Tsygankov, A. 2004. The effect of light intensity on hydrogen production by sulfur-deprived *Chlamydomonas reinhardtii*. *J. Biotechnol.* **114**, 143-151.

138. Laurinavichene, T.V., Fedorov.A.S., Ghirardi, M.L., Seibert, M., and Tsygankov, A.A. 2006. Demonstration of sustained hydrogen photoproduction by immobilized, sulfur-deprived *Chlamydomonas reinhardtii* cells. *Internat'l. J. Hydrogen Energy* **31**, 659-667.

139. Lee, C. M., Chen, P. C., Wang, C. C., and Tung, Y. C. 2002. Photohydrogen production using purple nonsulfur bacteria with hydrogen fermentation reactor effluent. *Intl. J. Hydrogen Energy* **27**, 1309-1313.

140. Lee, J.W. and Greenbaum, E. 2003. A new oxygen sensitivity and its potential application in photosynthetic H_2 production. *Appl. Biochem. Bioeng.* **105-108**, 303-313.

141. Lehman, L. J., and Roberts, G. P. 1991. Identification of an alternative nitrogenase system in *Rhodospirillum rubrum*. *J. Bacteriol.* **173**, 5705-5711.

142. Levin, D., Pitt, L. and Love, M. 2004. Biohydrogen production: prospects and limitations to practical application. Internat'l. J. Hydrogen Energy **29**, 173-185.

143. Li, H. and Rauschfuss, B. 2002. Iron carbonyl sulfides, formaldehyde, and amines condense to given the proposed azadithiolate cofactor of the Fe-only hydrogenaes. J. Am. Chem. Soc. **124**, 726-727.

144. Liu, H., Grot, S., and Logan, B. E. 2005. Electrochemically assisted microbial production of hydrogen from acetate. Environ. Sci. Technol. **39**, 4317-4320.

145. Liu, X., Ibrahim, S.K., Tard, C., and Pickett, C.J. 2005. Iron-only hydrogenase: Synthetic, structural, and reactivity studies of model compounds. *Coord. Chem. Rev.* **249**, 1641-1652.

146. Lutz, S., Jacobi, A., Schlensog, V., Böhm, R. Sawers, G., Böck, A. 1991. Molecular characterization of an operon (hyp) necessary for the activity of the three hydrogenase isoenzymes in *Escherichia coli. Mol. Microbiol.* **5**, 123-135.

147. Madigan, M., Cox, S. S., and Stegeman, R. A. 1984. Nitrogen fixation and nitrogenase activities in members of the family *Rhodospirillaceae. J. Bacteriol.* **157**, 73-78.

148. Maeda, I., Hikawa, H., Miyashiro, M., Yagi, K., Miure, Y., Miyasaka, H., Akano, T., Kiyohara, M., Matsumoto, H., and Ikuta, Y, 1994. Enhancement of starch degradation by CO_2 in a marine green alga, *Chlamydomonas* sp. MGA161. *J. Ferment. Bioeng.* **78**, 383-385.

149. Maeda, I., Chowdhury, W.Q., Idehara, K., Yagi, K., Mizoguchi, T., Akano, T., Miyasaka, H. Furutani, T., Ikuta, Y., Shioji, N., and Miura, Y. 1998. Improvement of Substrate Conversion to Molecular Hydrogen by Three Stage Cultivation of a Photosynthetic Bacterium, *Rhodovulum sulfidophilum. Appl. Biochem. Biotechnol.* **70-72**, 301-310.

150. Magalon, A., and Böck, A. 2000a. Dissection of the maturation reaction of the [NiFe] hydrogenase 3 from *Escherichia coli* taking place after nickel incorporation. *FEBS Lett.* **473**, 254-258.

151. Maier, T., Lottspeich, F., and Böck, A. 1995. GTP hydrolysis by HypB is essential for nickel insertion into hydrogenases of *Escherichia coli. Eur. J. Biochem.* **230**, 133-138.

152. Maness, P. C., Smolinski, S., Dillon, A. C., Heben, M. J., and Weaver, P. F. 2002. Characterization of the oxygen tolerance of a hydrogenase linked to a carbon monoxide oxidation pathway in *Rubrivivax gelatinosus. Appl. Environ. Microbiol.* **68**, 2633-2636.

153. Masters, R. A., and Madigan, M. 1983. Nitrogen metabolism in the phototrophic bacteria *Rhodocyclus purpureus* and *Rhodospirillum tenue. J. Bacteriol.* **155**, 222-227.
154. Masukawa, H., Mochimaru, M., and Sakurai, H. 2002a. Disruption of uptake hydrogenase gene, but not of the bidirectional hydrogenase gene, leads to enhanced photobiological hydrogen production by the nitrogen-fixing cyanobacterium *Anabaena* sp. 7210. *Appl. Microbiol. Biotechnol.* **58**, 618-624.
155. Masukawa, H., Mochimaru, M., and Sakurai, H. 2002b. Hydrogenases and photobiological hydrogen production utilizing nitrogenase system in cyanobacteria. *Internat'l. J. Hydrogen Energy* **27**, 1471-1474.
156. Mayer, S. M., Lawson, D. M., Gormal, C. A., Roe, S. M., and Smith, B. E. 1999. New insights into structure-function relationships in nitrogenase: a 1.6 A resolution X-ray crystallographic study of *Klebsiella pneumonia* MoFe-protein . *J. Mol. Biol.* **292**, 871-891.
157. McTavish, H., Sayavedra-Soto, L. A., and Arp. D. 1996. Comparison of isotope exchange, H_2 evolution, and H_2 oxidation activities of *Azotobacter vinelandii* hydrogenase. *Biochim. Biophys. Acta* **1294**, 183-190.
158. Melis, A., Niedhardt, J., Benemann, J.R. 1999. *Dunaliella salina* (Chlorophyta) with small chlorophyll antenna sizes exhibit higher photosynthetic productivities and photon use efficiencies than normally pigmented cells. *J. Appl. Phycol.* **10**, 515-525.
159. Melis, A., Zhang, L., Forestier, M., Ghirardi, M.L. and Seibert, M. 2000. Sustained photobiological hydrogen gas production upon reversible inactivation of oxygen evolution in the green alga *Chlamydomonas reinhardtii. Plant Physiol.* **122**, 127-135.
160. Melis, A. and Happe, T. 2001. Hydrogen production. Green algae as a source of energy. *Plant Physiol.* **127**, 740-748.
161. Melis, A. 2002. Green alga hydrogen production: progress, challenges and prospects. *Internat'l. J. Hydrogen Energy* **27**, 1217-1228.
162. Melis, A., Seibert, M. and Happe, T. 2004. Genomics of green algal hydrogen research. *Photosynth Res.* **82**, 277-288.
163. Meyer, T.J. 1989. Chemical approaches to artificial photosynthesis. *Acc. Chem. Res.* **22**, 163-170.
164. Miura, Y., Yagi, I., Shoga, M., and Miyamoto, K. 1982. Hydrogen production by a green alga, *Chlamydomonas reinhardtii*, in an alternating light/dark cycle. *Biotechno. Bioeng.* **24**, 1555-1563.
165. Misra, H. S., and Tuli, R. 2000. Differential expression of photosynthesis and N_2-fixation genes in the cyanobacterium *Plectonema boryanum. Plant Physiol.* **468**, 731-736.
166. Miyake, J., Mao, X. Y., and Kawamura, S. 1984. Photoproduction of hydrogen by a co-culture of a photosynthetic bacterium and *Clostridium butyricum. J. Ferment. Technol.* **62**, 531-535.
167. Miyake, J., Miyake, M., and Asada, Y. 1999. Biotechnological hydrogen production: research for efficient light energy conversion. *J. Biotechnol.* **70**, 89-101.
168. Miyamoto, K., Hallenbeck, P.C. and Benemann, J.R. 1979. Solar energy conversion by nitrogen-limited cultures of *Anabaena cylindrica. J. Ferment. Technol.* **57**, 287-293.
169. Montet, Y., Amara, P., Volbeda, A., Vernede, X., Hatchikian, E.C., Field, M.J., Frey, M. and Fontecilla-Camps, J.C. 1997. Gas access to the active site of Ni-Fe hydrogenases probed by X-ray crystallography and molecular dynamics. *Nat. Struct. Biol.* **4**, 523-526.
170. Moore, T.A., Moore, A.L. and Gust, D. 2002. The design and synthesis of artificial photosynthetic antennas, reaction centres and membranes. *Phil. Trans. R. soc. Lond. B* **357**, 1481-1498.
171. Müller, P, Li, X.P. and Niyogi, K.K. 2001. Non-photochemical quenching. A response to excess light energy. *Plant Physiol.* **125**, 1558-1566.

172. Murry, M.A., and Wolk, C.P. 1989. Evidences that the barrier to the penetration of oxygen into heterocysts depends upon two layers of the cell envelope. *Arch. Microbiol.* **151**, 469-474.

173. Mus, F., Cournac, L., Cardettini, V., Caruana, A. and Peltier, G. 2005. Inhibitor studies on non-photochemical plastoquinone reduction and H_2 photoproduction in *Chlamydomonas reinhardtii. Biochim. Biophys. Acta* **1708**, 322-332.

174. Nakajima, Y. and Ueda, T. 1997. Improvement of photosynthesis in dense microalgal suspension by reduction of light harvesting pigments. *J. Appl. Phycol.* **9**, 503-510.

175. Nakajima, Y. and Ueda, T. 1999. Improvement of microalgal photosynthetic productivity by reducing the content of light harvesting pigment. *J. Appl. Phycol.* **11**, 195-201.

176. Nakajima, Y. and Ueda, T. 2000. The improvement of marine microalgal productivity by reducing the light-harvesting pigment. *J. Appl. Phycol.* **12**, 285-290.

177. Nakajima, Y., Tsuzuki, M., and Ueda, R. 2001. Improved productivity by reduction of the content of light-harvesting pigment in *Chlamydomonas perigranulata. J. App. Phycol.* **13**, 95-101.

178. Nash, D., Takahashi, M. and Asada, K. 1984. Dark anaerobic inactivation of photosynthetic oxygen evolution by *Chlamydomonas reinhardtii. Plant Cell Physiol.* **25**, 531-539.

179. National Research Council. 2004. The hydrogen economy: opportunities, costs, barriers, and R&D needs. The National Academies Press, Washington D.C.

180. Neale, P.J. and Melis, A. 1986. Algal photosynthetic membrane complexes and the photosynthesis-irradiance curve: a comparison of light-adaptation responses in *Chlamydomonas reinhardtii* (Chlorophyta*). J. Phycol.* **22**, 531-538.

181. Nicolet Y., Piras, C., Legrand, P., Hatchikian, E.C. and Fontecilla-Camps, J.C. 1999. *Desulfovibrio desulfuricans* iron hydrogenase: the structure shows unusual coordination to an active site Fe binuclear center. *Structure* **7**, 13-23.

182. Nicolet, Y., deLacey, A.L., Vernede,, X., Fernandez, V.M., Hatchikian, E.C. and Fontecilla-Camps, J.C. 2001. Crystallographic and FTIR spectroscopic evidence of changes in Fe coordination upon reduction of the active site of the Fe-only hydrogenase from *Desulfovibrio desulfuricans. J. Am. Chem. Soc.* **123**, 1596-1601.

183. Niyogi, K.K. 1999. Photoprotection revisited: genetics and molecular approaches. *Annu. Rev. Plant Physiol. Plant Mol. Biol.* **50**, 333-359.

184. Nugent, J.H.A. and Evans, M.C.S. 2004. Structure of biological solar energy converters – further revelations. *Trends Plant Sci.* **9**, 368-370.

185. Odom, J. M., and Peck, Jr., H. D. 1984. Hydrogenase, electron-transfer proteins, and energy coupling in the sulfate-reducing bacteria *Desulfovibrio. Ann. Rev. Microbiol.* **38**, 551-592.

186. Ohkawa, H., Sonoda, M., Shibata, M., and Ogawa, T. 2001. Localization of NAD(P)H dehydrogenase in the cyanobacterium *Synechocystis* sp. Strain PCC 6803. *J. Bacteriol.* **183**, 4938-4939.

187. Osshima, H., Takakuwa, S., Katsuda, T., Okuda, M., Shirasawa, T., Azuma, M., and Kato, J. 1998. Production of hydrogen by a hydrogenase-deficient mutant of *Rhodobacter capsulatus. J. Ferment. Bioeng.* **85**, 470-474.

188. Ormerod, J. G., Ormerod, K. S., and Gest. H. 1961. Light-dependent utilization of organic compounds and photoproduction of molecular hydrogen by photosynthetic bacteria; relationships with nitrogen metabolism. *Arch. Biochem. Biophys.* **94**, 449-463.

189. Ort, D. and Yocum, C.F. 1996. Oxygenic Photosynthesis: the Light Reactions. Kluwer Academic Publishers, London.

190. Packer, L. and Cullingford, W. 1977. Stoichiometry of H_2 production by an *in vitro* chloroplast, ferredoxin, hydrogenase reconstituted system. *Z. Naturforsch.* **33c**, 113-115.

191. Paschos, A., Bauer, A., Zimmermann, A., Zehelein, E. and Böck, A. 2002. HypF, a car-bamoyl phosphate-converting enzyme involved in [NiFe] hydrogenase maturation. *J. Biol. Chem.* **277**, 49945-49951.

192. Paschos, A., Glass, R. S., Böck, A. 2001. Carbamoylphosphate requirement for synthesis of the active center of [NiFe]-hydrogenases. *FEBS Lett.* **488**, 9-12.

193. Pfennig, N. 1978. General physiology and ecology of photosynthetic bacteria. p. 3-14. In The Photosynthetic Bacteria. Clayton, R. K., and Sistrom, W. R. (eds). Plenum Press, New York.

194. Pierik, A.J., Hulstein, M., Hagen, W.R. and Allbracht, S.P. 1998. A low spin iron with CN and CO as intrinsic ligands form the core of the active site in [Fe]-hydrogenases. *Eur. J. Biochem.* **258**, 572-578.

195. Polle, J.E.W., Benemann, J.R., Tanaka, A. and Melis, A. 2000. Photosynthetic apparatus organization and function in wild type and a Chl b-less mutant of *Chlamydomonas reinhardtii*. Dependence on carbon source. *Planta* **211**, 335-344.

196. Polle, J.E.W., Niyogi, K.K. and Melis, A. 2001. Absence of lutein, violaxanthin and neoxanthin affects the functional chlorophyll antenna size of photosystem-II but not that of photosystem-I in the green algae *Chlamydomonas reinhardtii*. *Plant Cell Physiol.* **42**, 482-491.

197. Polle, J.E.W., Kanakagiri, S., Jin, E., Masuda, T. and Melis, A. 2002. Truncated chloro-phyll antenna size of the photosystems – a practical method to improve microalgal prod-uctivity and hydrogen production in mass culture. *Internat'l. J. Hydrogen Energy* **27**, 1257-1264.

198. Polle, J.E.W., Kanakagiri, S. and Melis, A. 2003. *tla1*, a DNA insertional transformant of the green alga *Chlamydomonas reinhardtii* with a truncated light-harvesting chlorophyll antenna size. *Planta* **217**, 49-59.

199. Posewitz, M.C., King, P.W., Smolinski, S.L., Zhang, L., Seibert, M. and Ghirardi, M.L. 2004a. Discovery of two novel radical S-adenosylmethionine proteins required for the as-sembly of an active [Fe] hydrogenase. *J. Biol. Chem.* **279**, 25711-25720.

200. Posewitz, M.C., Smolinski, S.L, Kanakagiri, S., Melis, A., Seibert, M. and Ghirardi, M.L. 2004b. Hydrogen photoproduction is attenuated by disruption of an isoamylase gene in *Chlamydomonas reinhardtii*. *Plant Cell* **16**, 2151-2163.

201. Posewitz, M.C., King, P.W., Smolinski, S.L., Smith II, R.D., Ginley, A.R., Ghirardi, M.L. and Seibert, M. 2005. Identification of genes required for hydrogenase activity in *Chlamy-domonas reinhardtii*. *Biochem. Soc. Trans.* **33**, 102-104.

202. Prince, R.C., and Kheshgi, H.S. 2005. The photobiological production of hydrogen: poten-tial efficiency and effectiveness as a renewable fuel. *Crit. Rev. Microbiol.* **31**, 19-31.

203. Qiang, H., Faiman, D., and Richmond, A. 1998. Optimal tilt angles of enclosed reactors for growing photoautotrophic microorganisms outdoors. *J. Fermentation and Bioengi-neering* **56**, 230-236.

204. Przybyla, A. E., Robbins, J., Menon, N., and Peck, Jr. H. D. 1992. Structural-function relationships among the nickel-containing hydrogenases. *FEMS Microbiol. Rev.* **88**, 109-136.

205. Rachman, M.A., Furutani, Y., Nakashimada, Y., Kakizono, T., and Nishio, N. 1997. Enhanced hydrogen production in altered mixed acid fermentation of glucose by *Entero-bacter aerogenes*. *J. Ferm. Bioeng.* **83**, 358-363.

206. Radmer, R. and Kok, B. 1977. Photosynthesis: limited yields, unlimited dreams. *Bios-cience* **29**, 599-605.

207. Rakhely, G., Colbeau, A., Garin, J., Vignais, P. M., and Kovacs, K. 1998. Unusual organ-ization of the genes coding or HydSL, the stable [NiFe]hydrogenase in the photosynthetic bacterium *Thiocapsa roseopersicina* BBS. *J. Bacteriol.* **180**, 1460-1465.

208. Rao, K.K., Gogotov, I.N. and Hall, D.O. 1978. Hydrogen evolution by chloroplast-hydrogenase systems: improvements and additional observations. *Biochimie* **60**, 291-296.
209. Rao, K.K., Hall, D.O., Vlachopoulos, N., Grätzel, M., Evans, M.C.W., and Seibert, M. 1990, Photoelectrochemical response of photosystem II particles immobilized on dye-derivatized TiO_2 films. *J. Photochem. Photobiol.* **5**, 379-389.
210. Raymond, J., Siefert, J. L., Staples, C. R., and Blankenship, R. E. 2003. The natural history of nitrogen fixation. *Mol. Biol. Evol.* **21**, 541-554.
211. Reade, J. P., Dougherty, L. J., Rodgers, L. J., and Gallon, J. R. 1999. Synthesis and proteolytic degradation of nitrogenase in cultures of the unicellular cyanobacterium *Gloeothece* strain ATCC 27152. *Microbiol.* **145**, 1749-1758.
212. Reissmann, S., Hochleitner, E., Wang, H., Paschos, A., Lottspeich, F., Glass, R. S., and Böck, A. 2003. Taming of a poison: biosynthesis of the NiFe-hydrogenase cyanide ligands. *Science*, **299**, 1067-1070.
213. Reüttinger, W., and Dismukes, G.C. 1997. Synthetic water-oxidation catalysts for artificial photosynthetic water oxidation. *Hem. Rev.* **126**, 1-24.
214. Rey, L., Fernández, D., Brito, B., Hernando, Y., Palacios, J. M., Imperial, J., and Ruiz-Argüeso, T. 1996. The hydrogenase gene cluster of *Rhizobium leguminosarum* bv viciae contains an additional gene (hypX) which encodes a protein with sequence similarity to the N10-formyltetrahydrofolate-dependent enzyme family and is required for nickel-dependent hydrogenase processing and activity. *Mol. Gen. Genet.* **252**, 237-248.
215. Roessler, P. and Lien, S. 1982. Anionic modulation of the catalytic activity of hydrogenase from *Chlamydomonas reinhardtii*. *Arch. Biochem. Biophys.* **213**, 37-44.
216. Roessler, P. and Lien, S,. 1984, Purification of hydrogenase from *Chlamydomonas reinhardtii*. *Plant Physiol.* **75**, 705-709.
217. Rossmann, R., Sauter, M., Lottspeich, F., and Böck, A. 1994. Maturation of the large subunit (HycE) of *Escherichia coli* hydrogenase 3 requires nickel incorporation followed by C-terminal processing at Arg537. *Eur. J. Biochem.* **220**, 377-384.
218. Rubio, L. M., and Ludden, P. W. 2005. Maturation of nitrogenases: a biochemical puzzles. *J. Bacteriol.* **187**, 405-414.
219. Rybtchinski, B., Sinks, L.E., Wasielewski, M.R. 2004. Combining light-harvesting and charge separation in a self-assembled artificial photosynthetic system based on perylenediimide chromophores. J. Am. Chem. Soc. **126**, 12268-12269.
220. Sakamoto, M., Kamachi, T., Okura, I., Ueno, A. and Mihara, H. 2001. Photoinduced hydrogen evolution with peptide dendrimer-multi-Zn(II)-porphyrin, viologen, and hydrogenase. *Biopolymers* **59**, 103-109.
221. Samson, G., Herbert, S.K., Fork, D.C. and Laudenbach, D.E. 1994. Acclimation of the photosynthetic apparatus to growth irradiance in the mutant strain of *Synechococcus* lacking iron superoxide dismutase. *Plant Physiol.* **105**, 287-294.
222. Sasikala, K., Ramana, C. V., Rao, P. R., and Subrahmanyam, M. 1990. Effect of gas phase on the photoproduction of hydrogen and substrate conversion efficiency in the photosynthetic bacterium *Rhodobacter sphaeroides* O.U. 001. *Internat'l. J. Hydrogen Energy* **15**, 795-797.
223. Scherer, S., Almon, H. Böger, P. 1988. Interactions of photosynthesis, respiration, and nitrogen fixation in cyanobacteria. *Photosyn. Res.* **15**, 95-114.
224. Schick, H. J. 1971. Substrate and light dependent fixation of molecular nitrogen in *Rhodospirillum rubrum*. *Arch. Microbiol.* **75**, 89-101.
225. Schmitz, O., Boison, G., Hilscher, R., Hundeshagen, B., Zimmer, w., Lottspeich, F., and Bothe H. 1995. Molecular biological analysis of a bidirectional hydrogenase from cyanobacteria. *Eur. J. Biochem.* **233**, 266-276.
226. Schmitz, O., and Bothe, H. 1996. $NAD(P)^+$-dependent hydrogenase activity in extracts from the cyanobacterium *Anacystis nidulans*. 1996. *FEMS Microbiol. Lett.* 135, 97-101.

227. Schmitz, O., Boison, G., Salzmann, H., Bothe, H., Schűtz, K., Wang, S. H., and Happe, T. 2002. HoxE – a subunit specific for the pentameric bidirectional hydrogenase complex (HoxEFUYH) of cyanobacteria. *Biochim. Biophys. Acta* **1554**, 66-74.

228. Schűtz, K., Happe, T., Olga, T., Lindblad, P., Leitão, E., Oliveira, P., and Tamagnini, P. 2004. Cyanobacterial H_2 production – a comparative analysis. *Planta* **218**, P. 350-359.

229. Seibert M., Flynn, T., Benson, D., Tracy, E. and Ghirardi, M. 1998. Development of selection/screening procedures for rapid identification of H_2-producing algal mutants with increased O_2-tolerance, in *Biohydrogen*, O. R. Zaborsky, ed., Plenum Publishing Corporation, New York, N.Y., pp. 227-234.

230. Seibert, M., Flynn, T., and Benson, D. 2001a. Method and apparatus for rapid biohydrogen phenotypic screening of microorganisms using a chemochromic sensor, U.S. Patent # 6,277,589.

231. Seibert, M., Flynn, T. and Ghirardi, M.L. 2001b. Strategies for improving oxygen tolerance of algal hydrogen production, in *BioHydrogen II*, J. Miyake, T. Matsunaga and A. San Pietro eds., Pergamon Press, Amsterdam, p. 65-76.

232. Seibert, M., Flynn, T., and Benson, D. 2002. System for rapid biohydrogen phenotypic screening of microorganisms using a chemochromic sensor, U.S. Patent # 6,448,068.

233. Serebriakova, L., Zorin, N. A., and Lindblad, P. 1994. Reversible hydrogenase in *Anabaena variabilis* ATCC 29413. *Arch. Microbiol.*, **161**, 140-144.

234. Shah, V. K., and Brill, W. J. 1977. Isolation of an iron-molybdenum cofactor from nitrogenase. *Proc. Natl. Acad. Sci. USA* **74**, 3249-3253.

235. Sofia, H.J., Chen, G., Hetzler, B.G., Reyes-Spindola, J.F. and Miller, N.E. 2001. Radical SAM, a novel protein superfamily linking unresolved steps in familiar biosynthetic pathways with radical mechanisms: functional characterization using new analysis and information visualization methods. *Nucleic Acids Res.* **29**, 1097-1106.

236. Steinberg-Yfrach, G., Rigaud, J.L., Durantini, E.N., Moore, A.L., Gust, D. and Moore, T.A., 2002. Light-driven production of ATP catalysed by F_0F_1-ATP synthase in an artificial photosynthetic membrane. *Nature* **392**, 479-482.

237. Stevens, S. E., and Nierzwicki-Bauer, S. 1991. The Cyanobacteria. In: Stolz JF (ed) Structure of phototrophic prokaryotes, pp15-47. CRC Press, Inc. Baco Raton.

238. Sun, L., Raymond, M.K., Magnuson, A., LeGourriérec, D., Tamm, M., Abrahamsson, M., Kenéz, P.H., Mårtensson, J., Stenhagen, G., Hammarström, L., Styring, S., Åkermark, B. 2000. Towards an artificial model for photosystem II: A manganese(II,II) dimer covalently linked to ruthenium(II) tris-bipyridine via a tyrosine derivative *J. Inorg. Biochem.* **78**, 15-22.

239. Sun, L., Åkermark, D., and Ott, S. 2005. Iron hydrogenase activity site mimics in supramolecular systems aiming for light-driven hydrogen production. *Coord. Chem. Rev.* **249**, 1653-1663.

240. Tamagnini, P., Troshina, O., Oxelfelt, F., Salema, R., and Lindblad, P. 1997. Hydrogenase in *Nostoc* sp. strain PCC 73102, a strain lacking a bidirectional enzme. *Appl. Enviorn. Microbiol.* **63**, 1801-1807.

241. Tamagnini, P., Costa, J. L., Almeida, L., Oliveira, M. J., Salema, R., and Linblad, P. 2000. Diversity of cyanobacterial hydrogenase, a molecular biology approach. *Curr. Microbiol.* **40**, 356-361.

242. Tamagnini, P., Axelsson, R., Lindberg, P., Oxelfelt, F., Wűnschiers, R., and Lindblad, P. 2002. Hydrogenases and hydrogen metabolism of cyanobacteria. *Microbiol. Mol. Biol. Rev.* **66**, 1-20.

243. Tard, C., Liu, X., Ibrahim, S.K., Bruschi, M., de Gioia, L., Davies, S.C., Yang, X., Wang, L.S., Sawers, G. and Pickett, C.J. 2005. Synthesis of the H-cluster framework of iron-only hydrogenase. *Nature* **433**, 610-613.

244. Thiel, T. 1993. Characterization of genes for an alternative nitrogenase in the cyanobacterium *Anabaena variabilis. J. Bacteriol.* **175,** 6276-6286.

245. Thiel, T., and Pratte, B. 2001. Effect of heterocyst differentiation of nitrogen fixation in vegetative cells of the cyanobacterium *Anabaena variabilis* ATCC 29413. *J. Bacteriol.* **183,** 280-286.

246. Torzillo, G., Carlozzi, P., Pushparaj, B., Montaini, E., and Materassi, R. 1993. A two-plane tubular photobioreactor for outdoor culture of *Spirulina. Biotechnology and Bioengineering,* **42,** 891-898.

247. Trebitsh, T. and Danon, A. 2001. Translation of chloroplast *psbA* mRNA is regulated by signals initiated by both photosystems II and I. *Proc. Natl. Acad. Scie.* **98,** 12289-12294.

248. Trüper, H. G., and Pfennig, N. 1974. Taxonomy of the Rhodospirillales. P. 19-26. In The Photosynthetic Bacteria. Clayton, R. K., and Sistrom, W. R. (eds). Plenum Press, New York.

249. Tsygankov, A.A., Borodin, V.B., Rao, K.K. and Hall, D.O. 1999. H2 photoproduction by batch culture of *Anabaena variabilis* ATCC 29413 and its mutant PK84 in a photobioreactor. *Biotechnol. Bioeng.* **64,** 709-715.

250. Tsygankov, A., Kosourov, S., Seibert, M. and Ghirardi, M.L. 2002. Hydrogen photoproduction under continuous illumination by sulfur-deprived, synchronous *Chlamydomonas reinhardtii* cultures. *J. Intern. Hydrogen Energy* **27,** 1239-1244.

251. Van der Linden, E., Faber, B. W., Bleijlevens, B., Burgdorf, T., Bernhard, M., friedrich, B., and Albracht, S. P. J. 2004. Selective release and function of one of the two FMN groups in the cytoplasmic NAD⁺-reducing [NiFe]-hydrogenase from *Ralstonia eutropha.* Eur. J. Biochem. **271,** 801-808.

252. Venter, J.C., Remington, K., Heidelberg, J.F., Halpern, A.L. 2004. Environment Genome Shotgun Sequencing of the Sargasso Sea. *Science* **304,** 66-74.

253. Verhagen, M.F., O'Rourke, T. and Adams, M.W. 1999. The hyperthermophilic bacterium, *Thermatoga maritima,* contains an unusually complex iron-hydrogenase: amino acid sequence analyses versus biochemical characterization. *Biochim. Biophys. Acta* **1412,** 212-219.

254. Vignais, P. M., Dimon, B., Zorin, N. A., Tomiyama, M., and Colbeau, A. 2000. Characterization of the hydrogen-deuterium exchange activities of the energy-transduing HupSL hydrogenase and H2-signaling HupUV hydrogenase in *Rhodobacter capsulatus. J. Bacteriol.* **182,** 5997-6004.

255. Vignais, P.M., Billoud, B. and Meyer, J. 2001. Classification and phylogeny of hydrogenases. *FEMS Microbiol. Rev.* **25,** 455-501.

256. Vignais, P. M., and Colbeau A. 2004. Molecular biology of microbial hydrogenase. *Curr. Issues Mol. Bio.* **6,** 159-188.

257. Volbeda, A., and Fontecilla-Camps, J. 2003. The active site and catalytic mechanisms of NiFe hydrogenases. *Dalton Trans.* **6,** 4030-4038.

258. Volbeda, A., Charon, M. H., Piras, C., Hatchikian, E. C., Frey, M., and Fontecilla-Camps, J. C. 1995. Crystal structure of the nickel-iron hydrogenase from *Desulfovibrio gigas. Nature* **373,** 580-587.

259. Volbeda, A., Montet, Y., Vernède, X., Hatchikian, E. C., and Fontecilla-Camps, J. C. 2002. High-resolution crystallographic analysis of *Desulfovibrio fructosovorans* [NiFe] hydrogenase. *Intl. J. Hydrogen Energy.* **27,** 1449-1461.

260. Voncken, F.G.J., Boxma, B., van Hoek, A.HA.M., Akhmanova, A.S., Vogels, G.D., Huynen, M., Veenhuis, M., and Hackstein, J.H.P. 2002. A hydrogenosomal [Fe]-hydrogenase from the anaerobic chytrid *Neocallimastix* sp. L2. *Gene* **284,** 103-112.

261. Wang, R., Healey, F.P. and Myers, J. 1971. Amperometric measurement of hydrogen evolution in *Chlamydomonas. Plant Physiol.* **48,** 108-110.

262. Wasielewski, M.R. 1992. Photoinduced electron transfer in supramolecular systems for artificial photosynthesis. *Chem. Rev.* **92**, 435-461.

263. Weaver, P. F., Lien, S., and Seibert, M. 1980. Photobiological production of hydrogen. *Solar Energy* **24**, 3-45.

264. Wenk, S.O., qian, D.J., Wakayama, T., Nakamura, C., Zorin, N., Rögner, M. and Miyake, J. 2002. Biomolecular device for photoinduced hydrogen production. *Int. J. Hydrogen Energy* **27**, 1489-1493.

265. Weissman, J.C. and Benemann, J.R. 1977. Hydrogen production by nitrogen-fixing cultures of *Anabaena cylindrica*. *Appl. Environ. Microbiol.* **33**, 123-131.

266. Winkler, M., Heil, B., Hei, B., and Happe, T. 2002. Isolation and molecular characterization of the [Fe]-hydrogenase from the unicellular green alga *Chlorella fusca*. *Biochim. Biophys. Acta* **1576**, 330-334.

267. Winkler, M., Maeurer, C., Hemschemeier, A. and Happe, T. 2004. The isolation of green algal strains with outstanding H_2-productivity, in: *Biohydrogen III*, J. Miyake, Y. Igarashi and M. Roegner, eds., Elsevier Science, Oxford, pp. 103-116.

268. Wolk, C. P. 1996. Heterocyst formation. *Annu. Rev. Genet.* **30**, 59-78.

269. Wünschiers, R., Stangier, K., Senger, H. and Schulz, R. 2001. Molecular evidence for an [Fe]-hydrogenase in the green alga *Scenedesmus obliquus*. *Current Microbiol.* **42**, 353-360.

270. Wykoff, D.D., Davies, J.P., Melis, A. and Grossman, A.R. 1998. The regulation of photosynthetic electron-transport during nutrient deprivation in *Chlamydomonas reinhardtii*. *Plant Physiol.* **177**, 129-139.

271. Yagi, T., Motoyuki, T., and Inokuchi, H. 1973. Kinetic studies of hydrogenases. *J. Biochem.* **73**, 1069-1081.

272. Yokoi, H., Mori, S., Hirose, J., Hayashi, S., and Takasaki, Y. 1998. Hydrogen production from starch by a mixed culture of *Clostridium butyricum* and *Rhodobacter* sp. M-19. *Biotechnol. Lett.* **20**, 890-895.

273. Zak, E., Norling, B., Maitra, R., Huang, F., Anderson, B., and Pakrasi, H. B. 2001. The initial steps of biogenesis of cyanobacterial photosystems occur in plasma membranes. *Proc. Nalt. Acad. Sci. USA.* **23**, 13443-13448.

274. Zhang, L. and Melis, A. 2002. Probing green algal hydrogen production. *Phil. Trans. R. Soc. Lond. Biol. Sci.* **357**, 1499-1509.

275. Zhang, L., Happe, T. and Melis, A. 2002. Biochemical and morphological characterization of sulfur deprived and H_2-producing *Chlamydomonas reinhardtii* (green algae). *Planta* **214**, 552-561.

276. Zito, F., Finazzi, G., Delosme, R., Nitschke, W., Picot,D., and Wollman, F.A. 1999. The Qo site of cytochrome b6f complexes controls the activation of the LHCII kinase. *EMBO J.* **18**, 2961-2969.

277. Zürrer, H. and Bachofen, R. 1982. Aspects of growth and hydrogen production of the photosynthetic bacterium *Rhodospirillum rubrum* in continuous culture. *Biomass* **2**, 165-238.

278. M. C. Posewitz, personal communication.

279. F. P. Healey, 1970. The mechanism of hydrogen evolution by *Chlamydomonas moewusii*. *Plant. Physiol.* **45**, 153-159

280. R. P. Gfeller and M. Gibbs. 1985. Fermentative metabolism of Chlamydomonas reinhardtii. II. Role of plastoquinone, *Plant Physiol.* **77**, 509-511.

281. M. W. W.Adams. 1990. The structure and mechanism of iron-hydrogenases. *Biochim. Biophys. Acta* **1020**, 115-143.

282. Curtis, C.J., Miedaner, A., Ciancanelli, R., Ellis, W.W., Noll, B.C., Rakowski DuBois, M., and DuBois, D.L. 2003. [Ni(Et$_2$PCH$_2$NMeCH$_2$PEt$_2$)$_2$]$^{2+}$ as a functional model for hydrogenases. *Inorg. Chem.* **42**, 216-227.

283. Wilson, A.D., Newell, R.H., McNevin, M.J., Muckerman, J.T., Rakowski DuBois, M., and DuBois, D.L. 2006. Hydrogen oxidation and production using nickel-based molecular catalysts with positioned proton relays. *J. Am. Chem. Soc.* **128**, 358-366.

284. Abdel-Basset, R. and Bader, K.R. 1998. Physiological analyses of the hydrogen gas exchange in cyanobacteria. *Photochem. Photobiol.* **43**, 146-151.

285. Kondo, T., Arakawa, M., Hirai, T., Wakayama, T., Hara, M. and Miyake, J. 2002. Enhancement of hydrogen production by a photosynthetic bacterium mutant with reduced pigment. *J. Bioscie. Bioeng.* **93**, 145-150.

9

Centralized Production of Hydrogen using a Coupled Water Electrolyzer-Solar Photovoltaic System

James Mason[1] and Ken Zweibel[2]

[1]Hydrogen Research Institute, Farmingdale, NY
[2]Primestar Solar Co., Longmont, CO

1 Introduction

This study investigates the centralized production of hydrogen gas (H_2) by electrolysis of water using photovoltaic (PV) electricity. H_2 can be used to power all modes of transportation. The logical first large-scale application of H_2 is as a replacement fuel for light-duty vehicles, light commercial trucks, and buses. Since H_2 is an expensive fuel compared to gasoline, consumer acceptance of H_2 is contingent on its use in advanced fuel economy vehicles such as fuel cell vehicles (FCVs), which lowers the cost of H_2 relative to the cost of gasoline used by conventional fuel economy vehicles.* The purpose of the study is to provide baseline projections of capital investments, levelized H_2 prices, and fuel cycle greenhouse gas (GHG) emissions of a centralized PV electrolytic H_2 production and distribution system. This is important in order to evaluate the economic and environmental impacts of utilizing PV electrolytic H_2 as a fuel source.

The use of PV electricity for electrolytic H_2 production is a means of storing solar energy and overcoming its limitations as an intermittent power source. However, the intermittency of solar energy reduces the utilization capacity factor of electrolysis plants, which increases H_2 production cost. The relevant question is whether the

---

* Examples of advanced fuel economy vehicles are fuel cell vehicles (FCV), hybrid electric vehicles (HEV), and plug-in hybride electric vehicles. Fuel cell vehicles (FCVs) are on the verge of being ready for mass production.[2] FCVs are an attractive first application of H_2 due to their enhanced fuel economy and superior driving performance. FCVs have powerful electric engines but do not require batteries to recharge since the H_2 running through the fuel cells produces electricity. The average fuel efficiency of FCVs is a factor of 2.2 greater than the fuel efficiency of conventional gasoline powered ICE vehicles.

production of electrolytic H_2 using PV electricity is economically viable. This study attempts to provide insight into this question.

In all cases, the analysis draws on the perspective of the *Terawatt Challenge for Thin Film PV* in terms of PV costs, efficiencies, reliability, and progress towards these goals.[1] The study assumes progress in PV technologies will occur. Then the important questions to be examined are: Does it matter? Will PV electricity be inexpensive enough to make electrolytic H_2 production practical? The study will answer these in the positive.

The organization of the study is as follows. In the first Section, a H_2 production and distribution system is described. Secondly, capital and levelized H_2 price estimates are investigated for each of the H_2 system components. Thirdly, a life cycle evaluation of primary energy and GHG emissions in the H_2 fuel cycle is performed. Sensitivity analyses are performed for the H_2 price and the life cycle energy and GHG emissions estimates. The study concludes with a summary of findings and suggestions for future research.

2 Description of a PV Electrolytic H_2 Production and Distribution System

The H_2 production and distribution system analyzed in this study is scaled to a quantity of H_2 for one-million FCVs. The components of the centralized H_2 system are: a PV power plant; an electrolysis plant; a pipeline compression station; 621 miles (1,000 km) of long-distance pipeline with nine booster compressors sited at 60 mile intervals; four city gate distribution centers; and 1,000 local filling stations. The local distribution of H_2 is by truck with metal hydride (MH) storage containers.[*] The H_2 system is completed with the inclusion of regional underground H_2 storage facilities designed to level seasonal variations in H_2 supply and demand.

Each PV electrolysis plant produces 216-million kilograms of H_2 per year. This H_2 production level is sufficient to support the annual H_2 consumption of one-million FCVs. In addition, the H_2 production level takes into account 3% H_2 distribution losses and the use of H_2 to power pipeline booster compressors, city gate compressors, and city gate distribution trucks.[†] One-million FCVs consume 202-million kilograms of H_2 per year, which is based on an average FCV fuel economy of 54.5

---

[*] A H_2 system requires a H_2 storage medium for delivery trucks, filling stations, and vehicles. The near-term choices for H_2 storage are metal hydrides, compression at 10,000 psia, and liquid at extreme low temperatures. This study chooses to use a metal hydride H_2 storage system as a baseline model because it is the least energy intensive means of storing H_2. Collaborative DOE and industry metal hydride research goals are to achieve 6% H_2/MH by weight storage ratio, a three minute recharging time, thousands of recharging cycles, and low cost by 2010.

[†] The projection of 3%-H_2 distribution losses is twice the natural gas distribution loss rate.[10]

mi/kg of H_2 and an average annual travel distance of 11,000 miles over the range of all FCV light-duty vehicles and light commercial trucks.

The H_2 from each PV electrolysis plant is transported to city gate distribution centers by pipeline. At the city gate distribution centers, the pipeline H_2 is stored in metal hydride (MH) containers, which contain 2,000 kg of H_2 at a H_2/MH storage ratio of 6% by mass, and the MH containers are loaded onto tractor-trailer trucks and delivered to 1,000 local filling stations. At filling stations, the MH containers are stored in above-ground, cast-iron frames for fast replacement by tractor-trailer, container trucks. With an average FCV fill-up rate of 4.5 kg H_2 per refueling stop, 330 FCVs can be refueled by one MH container with the MH container having a 75% of capacity discharge factor. The filling station MH containers are replaced on a two to four day cycle, and the empty MH containers are replaced and returned to the city gate distribution centers to be refilled with H_2.

Cost estimates, performance parameters, and operating life of the central components of a PV electrolytic H_2 system are listed in Table 1. All component cost estimates are based on an optimized manufacturing scale. PV cost estimates are from Zweibel[1] and Keshner and Arya.[3] The PV performance parameters of PV electrolysis plants are informed by studies of the solar hydrogen project at Neunburg vorm Wald, Germany.[4,5] The performance parameters of electrolysers are from the collaborative study of large, grid-connected electrolyser plants by Norsk Hydro and Electricité de France.[6*] The cost estimates for H_2 compressors are from Amos.[7] The energy consumption of compressors used to transport and distribute H_2 is estimated with an adiabatic compression energy formula provided by a Praxair representative[8] and includes Redlich-Kwong H_2 compressibility factors.[9] Land costs, site preparation work, engineering and design, labor, and dismantling costs are factored into the component cost estimates.

The pipeline cost of $2.0-million per mile is based on an average natural gas pipeline cost of $1.5-million per mile, without compressor cost, with the addition of a 33% premium to take into account the cost for extra-secure pipe welds. More research is needed to accurately assess the capital costs of an integrated long-distance pipeline design for large regions such as the U.S., Europe, etc. The metal-hydride (MH) H_2-storage container estimates are original to this study and are based on the assumption that some combination of metals such as magnesium, lithium, and boron

---

* The Cloumann et al.[6] study of electrolytic H_2 production is based on the use of grid-distributed electricity and the cost estimates include AC to DC rectifier/transformer units, which are not needed for electrolysis plants using dc electricity from PV power plants. The cost estimates also include compressors, H_2 drying/purification units, and pumps for water and KOH circulation. The electrolysis performance efficiency of 61%, lower heat value, from the Cloumann et al. study is a global efficiency and includes the energy to compress H_2 to a pressure of 33 bar, H_2 losses in the drying/purification phases, and the energy for pumping water and KOH. In contrast, this study models compression and pumping energy separately and assumes an electrolysis efficiency of 64.2%. Separate PV installations are dedicated to provide electricity for H_2 compression and water distillation and pumping. The assumed electrolyser efficiency of 64.2% is a conservative estimate and may prove to be closer to 66%.

Table 1. Cost and performance assumptions for future PV electrolysis H_2 systems.

	Parameters	Operating life (years)
A. PV power plant		
1. PV area cost ($/m^2)	$60/m^2	20, 30, 60
a. 2nd-generation PV area cost ($/m^2)	$50/m^2	30
b. Freight charges @ $142/short ton	$ 2/m^2	
2. PV module efficiency (1st generation)	10–14%	
a. 2nd-generation PV module efficiency	12–16%	
3. PV balance of system (BOS) costs	$50/m^2	60
a. 2nd-generation BOS (only labor costs)	$20/m^2	30
b. Freight charges @ $100/short ton	$ 2/m^2	
4. DC/DC converters	$75/kW$_{dc-in}$	30
5. PV system net efficiency (dc output per W_p installed)	85%	
a. losses from wiring, ambient heat, module mismatch, etc.	– 11%	
b. losses from dc/dc converters and coupling to electrolyzers	– 4%	
6. PV system availability (included in PV-system efficiency)	99%	
7. Average hours/day of peak insolation @ 271 W/m^2 insolation	6.5 hours/day	
8. O&M expenses including PV additions (% of capital)	1.0%	
9. Land cost ($/acre)	$1,000	
10. Insurance (% of Capital)	0.0%	
11. Property taxes (% of Capital)	0.5%	
B. Electrolysis plant		
1. Electrolysers (including dc-dc power conditioning)	$ 425/kW$_{dc-in}$	60
2. Electrolyser energy efficiency (H_2 out/electricity in, LHV)	64.2%	
3. Electrolyser availability	98%	
4. Electrolyser capacity factor	26.2%	
5. Compressors (low pressure, water injected, screw type)	$ 340/hp	30
a. compressor efficiency	70%	
b. energy to compress H_2 from 14.7 to 116 psi	1.37 kWh/kg H_2	
6. Water system (collection, pumping, purification)	$ 5,000,000	60
7. Administration, maintenance, and security buildings	$10,000,000	60
8. O&M expenses (% of capital)	2.0%	
9. Insurance (% of capital)	0.5%	
10. Property taxes (% of capital)	0.5%	
C. Other H_2 system components[a]		
1. Pipeline	$2,000,000/mile	60
2. Pipeline compressors (reciprocating)	$ 670/hp	40
3. Pipeline booster compressors (intervals)	60 miles	60
4. Metal-hydride (MH) H_2 storage capital cost	$ 30/kg MH	30
5. Insurance (% of capital)	0.5%	
6. Property taxes (% of capital)	1.5%	

[a]Other costs such as site preparation, engineering, legal, electrolyte replacement, etc. are included.

will be able to meet the assumed 6% H₂ by weight storage capacity standard. The assumed MH cost of \$30/kg is believed reasonable since magnesium production costs are < \$4/kg, lithium production costs are < \$2/kg, and boron production costs are < \$1/kg.[11] However, it needs to be emphasized that the MH cost and performance estimates are speculative and require additional analysis. The performance data for MH containers are from Chao et al.[12]

At present, the only PV technology clearly demonstrating the potential to meet the module cost (\$60/m²) and minimum performance (10% PV module efficiency) projections of this study is thin film PV.[1*] Other combinations of module performance and cost (e.g., those of wafer silicon) are not as economical at the system level. Over time, additional PV technologies are expected to meet the PV cost and performance projections, and existing ones are expected to continue their cost reductions and efficiency improvements. The baseline projections of this study assume a thirty-year PV module operating life. However, it is quite plausible, but not verifiable with present data, that the operating life of thin film PV will be sixty years with a 1%-annual degradation rate. Therefore, an analysis of H₂ production costs with sixty year PV module operating life is performed and the results presented in the sensitivity analysis section to provide a range in what can be realistically expected with future developments in thin film PV. A multi-MWₚ PV installation demonstrating the potential to achieve \$50/m² BOS costs, which includes land preparation, wiring conduit, electrical connection stations, PV system grounding, PV mounting hardware and installation, and union-scale labor, has been documented.[13†] The cost for dc/dc power conditioning equipment is categorized separately.

While a variety of electrolyser technologies are currently marketed, the type of electrolyser with a demonstrated ability to meet the cost and performance projections of this study are atmospheric, bi-polar, alkaline electrolysers.[4] Alkaline electrolysers have a long track record for dependability, low-cost maintenance, and long operating life. The operating life of electrolysers is affected by the utilization rate.[6] With a 26%

---

* It is assumed that 10% efficient thin film PV modules will be available for the near-term application of PV for large-scale electrolytic H₂ production. At present, the best efficiency for a thin film PV module being produced at the > 50 MWₚ/year scale is 9.4%. While some thin film PV modules with efficiencies > 12% have been produced on a small scale, there are numerous technical challenges in maintaining high efficiency levels while scaling-up PV manufacturing capacity. Therefore, it is assumed that 10% efficient PV modules will be available for the first large PV electrolysis plants, and over time PV modules with higher efficiencies will become available. In addition, reaching module costs in the \$50/m² range requires further innovation and economies of scale.

† Tucson Electric Power at the Springerville PV plant has achieved \$64/m²-BOS costs for MWₚ scale PV installations. With an increase to the multi-GWₚ scale installation, it is reasonable to believe that a 25% reduction in BOS costs can be achieved through the mass manufacture and purchase of standardized BOS components and through efficiency gains in the allocation of labor/machinery for PV plant installation.

capacity factor of PV electrolysis plants, the electrolyser operating life is 60 years.[14] At the low capacity factor of PV electrolysis plants, electrolyser maintenance will require nickel replating of electrolyser cells and electrodes only every twelve years rather than the normal seven year replating cycle with electrolyser capacity factors of 80% or greater. This reduction in maintenance cost almost entirely offsets the higher cost of low utilization factor electrolyser plants; an analysis that is expanded further in a later Section (see especially Fig. 3).

From the performance parameters in Table 1, the size of the electrolysis plant is 5.12 GW_{dc-in} of electrolysers coupled to a 5.69 GW_p PV power plant. An additional 0.15 GW_p of PV is required for the electrolysis plant compressors, water pumps, and water distillation plant. The compressors and water pumps at the pipeline compression station require another 0.13 GW_p of PV. The cumulative size of the PV power plant is 5.97 GW_p.

This study categorizes the costs of a PV power plant into:

1. PV modules;
2. dc/dc converters; and
3. balance of system (BOS) components, which include site preparation, PV mounting frames, wiring, and labor.

The operating life of a PV power plant has two distinct generations. The first generation is the initial construction of the PV power plant. While PV modules and dc/dc converters have a thirty-year operating life, many of the BOS components such as site preparation, mounting frames, underground wiring conduits, and PV array connection stations have a sixty-year operating life. With properly standardized module and BOS designs, capital investments in second generation PV power plants consists only in the costs of removing first generation PV modules and dc/dc converters and replacing them with new, second generation units without incurring the full range of BOS costs. Second generation BOS cost is reduced to labor for PV module mounting and inter-module wiring connection and is estimated at forty percent of first generation BOS cost. This study also investigates the economic impacts of second generation PV power plants with sixty-year PV module operating life.

The design of the PV power plant includes the annual addition of new PV to compensate for PV electricity output losses attributable to factors such as module soiling, PV module output degradation, and catastrophic PV module failures. The purpose of the PV additions is to maintain a constant level of electricity output to the electrolysers and compressors. Electricity losses from PV module soiling are assumed to be a constant 1.0% from year four to the end of the module operating life. The PV module degradation rate is assumed to be 1.0% per annum throughout the operating life of the PV modules. Catastrophic PV module failure, caused by factors such as manufacturing defects, glass stress fractures, and lightning strikes, is assumed to be 0.01% (1/10000) per annum. The financial accounting for the annual PV additions is treated as a normal O&M expense rather than as a capital investment.

To maximize the utilization capacity factor of PV electrolysis plants, it is assumed that PV electrolysis plants will be located at sites receiving high insolation (solar radiation) levels. Areas of the world with high insolation levels are presented in Fig. 1. This analysis assumes that PV electrolysis plants will be built at locations

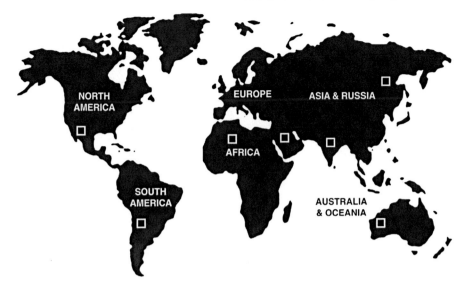

Fig. 1. Areas of world with high average solar radiation levels (boxes). Copyright permission granted by Encyclopedia Britannica.

with a minimum average insolation level of 271 W/m². This insolation level translates into 6.5 hours of average daily peak PV electricity and electrolyser H₂ production. PV installations are mounted at a fixed angle equaling the site's latitude. The application of tracking systems for large field PV plants has not yet demonstrated cost effectiveness. The rows of the PV arrays are spaced to prevent cross-shading of modules when the sun is low in the sky from 9:00 am through 3:30 pm on December 21. The total area of the PV installation is approximately a factor of 3.0 greater than the area of the PV modules, which provides a small safety margin for installation variances.* The actual spacing of PV array rows to prevent module cross-shading will vary according to the site's latitude.

The land area of a PV electrolysis plant to produce 216-million kg of H₂/year is a function of insolation level, PV module efficiency, the spacing between the rows of the PV arrays, and the land required for electrolyser, compressor, administration/maintenance/security buildings, water storage, water pumping and distillation facilities, PV for the pipeline compression station, and PV additions to compensate for PV degradation losses. A land area of 4 mi² is allocated for electrolysers, compressors, administration buildings and water storage, pumping and distillation facilities. The total land area for the 5.97-GW_p PV power plant is 94 mi² for 10% efficient PV modules, 79 mi² for 12% efficient PV modules, and 68 mi² for 14% efficient PV modules. The land area includes the addition of 1.9 GW_p of PV to compensate for PV output degradation losses. While this is a substantial land area, it is not prohibi-

---

* The row spacing estimate is based on 33° latitude and a sun altitude of 14.9° above the horizon at 9:00 am. The actual row spacing is a factor of 2.88 greater than the length of the modules and 0.12 is added as a safety buffer.

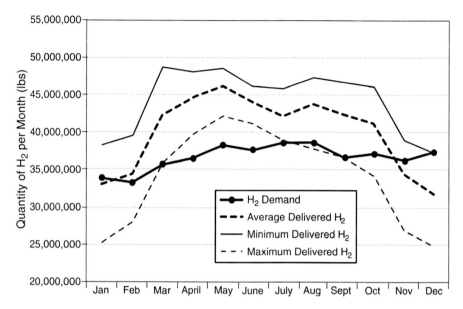

Fig. 2. Average monthly H₂ production and demand (one-million FCVs). Average fuel consumption is from U.S. Federal Highway Administration data;[16] and H₂ supply is estimated from average monthly insolation values for six locations in the southwest U.S. with data from NREL's Solar Radiation Data Manual for Flat-Plate and Concentrating Collectors.[17] Correlation between H₂ supply and H₂ demand is 0.61.

tive since the best locations for PV electrolysis plants are sparsely populated desert regions, which eliminates competition over competing land uses such as agriculture, forestry, grazing, mining, or other commercial land uses. For example, in Arizona only 17% of the total land area is privately owned, which indicates abundant sources of inexpensive land for PV electrolysis plants.

A water supply is required for H₂ production and to cool electrolysers and compressors. To produce a kilogram of H₂ requires 2.9 gallons of water. Hence, 216-million kg of H₂/year requires 630-million gallons of feed-water. Water is needed to cool electrolyser cells and compressors. Electrolyser cell cooling requires 93.7 gallons of water/kg H₂ produced,[15] and the cooling of compressors requires 13.2 gallons of water/kg H₂ compressed.[7] The quantity of water for electrolyser and compressor cooling is 792 million gallons per year, assuming cooling tower losses of 3% per hour.[18] The total quantity of water consumed by the electrolysers and compressors is 1.42-billion gallons of water per year.* The water can be economically supplied by

---

* In an interview with John Fortune of the Arizona Statewide Water Planning Unit, it was stated that the typical 90-acre golf course in Phoenix, Arizona consumes 400-acre feet or 130-million gallons of water per year. Therefore, eleven golf courses in Phoenix use the quantity of water for an electrolysis plant to produce H₂ for one million cars. The idea for an on-site water collection system was discussed with

either on-site water collection and storage systems or imported by train or truck. The quantity of water is one inch of rainfall over the PV plant area. Under no condition should the PV electrolysis plant draw water from underground aquifers, lakes, or rivers.

The effects of seasonal variation in insolation levels on seasonal H_2 supply/demand balances are presented in Fig. 2. The seasonal H_2 production profile is well suited to meet seasonal H_2 demand. The positive 0.61 correlation between monthly H_2 production levels by PV electrolysis plants and monthly H_2 demand reduces the required capacity of underground storage facilities.

The high H_2 output in the spring months insures that the underground H_2 storage facilities will have sufficient H_2 capacity to meet summer peak demand. The minimum and maximum H_2 production curves are based on the minimum and maximum insolation levels recorded for each month over a ten year record of insolation levels for six locations in the southwest U.S. from west Texas to east California. The curves for the minimum and maximum insolation levels represent the extreme case where all locations receive the historical minimum or maximum insolation level in the same month. It is highly unlikely that minimum or maximum insolation levels will occur in the same month at each of the locations distributed over such a large area. But the minimum H_2 production level estimate is useful as a yardstick in assessing the quantity of H_2 that should be stored in underground storage facilities as reserves to insure adequate H_2 supplies in the event of a variety of contingencies that could disrupt H_2 supply.

The pipeline transport of H_2 requires compression. The electrolysis plant uses low-pressure, water injected, screw-type compressors to compress H_2 from 1.02 bar to 8.0 bar to transport the H_2 a short distance ($\sim$ 10 miles) to a pipeline compression station. The energy to compress H_2 from 1.02 bar to 8.0 bar is 1.37 kWh/kg of H_2. There is no need for H_2 storage at the electrolysis plant. At the pipeline compression station, the H_2 is compressed from 8.0 bar to a pipeline pressure of 69.0 bar by high-pressure reciprocating compressors. The energy to compress H_2 from 8.0 bar to a pipeline pressure of 69.0 bar is 1.17 kWh/kg of H_2. PV electricity is used to power the compressors and water pumps at the compression station.

3   Capital Investment and Levelized Price Estimates

Capital cost estimates for the H_2 system are presented in Table 2. The PV power plant is the largest capital component. With 10% efficient PV modules, the PV power plant accounts for 59% of total capital investments. With cost reductions achieved by PV module efficiency gains, the proportion of capital for 14% efficient PV modules is reduced to 51%. The second largest capital investment component is the electrolysis plant. The electrolysis plant accounts for 18–22% of total capital for

---

Fortune. Fortune stated that Arizona is willing to work closely with companies and developers who build rain-runoff water collection and storage systems, and he stated that he believes an on-site rain-runoff water collection and storage system for electrolysis plants is feasible.

Table 2. Capital estimates for future PV electrolytic H₂ systems[a] (scaled to serve 1-million fuel cell vehicles).

	Capital Costs		
	10% PV Efficiency	12% PV Efficiency	14% PV Efficiency
1. PV power plant (5.972-GW$_p$)			
A. PV cost	3,702,390,460	3,085,325,383	2,644,564,614
B. PV BOS cost	3,105,230,708	2,587,692,257	2,218,021,935
C. DC/DC power conditioning	429,955,021	429,955,021	429,955,021
Subtotal	7,237,576,190	6,102,972,661	5,292,541,570
2. Electrolysis plant (5.121-GW$_p$)			
A. Electrolyser cost	2,176,288,879	2,176,288,879	2,176,288,879
B. Compressor cost	60,738,708	60,738,708	60,738,708
C. Water system cost	5,000,000	5,000,000	5,000,000
D. Administration buildings	10,000,000	10,000,000	10,000,000
Subtotal	2,252,027,586	2,252,027,586	2,252,027,586
3. Pipeline System (621 miles)			
A. Pipeline Cost	1,242,000,000	1,242,000,000	1,242,000,000
B. Compression Station Cost	103,260,246	103,105,428	103,105,428
D. Pipeline Booster Compressors	96,261,299	96,116,974	96,116,974
D. Underground Storage Facility	5,000,000	5,000,000	5,000,000
E. Administration Buildings	1,000,000	1,000,000	1,000,000
Subtotal	1,447,521,545	1,447,222,401	1,447,222,401
4. City Gate Distribution Centers (4)			
A. City Gate Distribution Centers	26,500,000	26,500,000	26,500,000
B. City Gate Compressors	28,957,500	28,957,500	28,957,500
C. H₂ Delivery Trucks	22,500,000	22,500,000	22,500,000
D. MH Containers	1,300,000,000	1,300,000,000	1,300,000,000
Subtotal	1,377,957,500	1,377,957,500	1,377,957,500
5. Refueling Stations (1,000)			
A. MH Container Stands	10,000,000	10,000,000	10,000,000
B. Filling Station Compressors	12,300,000	12,300,000	12,300,000
C. Filling Station Dispensers	30,000,000	30,000,000	30,000,000
Subtotal	52,300,000	52,300,000	52,300,000
Total Capital Costs of H₂ System	12,367,382,820	11,232,480,149	10,422,049,057

the 10% and 14% efficient PV module cases respectively. The pipeline system and city gate distribution centers are the next largest capital components and account for 12% and 11% of total capital respectively. The metal hydride H₂ storage containers are 94% of the capital investments for the city gate distribution centers. The remaining capital component is the local filling stations, which is less than 1% of total capital investments.

Hydrogen production and PV electricity prices are presented in terms of levelized prices. Levelized price is the constant revenue stream that recovers all capital investments (equity and debt) at the required rates of return and covers annual O&M expenses, insurance, property tax, and income taxes over the assigned capital recovery period. The levelized H_2 price estimates are based on a thirty-year capital recovery period and a 6.0% discount rate.

The discount rate is a weighted average cost of capital (WACC) and takes into account the capital structure of firms, cost of equity and debt, and income taxes. The capital structure of firms is assumed to be 30% equity and 70% debt. The cost of equity capital is 10%, the cost of debt is 7%, and the effective income tax rate is 39%. The debt instrument is assumed to be a 20-year, 7% coupon bond. The calculation of the discount rate is

$$\text{Discount Rate} = \text{WACC} = [0.7(0.07)(1 - 0.39)] + [(0.3)(0.10)] = 6.0\% \qquad (1)$$

The levelized prices of PV electricity and H_2 are derived by net present value cash flow analysis. The net present value cash flow method is described in Appendix A.1. A straight-line, ten-year depreciation schedule is applied with an annual depreciation rate of 9% of capital. The levelized PV electricity and H_2 prices are derived by choosing PV electricity and H_2 prices to generate a revenue level that results in a cumulative, net cash flow stream with a $0-net present value over the thirty-year capital recovery period. The annual net cash flow streams are discounted at the present value of the 6%-discount rate. Investment funds are allocated in year 1; construction occurs in year 2; and H_2 cash flow begins in year 3. The modular design of PV electrolysis plants and H_2 distribution systems enables the rapid initiation of H_2 marketing and cash flow.

The levelized H_2 and PV electricity price estimates are presented in Table 3. The PV electrolysis plant dominates H_2 production cost. The PV electrolysis plant component of the levelized H_2 pump price ranges from $3.75–$4.67 per kg H_2 contingent on PV module efficiency and PV area cost. The total levelized H_2 pump price ranges from $5.53–$6.48 per kg H_2.[*] The levelized H_2 pump price estimates do not include fuel use taxes. In the U.S., fuel use taxes typically range from $0.40–0.50/gallon of gasoline, which translates into a H_2 pump price of $6.52–$7.47/kg with tax.

While a kilogram of H_2 is a gallon of gasoline equivalent in terms of energy content, it is not a gallon of gasoline equivalent in terms of fuel cost when the H_2 is consumed by fuel cell vehicles (FCVs). The fuel efficiency of FCVs with their powerful electric engines is much greater than the fuel efficiency of internal combustion engines (ICE) vehicles. The average fuel economy of FCVs is 54.5-mi/kg H_2, whereas the average fuel economy of conventional ICE vehicles is 23.5-mi/gal gasoline. When H_2 is used to power FCVs, the gallon of gasoline equivalent price is $2.81–$3.22, which is comparable to high-end 2005–2006-U.S. gasoline prices.

The PV electrolysis plant cost components account for 68–72% of the levelized H_2 pump price. This can be seen by comparing the H_2 production costs listed in Table 3.B.1 to the levelized H_2 pump prices listed in Table 3.C. Of the PV electrolysis plant cost factors, the price of PV electricity is the dominant factor on H_2 production costs. The large effect of PV module efficiency on H_2 production costs is apparent by

---

[*] In terms of work energy, the energy content of a kilogram of H_2 is approximately equivalent to the energy content of a gallon of gasoline. Therefore, a kilogram of H_2 is considered to be a gallon of gasoline equivalent (gge) metric.

Table 3. Financial overview of a PV electrolytic H₂ system (scaled to serve 1-million fuel cell vehicles).

	Electricity price ($/kWh)	Capital investments (million $)	Annual revenues (million $)	PV additions expense (million $)	O&M expense (million $)
A. PV power plant					
with 10% Efficient PV	0.064	7,238	769	72	5
with 12% Efficient PV	0.054	6,103	649	60	5
with 14% Efficient PV	0.047	5,293	562	52	5

	H₂ price ($/kg)	Capital investments (million $)	Annual revenues (million $)	Electricity expense (million $)	O&M expense (million $)
B. H₂ production and distribution					
1. Electrolysis Plant		2,252			56
with 10% Efficient PV	4.67		1,013	753	
with 12% Efficient PV	4.12		894	635	
with 14% Efficient PV	3.75		813	553	
2. Pipeline Transport		1,448			54
with 10% Efficient PV	0.97		104	16	
with 12% Efficient PV	0.97		104	13	
with 14% Efficient PV	0.95		104	12	
3. City Gate Distribution Centers (4)	0.77	1,378	162	0	39
4. Local Filling Stations (1000)	0.07	52	11	6	1
C. Totals					
with 10% Efficient PV	6.48	12,367	2,059	775	144
with 12% Efficient PV	5.93	11,232	1,820	654	144
with 14% Efficient PV	5.53	10,422	1,652	571	144

reviewing Table 3. An increase in PV module efficiency, from 10% to 14%, lowers H₂ production costs by 20% and the levelized H₂ pump price by 15%.

The large effect of electricity price on H₂ production costs is readily apparent in Fig. 3, which breaks down H₂ production cost by electrolysis plant cost factors. The cost of electricity accounts for greater than 80% of H₂ production costs across the range of electrolyser capacity factors. One of the criticisms to the application of PV electricity to electrolytic H₂ production is its intermittent supply, which lowers the utilization capacity factor of electrolysers and increases H₂ production cost. The low electrolyser capacity factor cost penalty is evaluated in Fig. 3. Over the 25–95% range in electrolyser capacity factors presented in Fig. 3, the H₂ production cost of an electrolysis plant with a 25% capacity factor is approximately 11% higher than the H₂ production cost of an electrolysis plant with a 95% capacity factor.[*] In other

---

[*]From Fig. 3, it is obvious that the relationship between electrolyser cost and H₂ production cost across a 25–95% capacity factor range is non-linear. In this case, the appropriate method to evaluate the effect of electrolyser cost on H₂ production cost

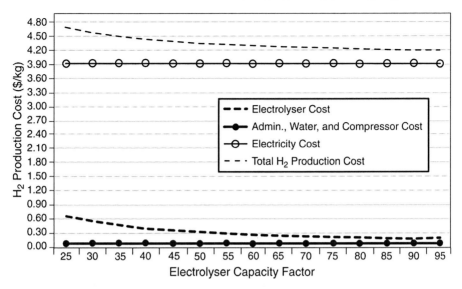

Fig. 3. Levelized H₂ production price as a function of electrolyser capacity factor.

words, the utilization rate of electrolysers is not a particularly important issue in terms of H₂ production cost because of the impact of offsetting factors such as electrolyser O&M expense and electrolyser operating life.

The critical element affecting the production cost of electrolytic H₂, over the range of electrolyser capacity factors, is electricity cost. While the 11%-H₂ cost penalty for the low electrolyser capacity factor from the use of PV electricity is significant, it is hardly prohibitive. In conclusion, based on the assumed progress in PV cost reduction, PV electricity can be an economically viable source of electricity for electrolytic H₂ production.

4 Sensitivity Analysis: H₂ Production and PV Electricity Prices

Sensitivity analyses are performed to evaluate the effect of changes in cost factor values on H₂ production and PV electricity prices. The cost factors for H₂ production are: PV electricity; electrolysers; electrolyser operating capacity factor; electrolyser efficiency (in terms of converting electricity energy input into H₂ energy output); electrolyser O&M expense; and the discount rate. The cost factors for PV electricity

---

across the range of electrolyser capacity factors is a log-linear regression model. A log-linear regression model transforms the non-linear dependent variable, H₂ production cost, into a linear variable by using its natural logarithm value. The log-linear regression result indicates that a 1% increase in electrolyser capacity factor reduces H₂ production cost by 0.16%. Hence, a 70% increase in electrolyser capacity factor decreases H₂ production cost by only 11.2% (0.16% x 70).

production are: PV modules; PV BOS; PV module efficiency (rated PV module conversion of sunlight into dc electricity under standard test conditions); average insolation level; and the discount rate. The assigned range of values for the electrolysis plant and PV power plant cost factors are presented in Table 4. The mean value for each of the cost factors is the value used to generate the baseline H_2 production and PV electricity price estimates reported in this study.

The sensitivity estimates for the effect of changes in cost factor values on levelized H_2 production and PV electricity prices are estimated by the least-squares, linear regression method. The regression results provide an estimate of the effect of unit changes in cost factor values on H_2 production and PV electricity prices. The sensitivity results are presented in Table 5, Fig. 4, and Fig. 5.

The appropriate unit change for each of the cost factors are presented in parenthesis in Table 5. The results in Table 5 report the increase/decrease (+/–) in H_2 production price (¢/kg) and in PV electricity price (¢/kWh) caused by a unit *increase* in cost factor values. The interpretation of the effect of a unit *decrease* in cost factor values requires changing the sign (+/–) of the estimated change in H_2 production and PV electricity price. Also, note that the regression sensitivity results can be applied to component values outside the range of component values presented in Table 4.

The sensitivity results reported in Table 5 for H_2 production price are as follows. A $0.01/kWh increase in electricity cost causes H_2 production price to increase by $0.55/kg. A $25 increase in electrolyser cost ($/kW$_{dc-in}$) causes H_2 production price to increase by $0.04/kg. A 1% increase in electrolyser capacity factor causes H_2 production price to decrease by $0.02/kg. A 1% increase in electrolyser efficiency (LHV) causes H_2 production price to decrease by $0.04/kg. A 1% increase in electrolysis plant O&M expenses, which includes water system and compressors, causes H_2 production price to increase by $0.09/kg.[*] A 1% increase in the discount rate causes H_2 production price to increase by $0.09/kg. To evaluate the effect of a decrease in cost factor values simply reverse the sign, positive or negative, for the change in H_2 production price.

The sensitivity results for PV electricity prices are as follows. A $5/m^2 increase in the area cost of PV modules causes PV electricity price to increase by $0.002/kWh. A $5/m^2 increase in the area BOS cost causes PV electricity price to increase by $0.002/kWh. A 1% increase in PV module efficiency causes PV electricity price to decrease by $0.004/kWh. A 42 W/m^2 increase in the average insolation level, which represents a 1.0 hour increase in the average daily peak insolation

---

[*] The linear regression estimates that evaluate the effect of electrolyser capacity factor on H_2 production cost need to be qualified. As previously noted in the footnote on page 275, the relationship between electrolyser cost and H_2 production cost over the full 25–95% range of capacity factors is non-linear. However, over the 25–29% range of electrolyser capacity factors applicable for PV power plants the relationship is approximately linear. Also, it should be noted that over the 25–29% capacity factor range the effect of change in capacity factor is greater than over the 25–95% range because the curve is steeper (greater change) at the low-end of the capacity factor range as can be seen in Fig. 3.

Table 4. Descriptive statistics: value ranges to generate regression estimates.

H_2 system components	Mean (Baseline value)	Minimum value	Maximum value
A. PV power plant			
Electricity Price ($/kWh)	0.054	0.044	0.064
PV Module Cost ($/m^2)	60	40	80
BOS Cost ($/m^2	50	40	60
PV Efficiency (%)	12%	10%	14 %
Insolation Level (W/m^2)	270	250	290
Discount Rate (%)	6.0%	5.2%	6.8%
B. Electrolysis plant			
Electrolyser Cost ($/kW$_{dc-in}$)	425	350	525
Electrolyser Capacity Factor (%)	27 %	25 %	29 %
Electrolyser O&M Expense (% of Capital)	2 %	0 %	4 %
Electrolyser Efficiency LHV (%)	64.2%	60.8%	67.6%
Discount Rate (%)	6.0%	5.2%	6.8%
C. Other H_2 system components			
Pipeline ($/mile)	2,000,000	1,500,000	2,500,000
Metal Hydride Containers ($/kg)	30	20	40

Table 5. Sensitivity of levelized H_2 pump price to change in component costs.

	Change in PV electricity price (¢/kWh)	Change in H_2 pump price (¢/kg H_2)
A. Effect of change in electrolysis plant values		
Electrolysis plant		
- Electricity cost (per ¢/kWh)		56.2
- Electrolyser cost (per $25/kW$_{dc-in}$)		4.4
- Electrolyser capacity factor (per 1.0%)		− 2.9
- Electrolyser efficiency LHV (per 1.0 %)		− 4.8
- Electrolyser O&M expense (per 0.5 %)		5.3
- Electrolysis plant discount rate (per 0.5 %)		4.2
B. Effect of change in H_2 distribution values		
- Pipeline (per $250,000/mile)		7.7
- Metal hydride containers (per $5/kg)		9.9
C. Effect of Change in PV Power Plant Values on Electricity Price		
- PV Cost $/m^2 (per $5/m^2)	0.2	
- BOS Cost $/m^2 (per $5/m^2)	0.2	
- PV Efficiency (per 1.0 %)	− 0.4	
- Insolation Level (per 0.5 average peak hours/day)	− 0.4	
- Discount Rate (per 0.5 %)	0.2	

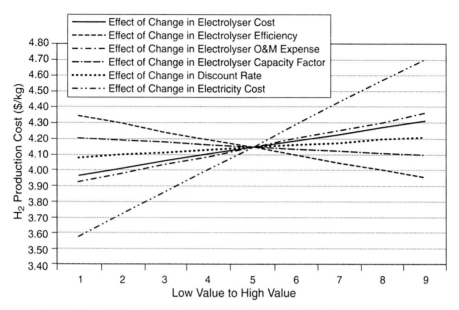

Fig. 4. Effect of change in electrolysis plant values on levelized H2 production cost.

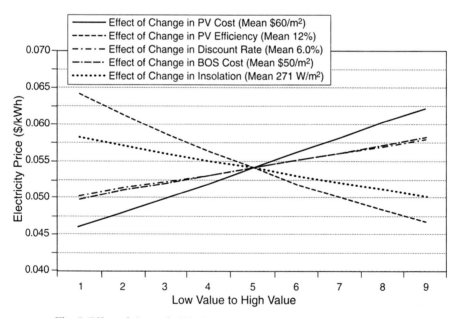

Fig. 5. Effect of change in PV plant values on levelized PV electricity price.

level, causes PV electricity price to decrease by \$0.008/kWh. A 1% increase in the discount rate causes PV electricity price to increase by \$0.004/kWh. And as previously stated, to evaluate the effect of a decrease in cost factor values simply reverse the sign, positive or negative, for the change in PV electricity price.

The slopes of the lines in Fig. 4 are a good demonstration of the relative impact of change in the cost factor values on H_2 production price. Consistent with previous findings, the sensitivity results clearly indicate the dominance of electricity cost on electrolytic H_2 production price. In decreasing order of effect are electrolyser efficiency, electrolyser O&M, and electrolyser cost. A degree of uncertainty exists regarding the cost of electrolysers. At present, large electrolysers are manufactured in small numbers and include ac/dc power conditioning equipment. It is possible that the cost of mass produced electrolysers (thousands of units per year) without power conditioning equipment will be lower than the $425/kW_{dc-in}$ cost estimate.

For PV electricity price, PV module efficiency and insolation level have the greatest impact. The variables having the next largest effect on PV electricity price are PV area cost and BOS area cost. A PV electricity price decrease associated with an increase in PV efficiency is contingent on holding area related PV manufacturing cost constant while achieving PV module efficiency gains.

Due to the large impact of insolation levels on PV electricity prices, a map of insolation levels for the U.S. is presented in Fig. 6. The map clearly indicates that the U.S. is endowed with a large land area with high insolation levels, i.e., insolation levels ≥ 271 W/m^2.

It is highly probable that PV module efficiencies will increase above the near-term 10% module efficiency, which implies that over time H_2 production price will decrease. A decrease in BOS cost is contingent on scale economies achieved through the bulk purchase of standardized BOS components and strict attention to the management of labor costs, i.e., design of tasks to maximize labor-time synergies and mechanization. In conclusion, it can be stated with a relatively high degree of confidence that the baseline cost estimates of this study are conservative and that over the long-term there is a reasonable expectation of decreases in PV electricity price, which translate into lower H_2 prices.

5 Economic Analysis of Second Generation (Year 31–Year 60) H_2 Systems

Many of the PV electrolytic H_2 production and distribution system components have an operating life that will exceed the assigned thirty-year capital recovery period. With the amortization of debt capital and the depreciation of equity capital assets, post-year-thirty H_2 production and distribution costs will decline. With the capital amortization of system components, H_2 production cost is reduced to O&M expenses for those system components. Therefore, it makes sense to evaluate both first and second generation H_2 production costs. First generation H_2 production is defined as the initial thirty-year capital recovery period, and second generation H_2 production is defined as the post-amortization, Year 31–Year 60 H_2 production period.

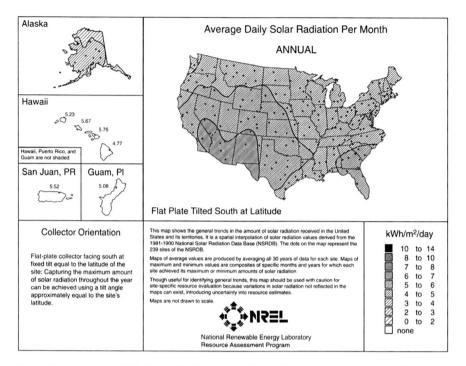

Fig. 6. Map of average U.S. insolation levels on a flat surface, tilted south at an angle equal to the site's latitude. The 250–290 W/m² range in insolation levels for the sensitivity analysis corresponds to solar radiation levels of 6–7 kWh/m²/day. This map was developed from the Climatological Solar Radiation (CSR) Model, developed by the National Renewable Energy Laboratory for the U.S. Department of Energy.

The system components with an operating life greater than thirty years are PV BOS infrastructure components, electrolysers, pipeline, underground H_2 storage facilities, and all buildings. Each of these system components has an operating life of sixty years. It is assumed that the pipeline reciprocating compressors will have a forty-year operating life since natural gas pipeline compressors have an operating life of forty or more years.

Because thin film PV is a relatively recent technology, there is a lack of data on long-term PV electricity production levels. The assignment of a twenty or thirty year operating life for PV is the standard method of economic analysis of PV power plants. However, it is plausible that PV modules will produce electricity for sixty years at a 1% average annual degradation rate. Because of uncertainty regarding the electricity production profile of thin film PV modules, three second generation PV scenarios are evaluated:

1. a 20-year PV module operating life model with PV module replacement at the end of twenty and forty years;

2. a 30-year PV module operating life model with PV module replacement at the end of thirty years; and

3. a sixty-year PV module operating life model with the PV modules left in place to degrade at an assumed 1%/year rate through Year 60.

It should be noted that the appropriate method to evaluate PV economic life is output degradation, and a greater than 30-year economic life is highly probable based on the observed life of silicon PV.

The twenty-year PV life model, which is the least probable model and is presented for comparison to other studies, provides the high case H_2 price estimates. The thirty-year PV life model supplies the intermediate case H_2 price estimates. The sixty-year PV life model gives the low case, second generation (Year 31-60), H_2 price estimates. The sixty-year PV life model is important, because unlike almost any other source of electricity, flat-plate, non-tracking PV has the unique attribute of very long life and very low O&M. For example, even a concentrating solar thermal system would not have this attribute. The closest parallel is hydroelectricity, which has demonstrated the clear value of a large initial investment followed by decades of low-cost generation.

The central financial assumption for the calculation of second generation levelized PV electricity and H_2 prices is the assignment of the depreciated 10% value of first generation assets as the second generation investment value for equity holders. All other second generation capital investments, revenues, expenses, depreciation, and taxes are entered into the net present value cash flow model in exactly the same manner as the first generation model. The capital structure of H_2 production and distribution firms is assumed to remain 30% equity and 70% debt. The rate of return on equity remains 10%, the rate of return on debt remains 7%, the income tax rate remains 39%, and the discount rate remains 6%.

The levelized H_2 pump price estimates for the second generation, thirty-year PV module life model are presented in Table 6.B. There is a 40% reduction in the levelized H_2 pump price of second generation H_2 compared to first generation H_2 pump price. The levelized H_2 pump price reduction is attributable to reductions in capital investments required for second generation H_2 production and distribution components. Second generation capital investments are 61% less than those for first generation H_2 systems. Three factors account for the large capital investment reduction of second generation H_2 production systems; reduced capital investments for the PV power plant and zero capital investments for electrolysers and pipelines.

Two factors account for the reduction in capital investments for the PV power plant. First, the electricity output from the first generation PV additions reduces the quantity of replacement PV from 5.971-GW$_P$ to 4.423-GW$_P$.* And secondly, the cost of

---

* The total quantity of PV additions to the first generation PV power plant is 1.888-GW$_P$. The weighted average PV output of the first generation PV additions is 82% of the rated output of the PV modules in Year 31. The de-rating of the first generation PV additions to 82% of rated output accounts for electricity output losses from PV module soiling, degradation, and catastrophic losses. Hence, the dc electricity output of the first generation PV additions is equivalent to 1.548-GW$_P$ of PV and reduces the quantity of PV replacements for the second generation PV power plant.

second generation PV modules is reduced by PV area cost reductions from $60/m^2 to $50/m^2, and BOS costs are reduced from $50/m^2 to $20/m^2. The BOS cost reduction is attributable to the sixty-year life of the BOS infrastructure components.

The levelized H_2 pump price estimates for the second generation, sixty-year PV module operating life model are presented in Table 6.C. Because the post-Year 30 electricity production profile of thin film PV is speculative at present, these findings are presented to establish the potential, low-end H_2 prices with future developments in thin film PV. The levelized H_2 pump price is 53% lower than the first generation levelized H_2 pump price for the sixty-year PV module operating life model. The capital investments for the second generation, sixty-year PV module operating life model are 83% less than the capital investments for first generation H_2 systems.

Another most important finding from the sixty-year PV module life model is the 59% reduction in the levelized electricity price. The levelized PV electricity price for the second generation, 60-year PV module life model is 35% less than the levelized PV electricity price for the second generation, 30-year PV module life model. At the low price of PV electricity produced by second generation, 60-year PV module life PV power plants, the levelized H_2 pump price is very attractive.

The findings for the 60-year PV module operating life model call attention to the importance of research into the factors that affect thin film PV module operating life with the goal to manufacture thin film PV modules with a sixty-year operating life. For example, it is currently the opinion that crystalline PV modules will produce electricity at an acceptable level for sixty years. Standard assessments of PV systems call attention to its high capital cost and low annual operating expense profile. With the development of 60-year PV life systems, second generation PV power plants will introduce a low capital cost and low annual operating expense model.

With the substantial price reduction for second generation H_2, it is interesting to investigate the application of H_2 as a fuel source for centralized, electricity production by combined-cycle steam turbine power plants. In essence, the use of H_2 produced by PV electrolysis to generate electricity at combined-cycle electricity generating plants is the transformation of PV electricity from an intermittent to a dispatchable source of electricity. This is an interesting case to explore because by the time that second generation PV electrolytic H_2 becomes available, 2040–2050 at the earliest, there are indications that the availability of fossil fuels for electricity generation will begin to be in short supply.

By 2040–2050, natural gas reserves will be in very short supply, and the production of coal will quite likely be approaching peak production levels.[19]. While nuclear power plants are a source of large-scale electricity generation, there exist major concerns regarding uranium supply (without breeder reactors), safety, waste disposal, and nuclear weapon proliferation. Therefore, it is prudent to explore the economic feasibility of other fuel sources such as PV electrolytic H_2 for centralized, electricity generating plants.

Therefore, only 4.423-GW_p of PV is required to replace the first generation 5.971-GW_p of PV.

Table 6. Levelized H₂ and PV electricity prices for first-generation (year 1–year 30) and second-generation (year 31–year 60) H₂ systems with 20-, 30-, and 60-year PV life.

A. First Generation H₂ Production			
	10% eff. PV ($/kg H₂)	12% eff. PV ($/kg H₂)	14% eff. PV ($/kg H₂)
a. PV electrolysis plant (20-year PV life)	4.89	4.34	3.95
b. PV electrolysis plant (30-year PV life)	4.67	4.12	3.75
Pipeline and compressors	0.97	0.97	0.95
City gate distribution center	0.02	0.02	0.02
City gate H₂ delivery trucks	0.13	0.13	0.13
City gate metal hydride containers	0.62	0.62	0.62
Filling station dispensing	0.07	0.07	0.07
a. Levelized H₂ pump price (20-year PV life)	6.70	6.15	5.74
b. Levelized H₂ pump price (30-year PV life)	6.48	5.93	5.53

B. Second generation H₂ production (20-year and 30-year PV life models)			
	12% eff. PV ($/kg H₂)	14% eff. PV ($/kg H₂)	16% eff. PV ($/kg H₂)
a. PV electrolysis plant (20-year PV life)	2.78	2.51	2.23
b. PV electrolysis plant (30-year PV life)	2.60	2.34	2.12
Pipeline and compressors	0.49	0.49	0.46
City gate distribution center	0.15	0.15	0.15
City gate metal hydride containers	0.52	0.62	0.62
Filling station dispensing	0.07	0.07	0.07
a. Levelized H₂ pump price (20-year PV life)	4.10	3.83	3.53
b. Levelized H₂ pump price (30-year PV life)	3.92	3.62	3.40

C. Second generation H₂ production (60-year PV life model)			
	12% eff. PV ($/kg H₂)	14% eff. PV ($/kg H₂)	16% eff. PV ($/kg H₂)
PV electrolysis plant	1.83	1.68	1.50
Pipeline and compressors	0.40	0.40	0.40
City gate distribution center	0.15	0.15	0.15
City gate metal hydride containers	0.62	0.62	0.62
Filling station dispensing	0.07	0.07	0.07
Levelized Pump Price of H₂	3.06	2.91	2.73

D. Levelized PV DC electricity prices			
	$/kWh	$/kWh	$/kWh
a. First generation H₂ system (20-year PV life)	0.072	0.061	0.053
b. First generation H₂ system (30-year PV life)	0.064	0.054	0.047
a. Second generation H₂ System (20-year PV life)	0.043	0.038	0.033
b. Second generation H₂ system (30-year PV life)	0.040	0.035	0.031
-- 2nd Generation H₂ System (60-Year PV Life)	0.026	0.023	0.021

E. H₂ system capital investments			
	$ billion	$ billion	$ billion
a. First generation H₂ system (20-year PV life)	12.367	11.232	10.422
b. First generation H₂ system (30-year PV life)	12.367	11.232	10.422
a. Second generation H₂ system (20-year PV life)	7.893	7.135	6.565
b. Second generation H₂ system (30-year PV life)	4.809	4.430	4.145
-- Second generation H₂ system (60-year PV life)	2.088	2.088	2.088

The delivered price of H_2 to centralized, electricity generating plants is lower than the delivered price of H_2 to filling stations. The lower delivered price of H_2 to centralized electricity generating plants is attributable to the fact that the H_2 can be transported by pipeline directly to the power plants, which eliminates city gate distribution and filling station costs. The levelized prices of grid-distributed electricity produced by H_2 fueled combined-cycle electricity generating plants are presented in Table 7. The assumed efficiency of combined-cycle, steam turbine, electricity generating plants is 55% in terms of converting H_2 energy into electricity.

From the results presented in Table 7.A, the levelized electricity price for electricity produced by combined-cycle power plants fueled with first generation H_2 is too expensive to be considered economically feasible. However, if the 60-year PV module operating life model proves relevant, then the levelized price of electricity generated by combined-cycle power plants using second generation H_2 as a fuel source could be as low as \$0.15–0.17/kWh. These electricity prices provide some assurance that if other options fail to meet electricity demand in the post-2040 period, dispatchable PV electricity will be a feasible option. Clearly, further progress in PV cost reduction, a near certainty by 2040, will reduce the price of electricity generated by H_2 fueled power plants.

6 Life Cycle Energy and GHG Emissions Analyses

6.1 Life Cycle Analysis Methods

This Section investigates life cycle energy and GHG emissions of a PV electrolytic H_2 system. The boundaries of the life cycle energy and GHG emissions analyses are *cradle to grave*. Five life cycle stages are evaluated:

Stage 1: materials production, which includes ore extraction, milling, part casting and machining, and transportation;
Stage 2: product manufacture and assembly;
Stage 3: product distribution;
Stage 4: product utilization; and
Stage 5: product disposal.

Construction, office facility utilization and employee travel to and from work are included. All components are scaled to a thirty-year operating life.

Life cycle primary energy estimation parameters are derived from published studies.[13,20,21,22] Recycling credits are allocated to the material production life cycle estimation parameters on the basis that 80% of materials are recycled at their end-of-life. The GHG emissions estimation parameters are generated with the energy software GREET1.6.[23] All energy values are reported in terms of Btu$_{prim}$/kg of delivered H_2, where prim is primary energy, and at the low heating value.

Primary energy is defined in this study as the total fuel cycle energy input per kg of H_2 energy delivered for consumption and accounts for the energy expended to extract, refine and deliver fuels. The primary energy estimates only include the fossil fuel energy from the use of system H_2 and PV energy. Electricity generation is based

Table 7. H_2 for electricity generation by combined-cycle power plants (efficiency = 55%).[a]

	Capital $/kWh	O&M $/kWh	H_2 Fuel $/kWh	Transmission $/kWh	Administration and profits $/kWh	Levelized electricity price $/kWh
			A. First generation H_2			
10% PV–H_2 @ $5.64/kg	0.011	0.0014	0.304	0.0029	0.03	0.349
12% PV–H_2 @ $5.09/kg	0.011	0.0014	0.274	0.0029	0.03	0.320
14% PV–H_2 @ $4.70/kg	0.011	0.0014	0.253	0.0029	0.03	0.299
			B. Second generation H_2 (30-year PV model)			
12% PV–H_2 @ $3.06/kg	0.011	0.0014	0.165	0.0029	0.03	0.210
14% PV–H_2 @ $2.78/kg	0.011	0.0014	0.150	0.0029	0.03	0.195
16% PV–H_2 @ $2.56/kg	0.011	0.0014	0.138	0.0029	0.03	0.183
			C. Second generation H_2 (60-year PV model)			
12% PV–H_2 @ $2.23/kg	0.011	0.0014	0.120	0.0029	0.03	0.165
14% PV–H_2 @ $2.08/kg	0.011	0.0014	0.112	0.0029	0.03	0.152
16% PV–H_2 @ $1.90/kg	0.011	0.0014	0.102	0.0029	0.03	0.148

[a]The data source for levelized costs for combined-cycle electricity generating plants is EIA, Annual Energy Outlook 2005, Market Trends – Electricity Demand and Supply, Fig. 71 – Data Table.

on a U.S. average fuel mix and power plant efficiency. Energy values are reported at the lower heating value. The GHG emissions are carbon dioxide, nitrous oxide and methane and are reported in grams of CO_2 equivalencies per kg of H_2 combusted.

A generalized analysis such as this produces only approximate life cycle energy and GHG emissions estimates because of cross-sectional variation in product and material production processes and local energy sources. Sensitivity analysis is an analytical tool to evaluate the effect of variances in life cycle estimation parameters on results. The sensitivity analysis performed in this study applies a 25% variance to each of the life cycle estimation parameters.

Energy and GHG emissions payback times are calculated to estimate the time it takes to recover the energy and GHG emissions embodied in the H_2 fuel cycle of FCVs compared to the energy and GHG emissions embodied in the gasoline fuel cycle of conventional internal combustion engine (ICE) vehicles. Payback time calculations are based on an average fuel economy for conventional ICE vehicles of 23.5 miles/gallon of gasoline and an average travel distance of 11,000 miles/year.

The primary energy content of a gallon of gasoline is 143,220 Btu, which is a factor of 1.24 greater than the 115,500 Btu energy content of a gallon of gasoline that is combusted in vehicle engines.[23] The fuel cycle GHG emissions from the combustion of a gallon of gasoline are 12.16-kg CO_2 equivalent. In comparison, FCVs have a fuel economy of 54.5 mi/kg of H_2 and an average travel distance of 11,000 miles/year.

Material resource issues associated with multi-GW_p scale PV manufacturing are evaluated by Zweibel[24] Material resource consumption for the other H_2 system components appears to be within sustainable bounds. The predominant resources for H_2 system components are iron, copper, and aluminum. The estimated 530,000-million metric tons of steel required for H_2 system components is only 0.1% of world annual steel production; the estimated 36,500-million metric tons of copper is only 0.3% of world annual copper production; and the 9,200-million metric tons of aluminum is less than 0.1% of world annual aluminum production.

6.2 Life Cycle Energy and GHG Emissions Analyses Results

The life cycle energy and GHG emissions findings are presented in Table 8. The total primary energy embodied in the life cycle of the H_2 production and distribution system is 35.8 MJ$_{prim}$/kg of delivered H_2. Of the total life cycle energy, the PV power plant accounts for 50%, filling stations account for 22%, the pipeline system accounts for 19%, the electrolysis plant accounts for 5%, and the city gate distribution centers account for 4%.

The total life cycle GHG emissions are 2.6-kg CO_2 Eq/kg of delivered H_2. The use of PV electricity to power the electrolysis plant compressors and pipeline compression station compressors, and system produced H_2 to power all other compressors significantly reduces H_2 fuel cycle CO_2 emissions. The high life cycle CO_2 emissions and primary energy use from the operation of filling station compressors, which are modeled to be powered by grid-distributed electricity with a U.S. average fuel mix, is one point in the H_2 system with potential for reductions in life cycle CO_2 emissions and primary energy consumption through the use of system H_2 or PV electric systems.

The primary energy payback time is 3.1 years, and the GHG emissions payback time is 3.1 years. With a thirty-year life cycle for all system components and the replacement of gasoline ICE vehicles with H_2 FCVs, the payback time estimates translate into vehicle operation with ~ 27 years of fossil fuel free energy use and zero-GHG emissions. The sensitivity results indicate that a ± 25% change in all life cycle estimation parameters change the primary energy payback time by ± 0.80 years and the GHG emissions payback time by ± 0.81 years.

The operation of H_2 powered vehicles results in energy savings of 90% and GHG emissions reductions of 90%. The analysis can be extended by including life cycle energy and GHG emissions embodied in the manufacture of FCVs and ICE vehicles. Research indicates that the life cycle energy and GHG emissions embodied in the manufacture of FCVs is basically the same as those embodied in the manufacture of current conventional gasoline ICE vehicles.[21] This finding lends support to the conclusion that H_2 powered FCVs reduce primary energy use and GHG emissions by 90%. Future growth in the quantity of renewable energy employed in the production of H_2 system components will lead to even greater reductions in the primary energy and GHG emissions profile of H_2 systems.

7 System Energy Flow/Mass/Balance Analysis

The compression energy estimates for electrolysis plant, pipeline, city gate, and filling station compression points are presented in Table 9. Total compression energy is 975 GWh, which is 13.5% of the energy content of gross H_2 production. However, the quantity of primary energy consumed for compression is less since the energy for

Table 8. Life cycle primary energy and CO_2 equivalent emissions.[a,b]

System components	Primary energy (MJ_{prim}/kg H_2)	CO_2 eq emissions (kg CO_2/kg H_2)	Payback sensitivity of energy to +/− 25% (Years)	Payback sensitivity of GHG emissions +/− 25% (Years)
PV power plant	21.26	1.5	0.48	0.47
Water system	1.01	0.1	0.02	0.02
Electrolysis plant	1.40	0.1	0.03	0.03
Pipeline	1.56	0.1	0.03	0.04
City gate distribution	1.35	0.1	0.03	0.03
Filling stations	9.25	0.7	0.21	0.22
Totals	35.82	2.6	0.80	0.81
Payback time (years)	3.1	3.1		
% reduction	89.7%	89.7%		

[a]Life cycle results are based on annual H_2 consumption of 203,613,391 kg H_2.
[b]The H_2 system payback times and % reductions are derived from the operation of one million-conventional ICE vehicles with a fuel economy of 23.5 miles/gal gasoline. The primary energy value of gasoline is 152 MJ_{prim}/gallon (LHV), and gasoline combustion carbon dioxide equivalent emissions are 10.83-kg CO_2 Eq per gallon gasoline.[23]

compressors is provided by PV electricity and H_2 from the pipeline. The total primary energy for all compression points is 706 GWh, which is 9.8% of the energy value of gross H_2 production. While the electrolysis plant and the pipeline compressors use the most energy, 58% of total compression energy, their contribution to primary energy consumption is only 6% because of the use of PV electricity and H_2 as the energy source to power the compressors. While filling stations account for only 20% of total compression energy, they contribute 84% of total primary energy because of the use of grid-distributed electricity.

Table 9. Energy consumption for H_2 compression.

Compression Points	Begin pressure (psi)	Final pressure (psi)	Compression energy (kWh/kg)	% of H_2 energy
Electrolysis plant compressors	14.7	116	1.37	4.1%
Pipeline compressor station	100	1000	1.25	3.8%
Pipeline booster compressors (9)	898	1000	0.54	1.6%
City gate compressors	798	1740	0.43	1.3%
Filling station compressors	363	1740	0.98	3.0%
Totals			4.57	13.7%

	H_2 Flow (kg/yr)	Compression Energy (MWh)	Compression primary energy (MWh)[a]
Electrolysis plant compressors	216,815,961	296,312	23,705
Pipeline compressor station	216,815,961	271,767	21,741
Pipeline booster compressors (9)	216,815,961	117,081	39,736
City gate compressors	210,311,614	90,422	30,688
Filling station compressors	202,006,772	198,959	590,112
Total compression energy		974,541	705,983
% of gross H_2 energy		13.5%	9.8%

A system energy flow chart is presented in Fig. 7, and energy mass and balance ratios are presented in Table 10. The mass efficiency is 94% and means that 94% of the H_2 produced at the electrolysis plant is available to vehicles at filling stations. The system energy efficiency is 77% in terms of H_2 energy output to total system energy inputs including H_2 and total primary energy inputs. Total system energy use is 44 MJ per kg of delivered H_2 of which 39 MJ is fossil fuel energy. The net energy ratio result indicates that 3.3 units of H_2 energy are produced for each unit of fossil fuel energy.

8 Conclusions: Summary of Results and Suggestions for Future Analysis

A summary of levelized H_2 pump prices, system capital investments, and levelized PV electricity prices are presented in Figs. 8–11 A summary of results for first generation (Year 1–Year 30) H_2 production are as follows. The levelized H_2 pump price, which does not include fuel use taxes, ranges from $6.48–$5.53/kg for 10% and 14% efficient PV modules respectively. With fuel tax, the H_2 pump price is $7.47–$6.52/kg, which is comparable to high-end 2005–2006 U.S. gasoline prices when the H_2 is for FCVs with a fuel economy 2.2-times greater than conventional ICE vehicles.

The capital investment for a PV electrolytic H_2 system to support one-million FCVs ranges from $12.4 billion for systems using 10% efficient PV modules to $10.4 billion for systems using 14% efficient PV modules. The PV power plant accounts for 59–51% of total H_2 system capital investments. The levelized PV electricity price ranges from $0.064/kWh to $0.047/kWh for 10% and 14% efficient PV modules respectively.

The most important findings of this study relate to the large price and capital investment reductions for second generation, Year 31–Year 60, PV electricity and H_2 production. Since electricity cost accounts for 80% of H_2 production cost, the reduction in Year 31–Year 60 PV electricity prices are summarized first. The long operating life of PV power plant BOS components causes a significant decrease in PV electricity prices and capital investments for Year 31–Year 60 PV electricity production. Second generation PV electricity prices are reduced to $0.040–$0.031/kWh for 10% to 14% efficient PV modules respectively in the case of a thirty-year PV module operating life and to $0.026–$0.021/kWh for 10% to 14% efficient PV modules respectively in the case of a sixty-year PV module operating life with 1% annual electricity output degradation. An overview of levelized PV electricity prices is presented in Fig. 8, and PV plant capital costs are presented in Fig. 9.

Hydrogen pump prices for second generation, Year 31–Year 60, electrolytic H_2 production are reduced to $3.90–$3.40/kg for 10% to 14% efficient PV modules respectively in the case of a thirty-year PV module operating life and to $3.06–$2.73/kg for 10% to 14% efficient PV modules respectively in the case of a sixty-year PV module operating life. A summary of H_2 pump prices is presented in Fig. 10.

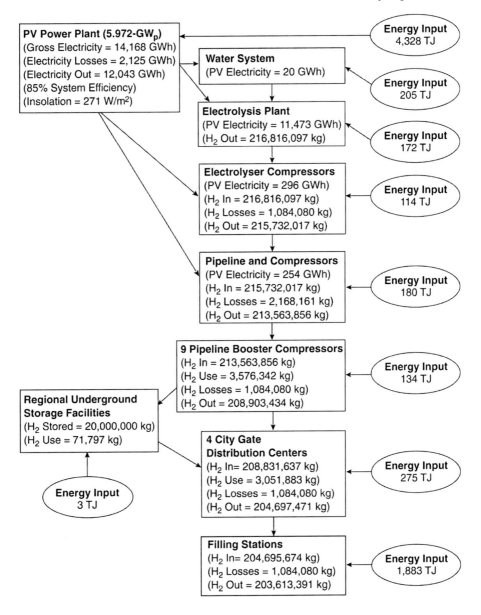

Fig. 7. H₂-system energy flow chart (lower heating values). Energy inputs in the right column are primary energy estimates for the system components.

Table 10. H_2 system energy mass and balance ratios (LHV).

		A. PV power plant			
	PV electricity supply (GWh)	PV electricity use (GWh)	PV electricity losses (GWh)	PV energy use (TJ)	PV life cycle primary energy use (TJ)
PV power plant	14,168		2,126	7,653	4,328
Delivered PV electricity					
Electrolysers		11,473	0	41,301	4,130
Electrolyser compressors		296	0	1,066	107
Pipeline compressors		252	0	907	91
Water system		21	0	76	1
Total	14,168	12,042	2,126	51,002	4,328

		B. H_2 system			
	H_2 flow (kg)	H_2 use (kg)	H_2 losses (kg)	Total H_2 use (TJ)	H_2 life cycle primary energy use (TJ)
H_2 production	216,816,097				
PV electrolysis plant	215,732,017	0	1,084,080	130	4,729
Pipeline	208,826,241	3,653,535	3,252,241	828	408
City gate distribution	204,697,471	3,044,690	1,084,080	495	275
Filling stations	203,613,391	0	1,084,080	130	1,883
Total	203,613,391	6,698,225	6,504,481	1,582	7,294

		C. H_2 system			
	Mass efficiency[a]	System energy efficiency[b]	System energy use (MJ$_{prim}$/kg H$_{2out}$)[c]	Net energy ratio[d]	System fossil fuel energy use (MJ$_{prim}$/kg H$_{2out}$)[e]
Electrolysis plant	99.5%	84.2%	23.9		23.2
Pipeline	96.8%	95.3%	6.1		2.0
City gate distribution	98.0%	97.0%	3.8		1.4
Filling stations	99.5%	92.4%	9.9		9.2
Total	93.9%	77.0%	43.6	3.3	35.8

a. Mass efficiency = H_2 out/H_2 in (does not include life cycle primary energy use).
b. System energy efficiency = H_2 energy out/H_2 in + fuel cycle primary energy.
c. System energy use = system energy use - H_2 use + primary energy (MJ) / kg H_2 out.
d. Net energy ratio = H_2 energy out/fossil fuel (primary energy) energy consumed in system.
e. Fossil fuel energy use = MJ fossil fuel (primary energy) energy/kg H_2 out.

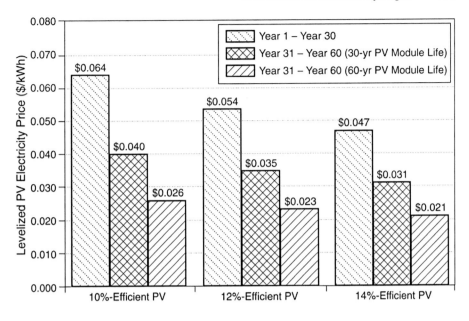

Fig. 8. Summary of levelized PV electricity prices ($/kWh).

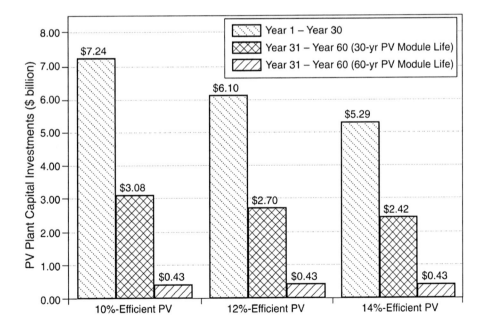

Fig. 9. Summary of PV power plant capital investments.

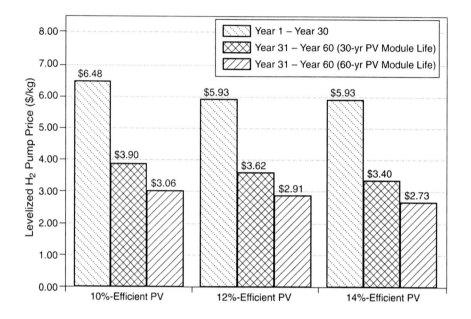

Fig. 10. Summary of levelized H₂ pump prices ($/kg H₂ = gallon of gasoline equivalent price).

Second generation H₂ system capital investments are reduced to $4.81–$4.15 billion for 10% to 14% efficient PV modules respectively in the case of a thirty-year PV module operating life and to $2.09 billion in the case of a sixty-year PV module operating life. A summary of H₂ system capital investments is presented in Fig. 11.

Since PV electrolysis plants are modular in design, it is possible to couple the expansion of PV electrolysis plants to growth in the FCV market. The creation of a H₂ production and distribution system is contingent on the development of a working partnership between PV, electrolyser, automobile, pipeline, metal mining and retail fuel companies. The capital investments required for the construction of a PV electrolytic H₂ production and distribution system is comparable to the capital investments in the construction of the cable and satellite infrastructure for the information technology industries in the latter part of the 20th century.

The total land area of the PV electrolysis plant ranges from 94 mi² to 68 mi² for 10% and 14% efficient PV respectively. The land area is not a problem since PV electrolysis plants will be located in sparsely populated desert regions. Annual water consumption is 1.47-billion gallons, which is a relatively small quantity of water and is easily supplied by either on-site, rain-runoff, collection and storage systems or water importation by train or truck.

The total life cycle primary energy is 35.8 MJ$_{prim}$/kg of delivered H₂. The life cycle GHG emissions are 2.6-kg CO_2 Eq/kg of delivered H₂. The primary energy and CO_2 payback times are 3.1 years respectively. The replacement of gasoline powered ICE vehicles with H₂ powered FCVs reduces primary energy consumption by 90% and GHG emissions by 90%.

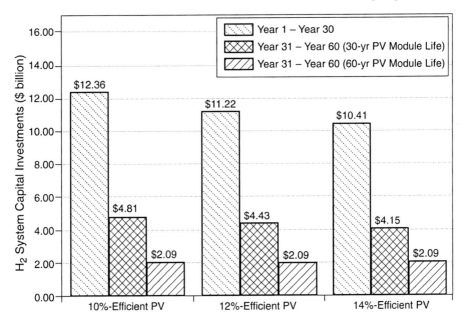

Fig. 11. Summary of capital investments (year 1–30 and year 31–60) H₂ production.

The PV manufacturing capacity to support the production of H₂ to power 250-million FCVs over a thirty-year timeframe, which is approximately 25% of the projected world fleet of light-duty vehicles and light commercial trucks, is presented in Table 11. The H₂ to power 250-million FCVs is 0.24-TW of energy, which replaces 0.52-TW of energy consumed by gasoline powered ICE vehicles. This level of H₂ production from PV electrolysis plants will require the annual manufacture of 50-GW_p of PV. The thirty-year cumulative quantity of installed PV is 1.735-TW_p.

The total capital investment for a PV electrolytic H₂ production and distribution system to deliver H₂ for 250-million FCVs ranges from $3.09-trillion to $2.60-trillion for H₂ systems with 10% and 14% efficient PV modules respectively. The annual capital investment to construct the H₂ system over thirty years is $103–$87 billion with 10% and 14% efficient PV modules respectively.

The PV technologies that currently demonstrate the potential to meet the cost and performance projections of this study are thin film CdTe and CIS PV, which raises questions regarding the resource availability of tellurium and indium to meet the required scale of PV production. The tellurium and indium production estimates of Zweibel[1,24] indicate that the tellurium and indium resource bases are likely sufficient to support the manufacture of 50-GW_p/year of CdTe and CIS PV. This conclusion is highly sensitive to assumptions about layer thickness and the availability and price of tellurium and indium. It needs to be emphasized that the tellurium and indium resource production projections are based on soft resource data analysis and substantial variation in assumed layer thicknesses and module efficiencies. An important devel-

Table 11. Installed PV and capital cost to produce H_2 for 250-million FCVs (with thirty-year PV life and first generation H_2 production assumptions).

	New PV installed/year (GW$_p$)	Total PV installed in 30 years (GW$_p$)	Number of PV manufacturing plants @ 3-GW$_p$/year capacity
PV electrolysis plants[a]	49.76	1,493	17.6
PV additions for output losses[b]	0.52	242	0.2
Total installed PV	50.28	1,735	23.0

	H_2 system with 10% efficient PV	H_2 system with 12% efficient PV	H_2 system with 14% efficient PV
Annual capital costs (billion $)	103	94	87
Total 30-year capital costs (billion $)	3,089	2,805	2,603

[a]Includes PV for electrolysers, compressors, water pumps, water distillation, and the pipeline compression station.

[b]The PV additions for the 49.76 GW$_p$ of PV installed in the first year are 0.52 GW$_p$ per year for thirty years. The PV additions increase each year by 0.52 GW$_p$. In the thirtieth year, the total quantity of PV additions is 15.73 GW$_p$. The PV manufacturing capacity of five PV manufacturing plants will be needed to supply the PV additions. A total of twenty-three PV manufacturing plants with an annual PV production capacity of 3 GW$_p$ each is required in year 30.

opment is the discovery of a very large source of economically recoverable tellurium in seabed ferromanganese crusts,[25] which will become available with growth in seabed mining in coming decades. On a final note, other analyses project that tellurium and indium resource constraints impose limits on PV production levels ranging from 20 GW$_p$/year to more than a 1,000 GW$_p$/year.

The primary challenges are: continued progress in thin film PV module efficiencies and cost reduction; the scale-up in the manufacturing capacity of PV and electrolyser components; and increasing the production of rare semiconductor metals.[1] The increase in tellurium and indium production will require timely investments for the addition of secondary metal production facilities, which will require coordination between PV manufacturers and metal mining and refining companies. Recycling processes for the full recovery of materials from retired PV modules need adopted to extend the long-term supply of rare semi-conductor metals. To hedge against the possibility that the supply of tellurium and indium falls short, further research on silicon based PV as well as new compound semiconductor thin films is important. Since the future supply of indium and tellurium is unpredictable, this research emphasis in PV is a necessary component of any strategy for the terawatt-scale application of PV.

The development of a PV electrolytic H_2 production and distribution system will provide substantial economic benefits. Growth in the PV, electrolyser, compressor and metal hydride industries will create millions of new jobs worldwide, which in turn will stimulate economic growth. The number of jobs created in the PV and electrolyser manufacturing industries will be many times the number of jobs lost in the gasoline production industry. The greatest economic benefits are the mitigation of global warming consequences and the development of sustainable energy systems to support global economic growth when fossil fuel production levels begin to decline over the course of the next several decades.

Areas for additional analysis are:

1. An economic evaluation of expanding secondary metal production facilities to support the timely growth in tellurium and indium production. This should also include further assessments of the economically recoverable tellurium and indium resource bases.

2. Analysis of the technical, material, and economic production parameters to manufacture PV modules with a 60-year operating life.

3. An evaluation of the daily PV electricity output profile to evaluate whether it matches the power requirements of H_2 compressors at the electrolysis plant and the pipeline compression station. In other words, can PV electricity be the sole source of power for electrolysis plant and pipeline compression station compressors?

4. Macro-economic analysis of labor market dynamics of multi-GW_p PV manufacturing plants.

In conclusion, the biggest challenge facing the use of PV for hydrogen production remains the carrying out of the research program to develop higher efficiency, lower cost PV; and assuring the interim market subsidies needed to keep investment in PV strong so that manufacturing scale-up continues. But the important conclusion of this paper is that the achievement of low-cost PV will then lead to cost-effective production of hydrogen for vehicular markets.

Appendices

Appendix 1. Energy Units and CO_2 Equivalent Emissions Estimates

See Table 12.

Appendix 2. Levelized Price Estimates Derived by Net Present Value Cash Flow Analysis

The levelized price of a product is the constant revenue stream that recovers all capital investments and covers all variable and fixed costs and taxes over the investment period. Therefore, the levelized electricity and H_2 prices presented in this study are derived by finding the electricity and H_2 price that generates a net cash flow resulting in a zero net present value for the sum of discounted annual net cash flows over the investment period. The net present value formula is

$$NPV = \sum_{t=1}^{N} \frac{NCF_t}{(1+k)^t} - I_0 \tag{2}$$

Table 12. Energy units and CO_2 equivalent emissions estimates[a]

A. Energy units		
	Low heat value	High heat value
Hydrogen (Btu/kg)	113,607	134,484
Conventional gasoline (Btu/gal)	115,500	125,000
Conventional diesel (Btu/gal)	128,500	138,700
Natural gas (Btu/scf)	928	1,031
Coal (Btu/short ton)	18,495,000	20,550,000

B. Primary energy and CO_2 equivalent emissions		
	Primary energy (MJ_{prim}/MJ_e)	CO_2 eq. emissions (g/MJ_e)
Electricity (US Fuel Mix)	2.96	220.2
PV electricity	0.10	7.1
Hydrogen (by PV electrolysis)	0.30	21.6
Gasoline (conventional)	1.24	89.3
Diesel (conventional)	1.19	92.7
Residual fuel oil (stationary boiler)	1.10	88.3
Natural gas (stationary boiler)	1.06	65.3
Coal (stationary boiler)	1.02	96.1

[a]The data source is GREET1.6 [23] except for the PV electricity and hydrogen by PV electrolysis primary energy and CO_2 equivalent emissions estimates, which are original to this study. The CO_2 equivalent emissions are carbon dioxide, nitrous oxide, and methane.

where NPV = net present value of the investment project, NCF_t = net cash flows per year for the project, k = cost of capital, which is a weighted average cost of capital (WACC), $(1 + k)^t$ = the discount rate to convert annual net cash flows to their present value, N = number of years, I_0 = shareholder investment in the project.

The definition of net cash flow (NCF) for capital budgeting purposes is after-tax cash flows from operations discounted at the present value of the cost of capital.[26] In net present value analysis the cost of capital is a pre-determined value based on the opportunity cost of capital. The cost of capital is defined as a weighted average cost of equity (WACC) and takes into account the firm's capital structure, the cost of equity and debt capital, and tax rates. The formula for the weighted average cost of capital (WACC) is

$$\text{WACC} = \text{Discount Rate} = \{[(\% \text{ equity}) (k \text{ equity})] \times [(\% \text{ debt}) (k \text{ debt}) (1 - \tau)]\} \quad (3)$$

where % equity is the percentage of the market value of the firm's market value owned by shareholders, k equity is the cost of equity capital, % debt is the percentage of firm's market value owned by creditors, k debt is the cost of debt, and τ is the tax rate.

Operating cash flows are revenues (Rev) minus direct costs that include variable costs (VC) and fixed cash costs (FCC):

$$\text{Operating Cash Flows} = \text{Rev} - \text{VC} - \text{FCC} \quad (4)$$

Since net cash flows are defined as the after-tax cash flows from operations, taxes have to be included:

$$\text{Taxes on Operating Cash Flows} = \tau (\text{Rev} - \text{VC} - \text{FCC} - \text{dep}) \quad (5)$$

Depreciation is defined as a non-cash charge against revenues in the calculation of net cash flows. Interest expenses and their tax shield are not included in the definition of cash flows for capital budgeting purposes. The reason is that when we discount at the weighted average cost of capital we are implicitly assuming that capital budgeting projects will return the expected interest payments to creditors and the expected dividends to shareholders. Meanwhile, the reduction in expenses from the tax shield is already counted in the term for the tax rate. Hence, the inclusion of interest payments or dividends as a cash flow to be discounted is double-counting.

Putting all of this together, the operational expression for the calculation of the net present value (NPV) of net cash flows is

$$NPV = \sum_{t=1}^{N} \left[\frac{(Rev - VC - FCC - dep)_t \, (1 - \tau)_t}{(1 + k)^t} \right] - I_0 \qquad (6)$$

which is equivalent to Eq. 2.

The levelized PV electricity and H_2 prices presented in this study are derived from Eq. 6 by choosing the electricity or H_2 price level for the revenue component that produces a zero net present value for the net cash flow streams over the investment period, which in this case is equivalent to the internal rate of return. The estimation of levelized PV electricity and H_2 prices by the net present value cash flow method insures that all creditors and shareholders receive their expected rates of return.

For this study it is assumed that the effect of inflation will be the same for cash inflows and outflows and rates of return. This inflation assumption implies that the inflation factor in Eq. 2 is the same in both the numerator and denominator, and hence, cancels out. Therefore, the net present value is both a nominal and real value. However, if the expected inflation rate for cash inflows, cash outflows, or rates of return are different, then inflation factors need to be added to the appropriate factors in Eq. 2 or equivalently in Eq. 6.

The application of net present value cash flow analysis to estimate levelized energy prices tends to provide estimates that are more conservative than almost any other estimation method, thereby making the analysis more robust.

Appendix 3. Adiabatic Compression Formula

Hydrogen compression energy is estimated with the adiabatic compression energy formula:

$$W_{J/kg} = \frac{\dfrac{y}{y-1} P_1 V_1 \left[(P_2 / P_1)^{(y-1)/y} - 1 \right] \dfrac{Z_1 + Z_2}{2Z_1}}{efficiency} \qquad (7)$$

where $W_{J/kg}$ = specific compression work; y = specific heat ratio (adiabatic coefficient); P_1 = initial pressure (PaA); P_2 = final pressure (PaA); V_1 = initial specific volume (m^3/kg); Z_1 = gas compressibility factor for initial pressure; Z_2 = gas compressibility factor for final pressure; and efficiency = efficiency of the compressors.[8] The gas compressibility factors are calculated by the Redlich-Kwon equation of state.[9] An average compressor efficiency of 70% is assumed over the 0.8–11.72-MPa range of pressures used in this study.

Appendix 4. Deviations from DOE H2A Assumptions

The U.S. Department of Energy's Hydrogen Program has developed a DOE H_2 Analysis tool for H_2 systems research. Researchers from the National Renewable Energy Laboratory and Argonne National Laboratory have constructed a techno-economic database known as the H2A guidelines to assist in the economic evaluation of a variety of H_2 delivery and forecourt scenarios.[27,28] Due to the widespread use of H2A guidelines it is believed appropriate to address some of the areas where the assumptions and methods underlying the results of this study deviate from the H2A default assumptions and values.

There are several differences in the financial assumptions. The H2A real after-tax discount rate is 10%, whereas in this study the real after-tax discount rate is 6%. The variation is attributable to differences in the capital structure for investments. The H2A uses a 100%-equity capital structure, whereas this study uses a capital structure of 30% equity capital and 70%-debt capital. The cost of debt is 10% for the H2A default value for 7% (30-year coupon bond) for this study. The tax rate is the same in both studies. The H2A assumptions include an inflation factor of 1.29%, while this study does not include an inflation factor, which is explained in Appendix 2. The net effect of these differences in financial assumptions is a lower levelized H_2 pump price estimate for this study compared to the levelized H_2 pump price under the H2A financial assumptions.

The assumptions of this study are premised on the commitment to a multi-trillion dollar, centralized H_2 production and delivery system in the U.S. over a thirty-year time period. Therefore, it is believed that the capital structure assumptions of 30%-equity capital and 70% debt are more realistic for the assumed scale of capital investments. In addition, there are cash flow benefits to financing capital budgeting projects with debt capital rather than equity capital because interest on debt is tax deductible whereas dividends payments are not. The 7% interest rate for 30-year coupon bonds is a reasonable assumption for the assumed scale of investments, particularly so if a national H_2 plan is adopted with government regulation and guaranteed bond issues.

Another major difference between this study and the H2A scenarios is the specification of a H_2 delivery and storage system based on metal hydride (MH) H_2 storage in this study, which is not included in the H2A scenarios. In the default H2A scenario for a compressed H_2 system, terminal and forecourt H_2 costs are estimated at $3.88/kg of H_2, whereas in this study city gate delivery and filling station costs are estimated at $0.84/kg of H_2. A review of the H2A database provides some answers as to why the H2A H_2 price estimates are higher than this study.

For one, the difference in financial assumptions explains part of the difference. Possibly the largest factor is a difference in the assumption regarding operating life of H_2 storage containers. H_2 storage containers, composite tube trailers for compressed H_2 storage or MH, are one of the most costly components in the H_2 delivery system. The default value for the H2A database is a ten-year operating life, whereas this study assumes a thirty-year operating life for MH containers. Also, the H2A compressed H_2 delivery and dispensing entails higher energy expenditures for H_2 storage and dispensing compared to MH systems. The compression energy for the H2A compressed H_2 delivery system is a factor of > 2.0 greater than for the MH system of this study, which translates into higher O&M expense for the H2A compressed H_2 scenarios. The higher energy levels for compression also mean larger compressors at higher capital investments per compressor.

Appendix 5. Summary of Reviewer Comments with Responses

The report was submitted to reviewers for comments. The comments addressed a variety of issues. A brief summary of the comments is provided and addressed by these categories:

1. PV power plant assumption;
2. the problem of creating a national H_2 supply sufficient to support the mass marketing of H_2 powered vehicles and the need for distributed H_2 production to address the short-term national H_2 problem problem; and
3. why the choice of a metal hydride H_2 storage system when an effective metal hydride storage system does not at present exist.

Responses to the reviewers' comments are offered in this Section.

(i) PV power plant assumptions

Reviewers state that while the PV power plant cost and performance estimates are optimistic, they are achievable based on the historical trajectory of PV development. The authors believe that the 10%-thin film PV power plant cost estimates are a legitimate baseline model to evaluate the near-term economic feasibility of using PV electricity for electrolytic H_2 production. 10%-thin film PV will be available for the near-term construction of the first 6-GW_p PV electrolysis plant. The projected PV cost of $60/m^2$ is premised on the assumption that the PV is manufactured at an optimized PV manufacturing plant with an annual production capacity of a 2–3-GW_p of PV. The size of the optimized PV manufacturing factory is based on the size of an optimized glass production facility, which is the single largest component of a PV module. Over time, PV electrolytic H_2 production costs will decline as PV technology advances to the 12% and eventually 14% efficiency levels.

Reviewers express some confusion regarding the thirty-year and sixty-year PV module operating life models for post-amortization second generation, Year 31–Year 60, H_2 production. The second generation H_2 production model is one of the most important concepts developed in this paper and is an area that deserves greater attention and research. If PV modules can achieve a sixty-year operating life, then PV will truly be an important technology. However, as the analysis demonstrates, even with a thirty-year PV module operating life, the sixty-year life of PV plant BOS infrastructure and the sixty-year operating life of electrolysers results in a 44% reduction in the second generation, Year 31–Year 60, H_2 production costs, which also is highly significant.

One reviewer raised the issue of the DOE target goal of \$0.04/kWh-electricity cost for the economic production of H_2 by electrolysis. While this is an achievable long-term goal for PV power plants, this analysis demonstrates that economical H_2, when used by advanced fuel economy vehicles such as fuel cell vehicles, can be produced by electrolysis with electricity costs as high as \$0.064 kWh. Electrolytic H_2 produced at this electricity price is comparable to 2006-gasoline prices when the H_2 is used by advanced fuel economy vehicles. This is a particularly attractive fuel price when the near-zero greenhouse gas emissions profile of PV electrolytic H_2 is factored into the cost assessment. With the likely near-term institution carbon taxes and increasing concern over the consequences of global warming, the near-term price of PV electrolytic H_2 is perceived as economical. And as the study indicates, the price of PV electrolytic H_2 will go down over time with the assumed progress in thin film PV technologies.

(ii) Questions related to national H_2 production and distribution issues

Reviewers asked why we did not consider distributed PV electrolytic H_2 production systems. For one, Ivy[29] conducted an excellent review of distributed PV electrolytic H_2 systems. But the most important reason is that the central issue to this study is the production and distribution of a sufficient quantity of H_2, which is to be widely distributed simultaneously and continuously to local markets throughout the nation, to support the mass marketing of H_2 powered vehicles in terms of millions of additional H_2 vehicles per year. This study attempts to establish the parameters for centralized PV electrolytic H_2 production and distribution to provide this scale of national H_2 production and distribution.

If the goal is to eliminate CO_2 emissions in the transportation energy use sector by mid-century, then this is the scale it will take. Two-hundred million vehicles consume approximately two-billion barrels of oil equivalent energy per year. By 2050, this will grow to more something on the order of three-hundred million vehicles. Today, we have approximately 150-oil refineries and a highly centralized oil production, refining, and distribution system. The development of a H_2 system to effectively replace oil use for transportation will also have to be highly centralized to produce and distribute the volume of H_2 required to make the replacement in a timely manner.

The logical solution to the national distribution of H_2 produced by centralized PV electrolysis plants in the southwest U.S. is to build an integrated national H_2 pipeline

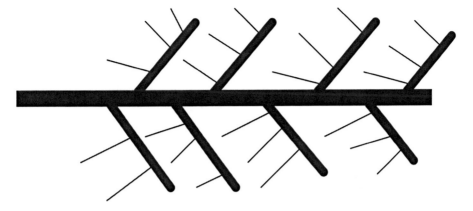

Fig. 12. H_2 pipeline system: trunk, regional lateral, and local spur pipelines.

with H_2 trunk pipelines built for the southern and south-central tier of the U.S. The national H_2 pipeline system can be augmented with trunk H_2 pipelines for the northern and north-central tier of the U.S. for the distribution of wind electrolytic H_2 produced in the upper Midwest (North Dakota, Wyoming). In addition, H_2 production by hundreds of gasification plants using biomass feedstocks can be distributed throughout the country to provide H_2 supplies to regions not readily served by either PV or wind electrolytic H_2 production.

The PV electrolytic H_2 model used in this study allocates 621 miles (1,000 km) of pipeline to each incremental increase in H_2 production to support one-million FCVs. In the first three plants are built in El Paso, Albuquerque, and CA/NV/AZ border, then 621 miles of pipeline in each location will transport H_2 to Houston, San Antonio, Austin, and El Paso from the El Paso plant; Los Angeles, Phoenix, San Diego, Las Vegas, San Bernardino, and Riverside from the CA/NV/AZ plant; and Albuquerque, Santa Fe, Colorado Springs, Denver, and Fort Collins from the Albuquerque plant. From there the pipeline networks keep extending with the construction of each additional PV electrolysis plant. The construction of five additional plants in west Texas enables the pipeline to reach markets on the East Coast and in the south-central U.S. The construction of five additional PV electrolytic H_2 plants on the CA/NV/AZ border enable the extension of H_2 pipelines throughout California and into the Pacific Northwest.

Two trunk pipelines can be constructed in existing pipeline corridors corresponding to the interstate highways I-10, I-20, and 1-40. Figure 12 presents a schematic of a trunk pipeline with lateral regional and local spur pipelines, which terminate at city gate distribution centers.

(iii) The choice of a metal hydride H_2 storage system

The authors acknowledge that metal hydride H_2 storage systems are still in the development stage, and a decision has not yet been made as to whether or not metal hydrides will be the final H_2 storage medium. Automobile manufacturers repeatedly state that low pressure, solid state H_2 storage is the preferred means for H_2 storage if the cost and performance criteria can be met. Therefore, it is important that H_2 systems research evaluate MH storage and delivery systems.

Also, to provide completeness to this study a H_2 storage medium is needed. It is beyond the scope of this study to evaluate all possible H_2 storage systems. The incorporation of a metal hydride system provides one legitimate model for the assessment of downstream H_2 distribution systems. By incorporating a MH storage and delivery system in this, two significant issues requiring additional research have been identified in the review process:

1. the need for a detailed comparative analysis of compression energy consumption and H_2 cost effects; and
2. determination of the operating life of MH storage containers since a 15,000-cycling life implies a thirty-year operating life.

References

1. K. Zweibel. The terawatt challenge for thin film PV, in *Thin Film Solar Cells: Fabrication, Characterization and Applications*, Jef Poortmans and Vladimir Archipov, eds., John Wiley, Hoboken, NJ (2005).
2. J. Motavalli, Commentary: hydride hurdles on the hydrogen highway. *Our Planet*, a weekly newsletter supplement to *E Magazine*, March 15, (2005).
3. M. S. Keshner and R. Arya, *Study of Potential Cost Reductions Resulting from Super-Large-Scale Manufacturing of PV Modules*, National Renewable Energy Laboratory, U.S. Department of Energy, Golden, CO. Final Subcontract Report, NREL/SR–520–36846, October (2004). http//:www.nrel.gov/docs/fy05osti/36846.pdf
4. A. Szyszka, Ten years of solar hydrogen demonstration project at neunburg vorm wald, Germany, *International Journal of Hydrogen Energy* **23**, 849-860, (1998).
5. T. Dietsch, Photovoltaics of the neunburg vorm wald solar hydrogen project, *Power Engineering Journal* **10**, 17–26, (1996).
6. A. Cloumann, P. d'Erasmo, M. Nielsen, B. G. Halvorsen, and P. Stevens, *Analysis and Optimisation of Equipment Cost to Minimise Operation and Investment for a 300 MW Electrolysis Plant*. A collaborative report prepared by Norsk Hydro Electrolysers AS and Electricité de France, Direction des Etudes et Recherches, Notodden, Norway and Moret sur Loing, France, (1994).
7. W. A. Amos, *Costs of Storing and Transporting Hydrogen*. National Renewable Energy Laboratory, NREL/TP-570-25106, US Department of Energy, Golden, CO, (1998).
8. Personal communication with Joseph Schwartz of Praxair (2004).
9. J. Peress, Working with non-ideal gases: here are two proven methods for predicting gas compressibility factors, *CEP Magazine* **March**, 39–41, (2003).
10. M. R. Harrison, T. M. Shires, J. K. Wessels, and R. M. Cowgill, *Methane Emissions from the Natural Gas Industry. Project Summary. United States Environmental Protection Agency*, Report No. 600/SR-96/080. National Risk Management Research Laboratory, Research Triangle Park, NC, 1997.
11. USGS, Minerals Information, United States Geological Survey (USGS), U.S. Department of the Interior, Reston, VA, 2006.
12. B. S. Chao, R. C. Young, V. Myasnikov, Y. Li, B. Huang, F. Gingl, P. D. Ferro, V. Sobolev, and S. P. Ovshinsky, *Recent Advances in Solid Hydrogen Storage Systems*, Texaco Ovonic Hydrogen Systems, LLC, Rochester Hills, MI, 2004.

13. J. E. Mason, V. M. Fthenakis, T. Hansen, and H. C. Kim. Energy payback and life-cycle CO_2 emissions of the bos in an optimized 3.5-mw PV installation, *Progress in Photovoltaics: Research and Applications* **14**(2) 179–190 (2006).
14. Norsk Hydro, personal communications with Harry Tobiassen. Norsk Hydro Electrolysers AS, Notodden, Norway, 2004.
15. J. M. Ogden, *Renewable Hydrogen Energy System Studies. Center for Energy and Environmental Studies*, Princeton University. Report prepared for National Renewable Energy Laboratory, DOE Contract No. XR-2-11265-1, Golden, CO, (1993).
16. Federal Highway Administration, *Highway Miles Traveled (Urban and Rural).* United States Department of Transportation, 2003.
17. W. Marion and S. Wilcox, *Solar Radiation Data Manual for Flat-Plate and Concentrating Collectors*, Manual Produced by the National Renewable Energy Laboratory's Analytic Studies Division. Contract No. NREL/TP-463-5607, DE93018229, April, 1994.
18. Portland, *Cooling Water Efficiency Guidebook,* Bureau of Water Works, City of Portland, Portland, OR, 2005.
19. J. E. Mason, World energy analysis: H_2 now or later? *Energy Policy* **35**(2) 841–853(2007).
20. V. Fthenakis and E. Alsema, Photovoltaic energy payback times, greenhouse gas emissions and external costs, *Progress in Photovoltaics: Research and Applications* **14** 275–280 (2006).
21. M. A. Weiss, J. B. Heywood, E. M. Drake, A. Schafer, and F. F. AuYeung, *On the Road in 2020: A Life Cycle Analysis of New Automobile Technologies,* Energy Laboratory, Massachusetts Institute of Technology, Cambridge, MA. Energy Laboratory Report # MIT EL 00-003, 2000.
22. L. Wibberley, *LCA in Sustainable Architecture (LISA)*, Developed by Sustainable Technology, BHP Billiton Technology, BlueScope Steel, Melbourne, Australia, 2002.
23. M. Wang, *GREET Version 1.6 (Excel Interactive Version)*, Center for Transportation Research, Argonne National Laboratory, University of Chicago, Chicago, IL., 2001.
24. K. Zweibel, *The Terawatt Challenge for Thin-Film PV: A Work in Progress* (January 2006 version). NREL Thin Film Partnership, National Renewable Energy Laboratory, Golden, CO, (2006).
25. J. R. Hein, A. Koschinsky, and A. N. Holliday, Global occurrence of tellurium-rich ferromanganese crusts and a model for the enrichment of tellurium, *Geochimica et Cosmochimica Acta,* **67**(6) 1117–1127 (2003).
26. T. E. Copeland, J. F. Weston, and K. Shastri, *Financial Theory and Corporate Policy*, 4th Ed., Addison-Wesley, Boston, MA, 2005.
27. M. Ringer, *Hydrogen Delivery Component Model*, National Renewable Energy Laboratory, Golden, CO, 2006.
28. M. Mintz, J. Gillette, A. Elgowainy, and J. Molberg, *Hydrogen Delivery Scenario Analysis Model (HDSAM)*, Argonne National Laboratory, University of Chicago, Chicago, IL, 2006.
29. J. Ivy, *Summary of Electrolytic Hydrogen Production*. Milestone report for the U.S. Department of Energy's Hydrogen, Fuel Cells, and Infrastructure Technologies Program, March 19, 2004. National Renewable Energy Laboratory, Golden, CO, 2004.

Index

Printed in the United States
109189LV00002B/195-232/P